VCR Troubleshooting
& Repair

VCR Troubleshooting & Repair

Third Edition

Gregory R. Capelo
Robert C. Brenner

Newnes

Boston • Oxford • Johannesburg • Melbourne • New Delhi • Singapore

Newnes is an imprint of Butterworth–Heinemann.

Copyright © 1998 by Butterworth–Heinemann

 A member of the Reed Elsevier group

Library of Congress Cataloging-in-Publication Data
Capelo, Gregory R.
    VCR troubleshooting & repair / Gregory R. Capelo, Robert C.
Brenner.—3rd ed.
       p. cm.
    Includes index.
    ISBN 0-7506-9940-X (alk. paper)
      1. Videocassette recorders—Maintenance and repair.   I. Brenner,
Robert C. II. Title.
    TK6655.V5C37   1997
    621.388¢337—dc21   97-27610
                                                     CIP

British Library Cataloguing-in-Publication Data
A catalogue record for this book is available from the British Library.

The publisher offers special discounts on bulk orders of this book.
For information, please contact:
Manager of Special Sales
Butterworth-Heinemann
225 Wildwood Avenue
Woburn, MA 01801-2041
Tel: 781-904-2500
Fax: 781-904-2620

For information on all Butterworth–Heinemann publications available, contact our World Wide Web home page at: http://www.bh.com

10 9 8 7 6 5 4 3 2 1

Printed in the United States of America

Contents

Preface

Seven-fifty for a movie? You've got to be kidding! Not really. Movie theater prices have increased steadily (almost in direct proportion to our kids' allowances). The five-cent Saturday matinee of the late forties has become the $7.50 "flick" of today.

However, an interesting phenomenon is occurring in the movie-watching consumer marketplace. Fewer people attend the expensive premiere showing, where long lines wrap lazily around the theater building. Instead, they prefer the comfort of home or apartment with a video cassette recorder or player, a large-screen television, a rented movie, handy snacks, and drinks.

From the moment the first commercial home video player came on the market, attendance at movie theaters began a slow but steady decline. At the same time, the number of users who were purchasing their own video cassette recorder (VCR) machines steadily increased. This increase became a torrent as VCR prices dropped from their original $1200 to less than $250 today. The standard features on machines today, such as wireless remote, multiday and multievent timers, Hi-Fi stereo, and direct cable capability, were either not technically available fifteen years ago, or simply too costly for the home video user.

GROWTH OF THE VCR MARKET

The video recorder market has steadily evolved over the last twenty years and is just now beginning to reach its prime. Today we can display fantastic color films on wide screens and with ear-rumbling sound. Stereo effects that could be experienced only in movie theaters are now possible at home. We can reproduce the same visual and sound effects from a rented copy of *Star Wars* that we saw and felt in the theater as a huge star cruiser roared out of the ceiling above and onto the screen. Sounds from a half dozen stereo speakers can fill our video world with the powerful engine rumble and action of this amazing space adventure. Today, far more films are seen at home on VCRs than in theaters, and this gap is widening.

In 1980, about 3 percent of U.S. households owned video cassette recorders. In 1995 over 90 percent of all U.S. households own a VCR! In Europe and Japan, the number of homes with VCRs is almost as high. The VCR is as common as an automobile or a telephone. Given this rate of acceptance and use, a new VCR is now being built every second. The number of home VCR owners worldwide is growing at an increasing rate.

A spokesman for RCA stated that ". . . while a number of households are still without the pleasures of home video, an ever-growing number of consumers are (now) buying their second VCR." Many of the VCRs sold in 1994 were second purchases . . . second units or upgraded replacements.

Often an upgraded model can be purchased for less than the cost of an original machine. Users find a second VCR convenient for taping from one set while watching a program on another. A second VCR is also indispensable for editing home videos and making copies of your masterpiece for relatives.

TROUBLESHOOTING AND REPAIR

Owning a VCR is fun, it's exciting, and it's entertaining. But when the machine begins to act strangely or the picture on your TV starts looking weird, the thrill quickly fades and a form of recreational panic sets in. This stress reaches a peak when the VCR owner sets the machine on the repair center counter and learns there is a $65 charge just to open the unit to find the problem.

To make matters worse, the problem can often be traced to operator error or to some simple action that would have restored normal operation. Nowhere is the adage "Read the manual if you want to save time and money" more important and repeatedly confirmed than with the ownership and use of a VCR. Yet the manual is usually the first thing that the owner misplaces or tosses away.

This book has been written to help all the users who either don't understand or can no longer find their operator's manual. It guides both novice and experienced VCR users through the magic of video recording. But more important, it gives valuable insights into those things that you can do yourself to restore correct operation or prevent failures from occurring.

OVERVIEW AND CHAPTER ORGANIZATION

This third edition has been revised to include troubleshooting and repair of current video cassette recorders. In addition to older models, it describes mechanical and electrical failures common to current machines, with techniques for determining the likely cause for malfunction. New circuit descriptions address 8mm, Hi 8, and Super VHS models.

This book is designed to provide information on VCR maintenance regardless of a person's repair experience and understanding of electronic and mechanical principles. The order in which the chapters appear segregates the book in two sections. Chapters 1 through 4 constitute the first section.

They provide a general introduction to VCR routine maintenance and repair for both the interested VCR owner and the novice repair person. No basic understanding of electronic theory or VCR operation is assumed. Basic troubleshooting skills are presented, followed by an introduction to VCR troubleshooting and routine maintenance. Additional useful information has been included in the appendix to supplement each chapter.

The first part concludes with a discussion of typical failures in Chapter 4. The troubleshooting flowcharts in this chapter are designed to aid novice and experienced technicians as well as VCR owners who want to diagnose and repair common failures in their own machines. At the end of each troubleshooting flowchart is a list of possible causes for a malfunction. Some VCR owners may elect to perform troubleshooting techniques learned in the first three chapters and directed by a flowchart.

When finding the failed component requires costly test equipment or a more experienced technician, the troubleshooting flowchart suggests taking your machine to a service center for final repair. Knowing when a repair requires service center action is helpful. By this point, you'll also know the type of repair necessary. This prevents unnecessary expense and faster corrective action.

The second part of this book consists of Chapters 5 through 7. These chapters are provided for those readers serious about understanding VCR operation and electronic troubleshooting. You are assumed to have knowledge of basic electronic theory and troubleshooting. This section guides you through the magnetic and electronic theory associated with VCR circuit operation and introduces you to the various tools and test equipment used by VCR service centers. This information is useful for service technicians new to the trade. The appendix contains additional information to aid the serious technician.

The following section describes the contents of each chapter and of the appendix and glossary at the end of this book.

CHAPTER SYNOPSIS

Chapter 1, "Introduction to VCR Maintenance," provides a quick overview of the history of the video machine and describes why VCR maintenance is so important. This preliminary chapter sets the stage for the rest of the book and shows both the novice and the experienced VCR user how each can gain maximum benefit from the book.

Chapter 2, "Basic Troubleshooting," covers the basic steps in analyzing problems and repairing your own VCR. It describes the troubleshooting process and gives insight into VCR components.

Chapter 3, "Routine Preventive Maintenance," describes in depth how you can prevent the breakdown of your machine and preserve precious video tapes.

In Chapter 4, "Specific Troubleshooting & Repair," you will find detailed flowcharts covering common VCR problems. These problems include a screen that won't display, noise lines dancing across the screen, a VCR that won't record, and a VCR that won't play back.

Chapter 5, "Magnetic Recording Theory," assists those who wish to better understand the principles behind video magnetic recording. This chapter covers video tapes, the process of magnetizing the particles on the tape, and the techniques for playing back and reproducing the video and audio information stored on the tape.

Chapter 6, "VCR Color Processing Theory," deals with the color information recorded and played back on a video tape recorder. It describes time-base error, TV broadcast signals, color-under in both record and playback, and symptoms of chroma record failure.

Chapter 7, "Luminance Operating Theory," explains how the FM luminance signal is recorded and played back, automatic gain control, clamps, filters, white and dark clip, limiters, compensation, de-emphasis, and current improvements to signal resolution.

In Chapter 8, "VCR Servo Control," you learn capstan and drum and direct drive servo operation in both record and playback. This includes signal phase, tracking, speed control, signal modulation, special effects and symptoms of servo failure.

Chapter 9, "VCR Audio Processing: Theory and Troubleshooting," is an in-depth look at audio recording and playback. This chapter covers linear track specifications, frequency modulation, companding, dropout, demodulation, de-multiplexing, signal processing, and pulse code modulation.

Chapter 10, "Miscellaneous VCR Circuits," is a catch-all chapter that incorporates additional subjects that most technicians want to know. Each section in itself is not sufficient to warrant its own chapter, so all are combined in this information-packed chapter. Specifically, Chapter 10 covers system control, tuners, RF modulators, demodulators, hook-up configurations, power supply theory and troubleshooting, and tuner, demodulator and RF modulator troubleshooting.

In Chapter 11, "Advanced Troubleshooting," you are introduced to the various types of tools and test equipment that become second nature to VCR repair technicians. After describing the "tools of the trade," the chapter explains how to use tools to find failures, how to work with solder and circuit board repair. It even deals with recommended equipment for the service technician's bench. Servicers of VCRs can use this chapter to familiarize new technicians with the equipment they will be using on the job.

Each chapter concludes with review questions to reinforce key points and aid in information retention. The questions are designed to enhance the use of this book in classroom situations. Answers to the chapter questions can be found in the appendix.

The appendix provides more meat-and-potatoes information related to the VCR. It covers selection criteria for VCRs and video tapes, editing and dubbing procedures, a suggested periodic maintenance schedule, and a preventive maintenance check-off chart. Finally, the appendix contains answers to the chapter review questions.

The glossary explains over 200 terms and expressions common to video cassette recorder servicing.

Included in this book are two comprehensive flowcharts covering the playback and record modes of the VCR. These charts show the big picture view of VCR operation.

VCR Troubleshooting & Repair concludes with a comprehensive index that references the subjects in this manual to a specific book page.

Gregory Capelo
Robert Brenner

Introduction to VCR Maintenance

Chapter 1 presents a brief history of the development of both the VCR and VTR. A basic overview of the internal operation of this magnificent machine is provided as an introduction to problem failure and analysis. The importance of learning to troubleshoot is also addressed.

"I still recall when I was a kid and my grandfather telling me that someday there would be an electronic device that could record pictures right off the television set. This machine would have the ability to play back both sound and picture immediately after it had been recorded. Then, during a later visit to his house, he proudly showed me his new Sony reel-to-reel half inch Portable Video Recorder Model AV3400. It was big and bulky, and was designed primarily for industrial use, but true to his word, there it was. It only recorded in black and white, and used large magnetic tape reels that held a half hour of program and had to be manually threaded in a fashion similar to an audio reel-to-reel tape recorder.

This marvel of that day included a black and white video camera and a special cable to record television programs. To record a TV program, he had to use a special monitor/receiver that could alter the TV signal so it could be recorded. The monitor/receiver connected to the recorder through an eight-pin connector and cable.

The possibilities for this new device excited me. My grandfather then told me that someday in the not-too-distant future there would be a device that could reproduce color video and be able to record a TV program using its own tuner without even having the TV on."—Capelo

For the past few years, futurists have been telling us that during our lifetime we will make three major purchases—a house, a car, and a computer. Lately, this expression has evolved into ". . . a house, a car, and AN ELECTRONIC SUPPORT AND ENTERTAINMENT CENTER."

The last purchase item includes the TV, the stereo, the computer, and the video cassette recorder.

As consumers, we owe much to the broadcast industry for its persistence in developing video tape recording. About twenty years before the introduction of the AV-3400, U.S. manufacturers such as Ampex and RCA busily worked on designs for video tape recorder (VTR) systems. Experiments were performed in 1951 to develop a rotating write-and-read head for recording and playback of video information.

In October 1952, the first barely visible video picture was demonstrated to an Ampex patent attorney and Ampex corporation founder Alexander Poniatoff. Also in that same year, Bing Crosby Enterprises demonstrated a broadcast VTR using fixed video heads and very high tape speeds. RCA introduced a longitudinal head VTR in 1953. It wasn't until January 2, 1955, that the first video tape recorder was produced using frequency modulation (FM) recording and special playback techniques. In that same year, a VTR using ½-inch tape was demonstrated in a closed circuit telecast from New York City to St. Paul, Minnesota. In 1958, sports history was made when the Los Angeles Rams football team used a video tape recorder to review the team's performance during a game's half time period.

In 1960, Toshiba introduced a process called *helical scan* that resulted in smaller and lighter broadcast tape recorders. Two years later, RCA announced that it had developed the first fully transistorized VTR. But it wasn't until 1965 that Sony introduced the first consumer video (CV) VTR. This reel-to-reel machine used ½-inch wide video tape on reels that were 7½ inches in diameter. It could record one hour of black and white video using a technique called *skip field* recording. By skipping every other field of video information, one hour of recording time could be placed on a 7½-inch reel of magnetic tape.

During playback, the recorded fields were played back twice reproducing an approximation of the original video image. But close observation revealed some loss of detail in the playback picture and a slight vertical jitter in the image.

Skip field technology did enable the recording of wider tracks of information on the tape which made it possible for the machine to tolerate some minor mechanical errors, with one major disadvantage. A normal TV cannot easily reproduce a playback picture without severe bending or flagging of vertical objects on the screen. Electronic synchronization of horizontal signals didn't work well when the same field was played twice, so special monitor/receivers were offered as optical accessories. These receivers had modified horizontal automatic frequency control circuits that produced a horizontally stable playback picture. These monitor/receivers don't have their own tuner for channel selection.

Other manufacturers followed shortly with the introduction of new ½-inch reel-to-reel recorders. However, there was no interchangeability from one manufacturer to another. As new models were introduced by the same manufacturer, users discovered that tapes recorded on one model could not be played back on a different machine.

In 1968, Japanese manufacturers organized the Electronic Industries Association of Japan (EIAJ) and established a standard for ½-inch VTRs. These standards included mechanical and electrical specifications. Several manufacturers in Japan such as JVC, Matsushita, and Sony produced VTRs meeting these new specifications. EIAJ developed specifications for color recording about a year later.

The ½-inch reel-to-reel VTRs became very popular in industry, but the consumer market had not yet discovered home video recording.

Under pressure by the EIAJ, skip field recording was discontinued, and a process called *full field recording* was developed.

THE FIRST CARTRIDGE MACHINE

In 1969, Ampex designed a cartridge machine based on the Japanese color recording format. It used a plastic cassette to hold the magnetic tape reels, but it was never commercially manufactured. Then, in 1971, Sony introduced the U-Matic ¾-inch tape format mounted in its own cartridge or cassette. U-Matic-capable video cassette recorders were not sold to the public until 1972, and many Japanese companies bought licenses from Sony to manufacture tape recorders based on this new format.

Although recent video cassette recorders (VCR) format developments threaten the future of machines based on the ¾-inch U-Matic format, these machines are still widely used in industry, education, and government. The large size and high cost of U-Matic recorders and U-Matic tape cartridges deterred consumers and kept this market extremely small.

In the early 1970s, an American company called Cartrivision, a division of Arco Industries, introduced a machine that used a cartridge containing two reels mounted one on top of the other. Using a modified form of skip field recording to conserve tape, the machine provided color picture information. This recording process skipped two fields for every single field it recorded (instead of skipping every other field like the Japanese machines did). Each recorded field was then played back three consecutive times. For example, field 1 was recorded, but fields 2 and 3 were skipped. Field 4 was recorded, the next two fields were skipped, and so on. During playback, field 1 would be played three times, then field 4 would be repeated three times.

Discontinuity in motion was visible in moving subjects, and this VCR required a modified TV to play back horizontally stable pictures. Cartrivision hoped to capture the consumer market by making available prerecorded movies and sports events. These machines were actually test-marketed in Chicago by Sears & Roebuck, but poor sales caused the project to be abandoned.

THE CONSUMER MARKET

The consumer market for VCRs really began in 1974 when Sony introduced the Betamax home VCR in Japan. Beta-format VCRs were not available in the United States until almost two years later.

In the meantime, JVC introduced another format called VHS for "Video Home System." The VHS format received wide publicity and was adopted by many other manufacturers. VCR sales in Japan grew rapidly and soon Japanese VCRs were being sold in the United States. Zenith, Sanyo, Sony, Sears, and Toshiba began selling Beta-format machines, while Quasar, Panasonic, RCA, and JVC sold VHS machines.

Today, VCRs from many other manufacturers and from countries other than Japan are reaching American markets. Korean companies, for example, are manufacturing VHS machines designed to the same specifications as the EIAJ units and there are estimated to be over 90 million VCRs in the United States today.

The choice of format is between Beta, VHS, and 8mm. VHS is by far the most dominant format today. The Beta format is still available but only through Sony. Betamax enthusiasts prefer the video reproduction quality of the Beta format over that provided by the more popular VHS. Although the 8mm format was introduced in 1985, it has not been widely accepted as a home taping format. Most of its popularity is found in the camcorder market.

Current Beta-format machines have three recording and playback speeds: Beta I, Beta II, and Beta III. The early Beta I speed produced excellent video but limited recording time to about an hour and thirty minutes. Although Beta I was superseded by Beta II and III, it was recently reintroduced for the consumer who is especially conscious of video quality. Many of Sony's VCR products can play all three speeds.

VHS machines are by far the most prevalent and have three speeds: SP (standard play), LP (long play), and EP (extended play) or SLP (super long play). The length of recording and playback time is associated with the speed in which the tape passes through the machine. Faster speeds produce higher quality video reproductions but short duration recording capability. Slower speeds enable long duration recording (such as an entire baseball game) but experience a fall-off in picture and normal (non-Hi-Fi audio) track sound quality. Everything's a trade-off.

Table 1.1 compares the current video tape formats with recording and playback times.

The possibilities and benefits associated with VCR use are endless. Favorite television programs can be viewed at convenient times. Prerecorded classics can be rented or purchased and played at home. Special entertainment and educational programs can be recorded and viewed repeatedly. With the use of a video camera, family events can be recorded and a visual history produced for future enjoyment. Many owners of home movie films are now having their 8mm and 16mm films reproduced on video tape for enjoyment on a VCR.

Table 1.1.
Comparison of Video Tapes and Run Time

Beta (L750 Tape)	
Beta I	90 minutes
Beta II	180 minutes
Beta III	270 minutes

VHS (T-120 tape)	
SP (standard play)	120 minutes
LP (long play)	240 minutes
EP (extended play)	360 minutes
SLP (super long play)	360 minutes

8mm (MP-120 tape)	
SP (standard play)	120 minutes
EP (extended play)	240 minutes

UNDERSTANDING THE VCR

Connecting a VCR to a television in basic configurations is relatively easy. The hook-up is similar to connecting your TV to the antenna cable system in your home or business. Record and playback operations are much like those of an audio cassette recorder in a stereo.

By understanding the electronic and mechanical theory and operation of a VCR, you will develop a genuine respect for these marvelous machines and a sensitivity for their proper care. You will also understand the machine's limitations.

The old expression "if you've seen one, you've seen them all" has relevance in the world of VCRs. In a sense the basic principles of operation have changed little since their invention; only the techniques or methods change. The most basic VCR currently available uses the same principles for magnetic recording as the stereo hi-fi VCR priced three times higher. No matter what type or size box is used to package the recorder, the basic workings inside are common.

There must be an input section, a recorder section, a storage medium, a reader section, and an output section as shown in Fig. 1.1. The input includes camera, microphone, another VCR, and cable and television signal paths. The recorder section converts all the input signals into a form suitable for saving, then writes or records that information on some form of storage medium.

The storage medium is the video tape itself. It's housed in a plastic cassette and comes in several sizes (formats). The cassettes are kept in special dust-preventing sleeves.

The reader section brings stored information, such as color television programs and sound signals, off the tape and converts it into signals suitable for the output unit. This output unit is actually the cable from your VCR to the television or to another VCR.

In all cases, the box, the television, and the "bells and whistles" (advanced features) can change, but the basic operation remains

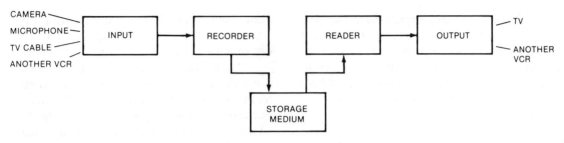

Fig. 1.1. *The basic components in every VCR system.*

the same. A $20 audio recorder uses the same principles for record and playback as the video recorder near your television. The input signal from the microphone or broadcast station is passed from the input to the record section. Here the combined effort of electronic circuitry and mechanical parts positions the magnetic tape and then writes information magnetically on the top layer of the tape.

The same thing occurs in your VCR, but the electronic and mechanical components are far more complex and operating requirements (specifications) far more stringent. Your VCR must be capable of handling both video and audio signals. A great deal of information must be stored on the tape and tolerances are rigidly controlled.

The magnetic tape in the plastic video cassette housing is similar to the brown magnetic tape in an audio cassette. It's constructed from a long strip of plastic or mylar wound onto a reel. On one side of the tape, the manufacturer glues a material that can easily be magnetized, usually a form of iron oxide mixed in an adhesive binder.

In the recorder is a cylinder called a *head cylinder* or *head drum*, around which the tape is pulled. Tiny electromagnets called *heads* are mounted in the surface of the drum. Electric signals from the video and audio part of the recorder section are passed through coils of wire in these heads.

The heads produce a tiny magnetic field that changes in direct proportion with the changing audio or video signals. The heads make contact with the oxide surface of the tape

passing around the drum, thus causing the head magnetic field to be felt by the tape. The oxide on the tape becomes magnetized in direct proportion to the signal in the head. Since several heads are on the drum, several electric signals are converted to magnetic fields applied to the tape, thus transferring the audio or video information onto the strip of tape where it's stored for later retrieval (playback).

When you insert a tape cassette into your VCR and begin to play back the stored information, complicated gears, arms, and other mechanisms pull the tape into the machine and place the tape against read heads located on the drum cylinder.

These heads sense the tiny energy fields stored in the magnetic layer of the tape and convert the changing magnetic fields (patterns of magnetic particles) into small electric signals. These signals are amplified to make them strong enough to send out on the coaxial cable which runs from your VCR to a nearby television set. The television set converts these electronic video and audio signals into light images on the screen and sound from a vibrating speaker system.

It's that simple . . . in basic theory, at least. But to accomplish these feats, very complex circuitry and arrangement of mechanical devices are required. Tolerances are extremely tight, and signal levels are strictly controlled to enable excellent quality video from the surface of a shiny brown tape. Ensuring that these parts combine to reproduce brilliant video and ear-pleasing audio is the subject of the following chapters.

The VCR is a complex, precision machine. It's expected to operate correctly and without fail. Care and proper operation are mostly ignored while everything is functioning properly. But when the machine begins to fail from months of neglect and misuse, you really learn to respect it.

WHY TROUBLESHOOT

One visit to a VCR repair center can leave a permanent impression on your mind and pocketbook. Most repair center costs average $125. In addition to charging for the replacement parts, many service centers charge a flat fee for labor, which ranges between $75 and $140. The average labor charge is $50 per hour; typically, two to three hours are spent troubleshooting and repairing a machine, so the repair costs quickly add up.

Some repair centers fix only the immediate problem. Others also replace worn belts and do alignment checks to extend the operational life of the machine.

The value of a book teaching troubleshooting and repair for the novice is more evident when you understand the causes for typical VCR failures—dirt buildup and the wearout of belts and pulleys.

Users don't intentionally damage a VCR, but lack of knowledge and proper preventive maintenance can be costly. This book provides understanding and appreciation for the VCR so you can give it the tender loving care it needs. It will also help you identify problems quickly, fix failures, and get on with the show. VCRs can serve you well for many months; this guide will show you how and why.

CHAPTER REVIEW QUESTIONS

1. In what year was the first consumer video tape recorder introduced?

2. What Japanese organization addressed the concern about tape interchangeability between manufacturers?

3. What are five disadvantages of the early skip field recorders?

4. What does the acronym VHS mean?

Chapter 2

Basic Troubleshooting

Like automobiles, VCRs break down after lots of use. Some break down sooner than others. Finding the problem can be easy or difficult, depending on your understanding of how to analyze a problem, identify the failed part, and step toward the correct repair. This chapter will show you how to find problems in your VCR in the shortest amount of time.

INTRODUCTION TO TROUBLESHOOTING

Imagine for a moment that you're in the midst of watching a video tape when suddenly the TV screen goes blank and the VCR stops working. What do you do? What failed?

This chapter is devoted to something we often wish we could pass off or ignore—trouble. Trouble is like a flat tire: no one wants it, but when it happens we want to fix it quickly and get the experience behind us. Knowledge and action overcome trouble.

You know from reading the owner's manual that your VCR is an electromechanical machine; it operates using analog and digital principles to record and play back audio and video signals.

A VCR generally doesn't break down slowly, with graceful degradation (at least you can't see it). If it fails, it's usually with a hard, consistent failure. In addition, the electronic devices that make up your VCR function within strict rules of logic. The most effective way to respond to a failure in these devices is to think the problem through just as the machine operates, logically. Understand what should happen and compare the "shoulds," one by one, with what is really happening.

A deductive technique called troubleshooting is particularly appropriate for solving VCR failure problems. Troubleshooting could be really frustrating if you were left to struggle through the process by yourself. This book provides you with the techniques for quick and easy troubleshooting and repair.

Before examining troubleshooting steps, let's review some important safety precautions.

SAFETY PRECAUTIONS DURING TROUBLESHOOTING AND REPAIR

As you would with any electrical device, you must observe certain precautions to prevent injury to yourself or damage to the VCR. Observing these precautions can save you time, money, frustration, and possibly injury.

1. Stay out of the power supply.

2. Observe the manufacturer's service safety precautions printed in the service manual.

3. Turn the power off and unplug the VCR whenever possible.

4. Ground yourself against static electricity.

5. Handle video tapes carefully.

6. Don't cycle the power quickly.

7. Keep liquids away from the VCR.

8. Handle components with care.

Don't troubleshoot the VCR power supply. These circuits convert the 115-volt line power in your home or office into the 5–10 volts used by the circuit boards. That 115 volts of electricity can be painful! It could be deadly! Some VCR power supplies are connected to one side of the AC power line for their ground reference. If you touch this type of power supply without the proper isolation, you could get shocked. Limit power supply repairs to fuse replacement, and only replace fuses with the AC cord disconnected. Be sure the replacement fuse has the same current rating as the old one.

Familiarize yourself with the safety precautions outlined in the service manual before opening the VCR.

Always turn the power off, and then pull the power cord out before touching anything inside the VCR. Touch a grounded metal object like a desk lamp to discharge any static electricity present in your body. Many failures are caused by static electricity on people who don't follow this rule.

Handle your video tapes carefully. Don't lay tapes on a very dusty, dirty surface. Keep cigarette ash away from the tapes and VCR. Don't touch the tape surface. Don't set your tapes on, or in front of, a TV or color monitor. Magnetic energy generated by the TV will weaken and possibly even erase the magnetic signals on the tape.

Don't cycle the power on and off quickly. Wait 7 to 10 seconds for the capacitors in the power supply to discharge fully and the circuits to return to a stable (quiescent) condition.

Keep liquids away from the unit. It's amazing how sticky soda pop (soft drink) becomes after frying components all over the inside of the unit.

Handle components with care. Don't let integrated circuits (IC chips) lie around. The pins will get bent. Watch out for static electricity—IC chips may need special handling.

STEPS TO SUCCESSFUL TROUBLESHOOTING

Effective and efficient troubleshooting requires gathering clues and applying deductive reasoning to isolate the problem. Once you know the cause of the problem, you can analyze, test, and substitute (good components for suspected bad components) to find the particular part that has failed.

The use of special test equipment such as an NTSC pattern generator and an oscilloscope can speed the analysis, but for many failures, good old brain power can suffice. Once you determine whether the problem is electronic or mechanical, deductive analysis changes to intelligent trial-and-error replacement. Reducing the number of suspected components to just a few and using intelligent substitution is the fastest way to identify the faulty device.

In general, follow these steps when your VCR fails:

1. Don't panic.

2. Obtain the service manual for the VCR model.

3. Observe the conditions in which the symptoms appear.

4. Make note of as many of the symptoms as possible.

5. Use your senses to locate the source of the problem.

6. Retry.

7. Document your findings and test results.

8. Assume one problem. (When multiple symptoms are present, troubleshoot the easiest one first.)

9. Diagnose to a section (fault identification).

10. Consult the troubleshooting charts (in Chapter 4).

11. Localize to a stage (fault localization).

12. Isolate to a failed part (fault isolation).

13. Repair.

14. Test and verify.

The following pages discuss the steps to troubleshooting success.

When something goes wrong, the first step is to determine whether the trouble results from a failure, a loose connection, or human error. Once you're sure a failure has occurred, the next step is to determine which portion of the VCR is not operating—mechanical or electrical.

Then, step by step, partition each section into stages and try to track the trouble to a single component. If one function isn't working, for example, the problem could be in the acti-

vation button itself, the connecting wires, or the electronic circuitry of the VCR.

Next, you need to understand the mechanical actions of the machine and the interaction of the mechanical and electronic parts of the unit.

COMPONENT RECOGNITION

What's a VCR made of? The housing or case is made of high-strength, molded plastic or metal, depending on your unit. These cases are not likely to fail under normal use.

MAKE SURE THE POWER IS OFF AND THE UNIT IS UNPLUGGED, then open your VCR. Use the disassembly instructions found in the service literature.

There are a number of subassemblies inside your VCR including the VCR (circuit) boards, levers, gears, tape drive motors, tuner, and power supply. A high percentage of VCR problems results from mechanical failures, so unless you are experienced at replacement of electronic components, the introduction to IC chips, capacitors, resistors, diodes, and transistors will serve only to familiarize you with what is on your VCR circuit boards. These devices are soldered onto the board and should be replaced only by those experienced in repair. Most home users will let a technician replace soldered components.

To help you understand more about VCR circuit boards and their components, refer to Fig. 2.1. Circuit boards, like the one shown in Fig. 2.1, are made of fiberglass and have many

Fig. 2.1. A typical VCR circuit board.

colorful items mounted on them. These items include sockets, connectors, wire traces embedded into the boards, integrated circuits (also known as ICs or chips), resistors, capacitors, and transistors, as well as variable controls which align the unit.

Figure 2.2 compares the size of some of the devices you will find mounted on a circuit board relative to a one-cent coin.

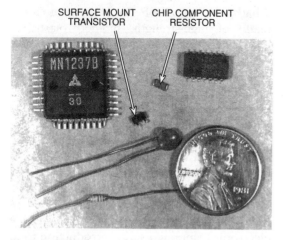

Fig. 2.2. Typical circuit board devices.

Chips

Those black-case, centipede-looking things are the chips or ICs. They serve the function of hundreds of transistors (or vacuum tubes, the predecessors of transistors), and cause the VCR to work logically. Much of the unit works on the principles of digital logic. The system control works like a Von Neumann machine, a computer-like device that works in binary digits (bits). All conditions are either ON (logic 1) or OFF (logic 0), and all operations occur in sequence. Dr. John Von Neumann first described his idea of a binary computer at a conference at the Moore School of Electrical Engineering in 1945.

The analog part of the machine operates more like a radio or TV, receiving signals and amplifying and processing them for recording and playing back as needed. ICs are used in this part of the VCR too. They act to amplify, shape, and switch the signals in the unit.

There are many configurations of IC chips on your circuit board: 8-pin, 14-pin, 16-pin, 24-pin, 40-pin, and 64-pin. Many of the ICs in a VCR are custom chips that are not inserted in sockets. So repair isn't quick and easy like it was in the days of tubes. In fact, special soldering skills are needed.

Notice that each IC chip has a notch or groove at one end, as shown in Fig. 2.3. This notch marks the end of the chip where pin 1 can be found. Pin 1 is to the left of the groove as you look down at the top of the chip with the groove pointed away from you. The pins are numbered counterclockwise starting from pin 1, so that the highest numbered pin is directly across from pin 1. As you'll learn later, in chip replacement, you must insert the new chip into the holes with pin 1 in the right place.

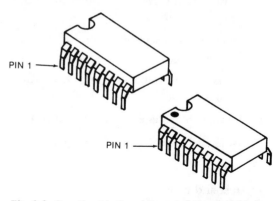

Fig. 2.3. Locating pin 1 on integrated circuits (chips).

ICs have special markings that tell a lot about what's inside. Look at the printing on the top of the chips on your VCR's circuit boards. First, you'll notice that many different companies make chips, and that most of these companies are outside the United States—Korea, Japan, and Taiwan, for example. Most companies place their logo on the chip.

You'll also notice letter-number combinations on the chips. Some ICs have two sets of letter-numbers. One set identifies the type of device, and the other set tells when the chip was made.

The prefix ("MN" in MN1237B) usually identifies the manufacturer, although sometimes it identifies the device family associated with a manufacturer. The prefix is sometimes omitted.

The core number is three or more digits long. It indicates the basic logic family or the specific application in the circuit. Many of the IC chips in the VCR are specifically designed for applications only in that model.

Resistors

Resistors are used to restrict or limit the flow of electrical current through the board's circuitry. They are commonly constructed using a carbon or metal film to resist the flow of electrons. Many types and sizes of resistors can be found in a VCR. The larger they are, the more heat (wattage) they dissipate.

Study the two resistors shown in Fig. 2.2. The surface mount resistor is about an eighth of an inch long. These resistors are typically rated at one-tenth of a watt. Surface mount resistors have leadless solder terminals on each end so they can be connected directly to the surface of a printed circuit board without using wire leads.

A surface mount resistor can be distinguished from other surface mount devices by its black epoxy center section and silver end caps. Usually a three-digit number is printed on the black section to indicate the part's resistance value.

The axial lead resistor in Fig. 2.2 is rated at one-quarter watt. It is mounted on a VCR circuit board using the wire leads that extend from each end. The leads are pushed through tiny holes in the circuit board and then soldered to a conductive pad surrounding the hole. The pad around the lead hole connects to copper traces on the circuit board.

Figure 2.4 shows axial lead resistors in their mounted position. Notice how the leads are bent so that they can push through small holes in the board. The resistors on this board are labeled with an "R" prefix and a number.

Fig. 2.4. Close-up of a typical VCR circuit board.

Figure 2.5 is a drawing of three common types of resistors with pins that are soldered into specific holes of VCR circuit boards.

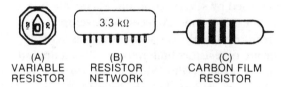

(A)	(B)	(C)
VARIABLE	RESISTOR	CARBON FILM
RESISTOR	NETWORK	RESISTOR

Fig. 2.5. Three types of resistors.

Notice the bands on the axial lead resistors in Figs. 2.2 and 2.5C. The value of resistance is given in ohms and can be determined by comparing the color bands with the colors in Table 2.1.

Table 2.1. Color Code Chart

Color	Digit	Multiplier
Black	0	1
Brown	1	10
Red	2	100
Orange	3	1000
Yellow	4	10,000
Green	5	100,000
Blue	6	1,000,000
Violet	7	10,000,000
Gray	8	100,000,000
White	9	1,000,000,000
Gold		±5% tolerance
Silver		±10% tolerance
None		±20% tolerance

For example, imagine that the color stripes around the resistor are brown-gray-orange-gold, in that order. The first two bands describe the primary number (18). The third band represents the multiplier, the numbers of zeroes to add to the primary number—in this case three. The last band is the tolerance value, or how close to the indicated value the actual value must be. As Table 2.1 shows, the brown band stands for 1, the gray for 8, and the orange for three zeroes after 18. By using Table 2.1, this resistor's value can be found to be 18,000 ohms (18 kΩ). The gold band represents a 5 percent tolerance value which means the actual resistance value can be plus or minus 5 percent away from the 18 kΩ designation (from 17,100 to 18,900 ohms).

Figure 2.5B shows a single-in-line package (SIP) network resistor. Resistor networks are used by several components on the board at the same time. A recently developed electronic device, the resistor network is actually a group of resistors built into an SIP or a dual-in-line package (DIP). Several SIP resistor networks are mounted on the boards.

The resistance designation of network resistors is printed on the side of each package. Assume that one of these devices was marked 8X-1 2 0 2. The 202 is a key to the resistance value in ohms. The first two numbers (2 and 0) indicate the significant figures. The third number (2) tells how many zeroes to add to the significant figures. Thus, 202 means 20 plus 2 zeroes or 2000 ohms (2 kΩ).

Some network resistors are marked directly with the value of resistance. One marking for example could be 898-1-R8.2K. The 8.2K labels this resistor package as a network of 8.2 kΩ resistors.

The device drawn in Fig. 2.5A is a variable resistor. These are used to align the electronics in the VCR. They are used throughout the unit to adjust the various circuits for proper operation. The volume control on the outside of a portable radio is one example of a variable resistor.

Capacitors

In addition to chips and resistors, circuit boards have a number of capacitors mounted on them. A capacitor builds up and holds an electric charge for a period of time determined by the design of the circuit. They are used to couple changing (AC) signals between circuit sections, to smooth ripples in waveforms, and to store potential for future use by other electronic components in the circuit.

Capacitors come in many varieties. Some of these include (1) electrolytic, (2) tantalum, (3) ceramic film, (4) mylar, and (5) variable. These five types of capacitors can be found in the VCR circuit boards all over the unit. Figure 2.4 also shows several different types of capacitors mounted on the board.

Surface mount capacitors are popular devices in VCRs. The smallest surface mount capacitors are made of ceramic. They are identical in size and shape to the surface mount resistors previously described with one exception—the center section is brown. The end caps are silver like the end caps on the surface mount resistors.

Figure 2.6 shows several different types of surface mount capacitors that you'll find on VCR circuit boards. The electrolytic surface mount capacitor is shaped like a metallic canister. Notice the size difference between the ceramic, electrolytic, and tantalum surface mount capacitors.

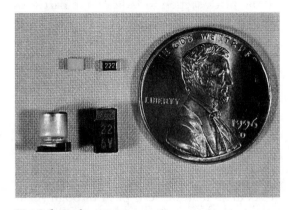

***Fig. 2.6.** Surface mount capacitors.*

Capacitors are measured in fractions of farads. You'll see values listed in F for microfarad and pF for picofarad. Micro means "to the sixth decimal place" or 0.000001 (one millionth) and pico means "to the twelfth decimal place" or 0.000000000001 (one trillionth). Thus .022 microfarad means 0.000000022 farad and 47 picofarad means 47 trillionths of a farad.

Capacitor value identification is challenging because most companies like to use their own identification standards. The designation "10 F" and "16 V" on the side of an electrolytic capacitor means that this is a 109-microfarad capacitor rated at 16 volts.

A ceramic film capacitor with the value ".047 ± 10%" stamped on it is a .047 microfarad capacitor. Other ceramic film capacitors are color-banded and can be decoded using the color chart given in Table 2.2. They have two additional color bands which refer to the capacitor's temperature dependence and tolerance. Capacitance varies with temperature (and to some extent with capacitor size and applied voltage). The temperature dependence of the capacitance value is given at the coefficient parts-per-million-per-degree-centigrade (PPM/C). Tolerance defines how much variation is permissible between the actual value of the capacitance and the nominal capacitance value.

Table 2.2. Capacitor Color Codes Indicating Capacitance in Picofarads

Color	Digits	Multiplier	Tolerance
Black	0	1	
Brown	1	10	
Red	2	100	
Orange	3	1000	
Yellow	4	10,000	
Green	5	100,000	
Blue	6	1,000,000	
Violet	7		
Gray	8		
White	9		±10%
Gold			±5%
Silver			±10%

All capacitors with wire leads have their capacitance value printed on their body. The tantalum and electrolytic surface mount capacitors also have values printed on them, but the ceramic surface mount capacitors have no such printed markings. Each technician must identify the value of these unmarked components by referring to the schematic for the model of VCR being analyzed. When replacing these small devices, the value of the replacement part is usually printed only on the packaging that contains the new part.

Both the electrolytic and tantalum capacitors have polarity markings. These markings must be followed to insert a replacement part in the circuit in the same way that the old part was installed. The polarity in the circuit must be maintained. Electrolytic capacitors often have an arrow printed on the negative polarity end of the case. Tantalum capacitors frequently have a plus mark printed on the positive end of the case.

Variable capacitors are not common in a VCR. They are used most often in the adjustment of high-frequency oscillations for the microprocessor timing clock pulse or in the chroma 3.58 MHz oscillators. A capacitor can be used to adjust the color signal. Variable capacitors are usually measured in the picofarads. The larger leaded-variable capacitors are being replaced by surface mount variable capacitors in new VCR models.

A mylar capacitor usually has a colorful, shiny, hard shell. Figure 2.5 shows a mylar capacitor. It has 472j printed on it and underneath that, you can see 100e. This is a .0047 microfarad capacitor at 100 volts. The 47 represents the number of microfarads and the 2 represents the number of zeroes that precede the first two numbers. The "j" label represents the tolerance of the part capacitance rating.

Inductors

Figure 2.7 shows two types of inductors found in a VCR. Inductance values are measured in microhenries. While we must determine the

INDUCTOR   INDUCTOR

(A) COLOR BANDED   (B) UNMARKED

Fig. 2.7. Two types of inductors.

value of unmarked inductors by referring to the schematic, some inductor values can be determined by reading the colors on the device and comparing them with the color code chart in Table 2.1.

Diodes

Diodes are tiny, often glass devices shaped like resistors. They are marked with printing on the side, although you'll often have a tough time reading it. There are several diodes on the circuit boards of a VCR. To determine if you are looking at a diode or at something else, look for glass construction and a label on the side. A "1Sxxxx" label denotes a diode made in Japan. U.S. manufacturers identify diodes with a "1Nxxxx" label. Look at D509 located at the bottom right corner of the VCR board in Fig. 2.4. This is typical of many of the diodes found in your unit.

Surface mount diodes look identical to surface mount transistors. These diodes even have three leads like the transistors. The only ways to distinguish a surface mount diode from a surface mount transistor is to look at the identifying letters and numbers printed on the device.

Transistors

The small rectangular devices on the board in Fig. 2.4 are transistors. There's one in the lower left corner of the picture. It has three leads and is identified by a "Q" prefix followed by numbers. Transistors can be found in different sizes and shapes in the VCR. Figure 2.2 includes a

surface mount transistor. The key to recognizing a non-surface-mount transistor is its designator—2SAxxx, 2SBxxx, 25Cxxx, and 2SDxxx. Surface mount transistors are identified using a code. For example, a surface mount transistor with the marking "1ZR" is actually a 2SD1030R made by Matsushita. Manufacturers supply cross-reference lists to correctly identify surface mount transistors. Table 2.3 lists some common markings with their cross reference for some Japanese transistors.

Table 2.3. Surface Mount Transistor Markings Cross-Referenced to Specific Transistor Types

Marking	Transistor	Manufacturer
A	2SB709	Matsushita
B	2SB709A	Matsushita
C	2SB710	Matsushita
FB	2SC1009AFA3	Mitsubishi
FC	2SC1009AFA4	Mitsubishi
LY	2SC2710Y	Toshiba
SO	2SA1162-0	Mitsubishi
ZK	2SB1114	NEC
1ZR	2SD1030R	Matsushita
2N	2SD1478	Matsushita

COMPONENT FAILURES

While the use of troubleshooting equipment is essential to analyze and isolate different VCR problems, many failures can be found without expensive equipment. In fact, troubleshooting and repair can be relatively simple if you understand how electronic components fail. Failures generally occur in the circuits that are used or stressed the most. These include the motor drive transistors and ICs, power supply transistors, and mechanical switches. The microprocessors are highly reliable devices and seldom fail. Most electronic failures involve other components such as the transistors and diodes.

Often a VCR will exhibit strange symptoms when it starts to fail. The system control electronics will cause the unit to do things it never has done before, such as rewinding in the middle of a recording. The system control microprocessor is the first component suspected by inexperienced people who attempt repair. After spending hours trying to prove that the system control microprocessor is the failed part, a novice discovers that the end-of-tape sensor has failed and the system control microprocessor is doing just what it was designed to do at the end of tape movement—rewind.

Most manufacturers claim there are far more microprocessors replaced than is necessary. The inexperienced technician often starts by replacing the microprocessor to correct a system control problem. After waiting for the part to arrive from the manufacturer, the technician replaces the IC only to discover that the problem still exists. Usually the culprit is a device which passes a status report to the microprocessor from another part of the machine. The only problem is that now someone has to pay for a microprocessor IC that costs $20 to $50. The manufacturer won't take it back because it's been installed. Guess who foots the bill for poor troubleshooting and repair.

Microprocessors do fail, but not often. The moral is to look for the cause elsewhere first and to check all signals, voltages, and waveforms before blaming the failure on the main system control IC.

Transistors and diodes can fail by internally opening (which causes an open or break in the circuitry), or having a short in the output. Either situation causes total loss of signal.

Failures also result from *leaking*. A diode, for example, is designed to conduct current in one direction only. When it becomes leaky, it will allow some current to flow in the wrong direction. Transistors also leak under a partial failure operation. If either part leaks too much we say it has *shorted*.

Capacitors fail when they short internally or when one of the leads disconnects, causing an open. Again, there is a loss of signal.

Resistors can absorb too much current and actually bake in the circuit. The result is usually an open circuit with shorting during

the "meltdown." All of the devices mentioned so far are solid state. They are constructed of materials (metals, plastics, oxide, etc.) that change as the components age or are subjected to severe temperatures or high voltages. Such a change can cause the device and the circuit or VCR to behave strangely.

Fortunately, VCR circuit boards are not normally subjected to high voltages. But they can get pretty warm if the natural air flow is restricted, and this affects the operation of the components. When we use the VCR, we place the circuitry under stress. It heats up when we turn it on and use it, cools down when we turn the machine off, and reheats when the machine is turned on again. This hot–cold–hot effect cause thermal stress. The result can be a break in the connection of a wire leading from inside the chip to a pin. This produces an *open circuit*, which requires chip replacement. Thermal stress can also create breaks in board traces and solder connections.

Even if there is no break in the IC or lead connection, after exposure to high voltages or temperatures the operating characteristics of a device can change. A chip may work intermittently or simply refuse to work at all. Theoretically, a wearout failure like this won't occur until after several hundred years of use. But we shorten the life span of the components by placing them in high-temperature, high-voltage, or power-cycling environments that cause them to fail sooner.

Other electronic problems occur outside the chip—between the chip leads and the support structure pins which connect the device to the rest of the VCR. Such failures include inputs or outputs shorted to ground, pins shorted to the +5 volt supply, pins shorted together, open pins, and connectors with intermittent defects. Most commonly, trouble results from opens or shorts to ground. Because they produce more heat, chips, transistors, and diodes fail far more often than capacitors, resistors, and inductors.

Chapter 3, "Routine Preventive Maintenance," tells more about the effects of heat. If you keep the VCR cool and clean, it should work well for many years.

Some electronic components fail more frequently than others. It might be helpful to know which parts to suspect first. You've already learned that devices that get warm, such as transistors, chips, and diodes experience thermal wear. These parts are high on the list of suspects. Also you should consider fuses, electronic sensors, and mechanical switches.

Defective capacitors would be next in the order of suspected components after the semiconductors (transistors, diodes, and ICs). Last on the list of most likely suspects are inductors and resistors.

Electronic failures in a VCR most commonly occur in one of these parts:

- Mechanical switches and contacts.
- Fuses.
- Diodes, transistors, and ICs.
- Sensors, including hall effect ICs, light-emitting diodes (LEDs), and photo sensors.

Second, you should consider:

- Capacitors.

Finally, you should check:

- Resistors.
- Inductors.

This grouping is by no means perfect, but it will provide some order of failure and tell you which parts to suspect first. The items within each grouping are not necessarily in order. Just because switches are listed first doesn't mean that they fail more often than ICs, but each of the items in the first list should be suspected before those in the following groupings.

Other Electronic Failures

Electronic failures during repair can be caused by careless mistakes at home, or in the shop by overzealous or undertrained technicians. Here are some possible repair-generated failures.

Devices "blown-up" in handling

This problem occurs when someone picks up static-sensitive ICs like the microprocessor chip without first grounding any static electricity that a person might be carrying.

Bent or broken pins

Watch the way you put those ICs in. You can only straighten a pin so many times before it breaks off completely (Fig. 2.8A).

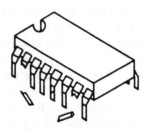

(A) BENT AND BROKEN PINS   (B) SOLDER SPLASH

Fig. 2.8. Failures can be caused by overzealous or undertrained repair people.

Solder "splashes"

These are caused by dropping tiny balls of solder from the end of the soldering pencil right on top of the board, shorting out some of the circuit as in Fig. 2.8B.

Liquid fry

This occurs when someone holds or sets a liquid on top of or too close to the VCR and then accidentally spills the liquid into the top of the machine while the VCR is running. It's a real mess to clean up, and you usually have to replace lots of components.

Component failure by asphyxiation

Blocking the VCR vent openings or stuffing your VCR into a tight cabinet where it has no room to breathe "kills" components.

The connector that doesn't

This can be an improper hook-up or poor quality cables. Plugging cables into the wrong pin could blow some components. If cable connectors are badly corroded or poorly soldered, no signal can get through the cable.

The message is clear. If a mistake can be made, someone has probably made it.

How VCR Mechanisms Fail

Seventy-five percent of all VCR failures are a result of a mechanical defect or mechanical wear. Motors may not rotate at an even speed, mechanical levers may lose traction. Bearings wear and may become noisy or cause uneven rotation. Friction generated by pulling video tape coated with oxide particles along the tape path can cause guides and heads to wear. In time, the video tape grinds its own track into the tape path causing tape alignment error.

Mechanical failures can appear to be a defect in the electronics section. A worn tape path, for example, can cause the audio to record and play back at very low volume levels. When the problem is diagnosed, the audio amplifier may seem defective. But after replacing several components in the audio circuitry, you may discover that the signal coming from the audio head is insufficient because the tape is not traveling across the head as it should be.

Other mechanical defects may cause the system control to act strangely. Bent levers which do not fully engage sensing switches can cause the microprocessor to misinterpret signals and operate strangely. The microprocessor depends upon information from many switches and sensors to keep the system control electronics informed of system status. If this information is missing from any of its inputs, the microprocessor may command the machine to shut down.

A defective capstan motor will cause the tape to be pulled through the machine at an uneven speed. When listening to the reproduced sound from the normal audio track, we hear this as *wow* and *flutter*—a wavering in the sound produced by the VCR. When listening to the Hi-Fi audio playback, we won't normally hear wow and flutter since this form of sound reproduction is somewhat immune to this problem. When wow and flutter are present in the normal audio reproduction, music seems to warble in pitch and in severe cases even normal voice playback seems to flutter. The *servo circuit* attempts to correct this problem by making rapid voltage changes to compensate

for motor imperfections, but the correction lags behind the errors generated by a defective motor. If you examine the servo circuitry with test instruments, you may suspect a defective electronic component when the problem is a mechanical defect with the capstan motor.

Isolating a defect within a VCR is easier if you can differentiate between mechanical and electronic failures. Mechanical devices fail much more frequently than electronic components, so most VCR failures are due to mechanical defects or wear.

The mechanical aspect of a VCR is quite reliable if you operate the unit carefully and conduct periodic routine maintenance. But sometimes you forget. You operate the VCR in an area where someone taps cigarette ashes into a tray on top of the unit. Or you slam a cassette into the VCR's front load cassette tray or force the tape in at an angle. Rough treatment of a sensitive video machine causes rough display and operational performance.

One day you are in the middle of enjoying a rented movie and the machine quits. Not only does it stop playing the tape, but it refuses to release that expensive movie.

HOW TO LOCALIZE FAILURES

Isolation of a VCR failure to a particular area involves localizing the defect to a mechanical or electronic malfunction. If the VCR will accept a tape, record, play back, fast forward, and rewind, your job will be easier.

There are two ways to localize failures; the technical (electronic) approach and the mechanical approach.

Technical (Electronic) Approach

This method uses several troubleshooting tools to measure voltage and waveform levels in the circuitry of the VCR. The tools include an audio test signal generator, an NTSC video waveform generator, an oscilloscope, a multimeter, a monitor/receiver, and the correct service literature.

The technical approach requires a knowledge of electronics, a basic understanding of how the unit functions, and special test equipment. It is usually a last resort, so the technical approach has been reserved for Chapter 11, "Advanced Troubleshooting."

Mechanical Approach

An important part of mechanical troubleshooting involves careful observation of unit operation. Check for broken belts, bent levers, and foreign objects lodged in the mechanism. If the video cassette will not insert properly into the tray, carefully check that area for misaligned gears or bent levers. Figure 2.9 shows a typical cassette tray. Check for bent levers, misaligned gears, or switches that have failed.

Fig. 2.10. A typical VCR mechanical assembly.

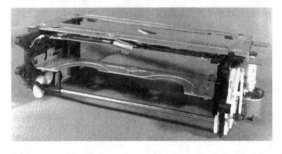

Fig. 2.9. Typical cassette tray assembly.

It's possible that an occasional plastic toy can mysteriously sneak into the tape load mechanism and wreak havoc on the mechanical operation. Figure 2.10 is a typical mechanical chassis assembly. Check for foreign objects that could obstruct tape or mechanical part movement or break belts or other components.

If the VCR mechanism does not operate normally, a careful check of both the top and bottom portions of the mechanical chassis can provide helpful clues. Be sure to disconnect the power. Then expose the top and bottom chassis by removing the cabinet case and swinging out the circuit boards. If all appears to be normal, perform an operational check next.

Inspection of the mechanical components includes observing the operation of the machine during load, eject, play, fast forward, and rewind operations. **Be especially careful because the AC power is applied to the unit during these tests.**

Use a video tape (with little or no value) to test for proper mechanical operation. Valuable tapes are never used for testing, especially if the mechanical condition of the unit is unknown. It would be a shame to have the VCR eat a priceless tape.

Most VCRs can be operated without a video tape in the unit by following the repair instructions in the manufacturer's service literature. End-of-tape sensors are defeated to prevent light from entering when no tape is in place.

The machine uses sensors called *cassette-down sensors* or *cassette-down switches* to tell when a tape is in place. The sensors can be defeated for troubleshooting purposes to make the unit think a tape is in place.

Once the VCR has been fooled into thinking that a tape is in the unit, you can engage the play, fast forward, and rewind functions. During this test procedure, check the unit for proper reel table, capstan, and head drum rotation as well as proper tape load and unload. If all appears normal, return the defeated tape end sensors and cassette-down switch to normal and place a tape in the machine.

Load the test tape into the VCR and attempt to play, record, and fast forward. Carefully check the tape path to insure that it is not damaging the tape.

Check the condition of the video tape where each of the guides contacts the tape surface. Look for any creasing of the video tape at all points of contact. The tape is examined both where it enters and where it exits the capstan shaft. At this point a defective rubber pinch roller can damage the tape.

During this test, load and unload the video tape several times. Observe the video tape as the arms pull it out of the cassette during load and allow it to go back into the cassette during eject. These are critical operations that the machine must perform smoothly to prevent tape damage.

Inspect the levers and arms responsible for positioning the tape in the tape path during the various modes of operation. Be sure the placement is correct. These arms swing in and out of position at various times depending upon the function you select.

During play, forward hold back tension is measured at a point on the tape just before it enters the video head drum. Proper tape tension is critical. Too little tape tension reduces the picture quality; too much tension reduces the life of the video tape and increases the wear on the video heads.

If the VCR passes these checks, remove the tape again and fool the machine once more by defeating the end-of-tape and cassette-down sensors. Measure fast forward, rewind, and brake torques and check the play take-up torque. Improper torques indicate that cleaning or replacement of worn parts is necessary.

Some manufacturers sell test cassettes with built-in sensors that monitor and indicate the tension in the mechanical system. These cassettes are convenient but the price ranges from $150 to $300 per cassette.

VCR Troubleshooting Approach

Usually when an electronic or mechanical part comes to the end of its useful life, a catastrophic failure occurs. While you can't always see an electronic or mechanical defect, you can find the problem without much effort. (But, don't think that every time your VCR quits working, you've just had a catastrophic failure.) In most cases, the troubleshooting flowcharts in this book will help you locate and correct the trouble quickly; but for problems that are not as easy to identify, follow these guidelines.

Don't panic. You now have a troubleshooting book that will help.

Obtain the service manual for the VCR model.

Observe the conditions under which the symptoms appear. What conditions existed at the time of failure? What function was the unit in at the time of failure? What was on the TV screen? Did the VCR make any unusual mechanical sounds before it failed?

Make note of as many of the symptoms as possible. After you compile your list of the various symptoms, you may find that, collectively, the symptoms all point to a common cause.

Use your senses to locate the source of the problem. Is there any odor from overheated components? Does any part of the VCR feel overly hot? Do you see anything that is burnt or cracked from overheating?

Retry. If the VCR has no power, check the power plug and the power cord. Is the plug snug in the socket? Is the wall socket working? If not, correct the situation and try again.

If your unit has power, try ejecting the tape. If the tape ejects, replace it with a "working tape" (not a valuable one) and try to operate the VCR again.

If the unit has power, and appears to operate mechanically, reseat all the connector cables that may be associated with the failure symptom. Cables have a habit of working loose if they aren't clamped down. Once you've checked all the connections, retry.

If a tape won't load or unload, check the chassis for any obvious problems like broken belts or foreign objects. Repair techs often laugh about the times when they've lifted carrots, cereal, toast, doll clothes, toy soldiers, and coins out of the delicate mechanisms inside.

One woman chased an elusive cockroach into the VCR and promptly emptied a can of insect spray into the cassette tray opening. The cockroach died—so did her VCR. The spray destroyed the friction surfaces on the belt pulleys, and caused an acidic reaction that ate into the top layers of the electronic circuit boards. Getting rid of that single cockroach cost her over $300 (a new VCR and a replacement can of bug spray).

If the VCR has no power, unplug the unit and check the power supply for blown fuses. Fuses blow when too much current tries to pass through the wire. Once in a while they seem to blow on their own from age or some strange malady, but normally a blown fuse indicates a failure somewhere in the system.

Document your findings and test results. Write down all that you see and sense. Write down all the conditions you observed at the time of failure and the conditions that exist now that failure has occurred. Answer these questions:

- What is your unit doing?

- What is it not doing?

- What is being displayed on the function panel and on the TV screen?

- What is still operating?

- Is power still indicated by the clock and other parts of the VCR?

- Will the unit load and eject the tape?

- Is it damaging the tape?

Assume one problem. In electronic circuitry, the chance of multiple simultaneous failures is low. Usually, a single component malfunctions, causing one or more symptoms. If the symptoms observed lead you to believe that you have more than one failure, troubleshoot the symptom that seems to be the least complicated. Chances are, when you locate the cause of the first symptom, the source of the other failure symptoms will either be solved or become apparent.

Diagnose to a section. You must determine if the malfunction is electronic or mechanical.

Consult the troubleshooting charts. Chapter 4 includes flowcharts of the most common VCR troubles. If the symptoms that you see match a problem described in the charts, follow the instructions under "Troubleshooting Procedure."

CAUTION

Any time you open the VCR:
1. *Be sure the power is off.*
2. *Unplug the VCR from the AC outlet whenever possible to avoid electrical shock.*
3. *Touch a metal lamp or other grounded object to remove any stray static electricity.*

Localize to a stage. Turn off the power to the VCR and disconnect the power plug. Disassemble the VCR as shown in the service manual. Follow the troubleshooting steps and procedures in Chapter 4, "Specific Troubleshooting and Repair," to localize the failed stage.

Isolate to a failed part. Closely following the procedures in Chapter 4 should guide you to the general area. But beware. There are many possibilities and potential pitfalls.

Take repair action. First, replacement of an IC is not easy. Those fragile pins on the ICs bend easily, and it takes very few straightening actions to break a pin completely off.

Removing and reinstalling electronic components that are soldered into the circuit boards are difficult actions and require more than a passing knowledge of soldering techniques. Only attempt this part of the test-repair procedure if you have experience soldering and desoldering delicate, printed circuit boards. Otherwise, bite the bullet and take your machine to someone who has the knowledge, experience, and equipment to successfully make this repair.

Sometimes a problem is caused by noise. Not audible noise, but electrical noise, the kind that produces static on the radio. This noise also affects VCRs. Noise in the VCR can cause very strange symptoms on the screen or in the sound. Figure 2.11 shows how automobile ignition noise can cause interference in a TV presentation. Notice the jumping picture.

NOTE

To avoid noise problems, keep cables clear and away from power cords, especially coiled power cords.

Fig. 2.11. Interference caused by automobile noise.

Sooner or later you're going to be confronted with those once-in-a-while failures called *intermittents*. These can be really frustrating.

Unlike a hard (constant) failure, an intermittent problem shows up randomly, or only at certain times (usually when you expect it least). Intermittent failures are difficult to locate using standard troubleshooting methods.

Since intermittent failures can be caused by shock, vibration, or temperature change, these conditions can be set up to find and correct the problem. Here are some helpful hints regarding intermittent failures. Be careful not to short out any connections or pin leads. Be sure to use only a nonmetallic or wooden object to probe components inside an energized VCR.

CAUTION

The following steps are conducted with the VCR open and operating. Dangerous voltages are present inside the VCR.

First, check, clean, and reseat all connector boards and cable plugs.

Second, tap gently at specific components on the suspected board using a nonmetallic rod or screwdriver.

Third, heat the suspected area with an infrared lamp or hair dryer. Don't overheat it.

Fourth, spray canned coolant on a suspected component. Service technicians sometimes use this technique to find a component that fails intermittently. Several companies sell pressurized cans of coolant spray that have long plastic extender nozzles for pinpoint application on the top of a suspected chip. By cooling the device with the VCR energized, and operating the VCR, you can identify components that are on the verge of total failure. The VCR works for a few moments until the part heats back up and starts causing problems again.

Finally, after you've found the problem area, make sure the power is off, then use a strong light and a magnifying glass to look for small cracks in the circuit board, defects in the wiring, or poor solder connections.

If the problem is caused by a marginal part, replace the pesky rascal. Be sure your replacement is of the same type and specification as the original (i.e., replace CX193A with another CX193A).

A large section of Chapter 3 is devoted to identifying and solving the intermittent problem. For now, let's say that good cleaning, cable and board reseating, and inside-the-case temperature control will prevent the occurrence of most random failures.

The final method for fault isolation to a component is signal tracing. This technique will be covered in the advanced troubleshooting chapter (Chapter 11).

Removing and replacing a chip. A disassembly and reassembly guide is located in most service manuals. Look for yours before you begin.

It takes a little practice before you can remove a chip without it jumping out, flipping in midair, and sticking you right in the thumb or index finger with that double row of tooth-like pins. An IC is the most difficult component to remove on any VCR circuit board.

But getting the chip out is only part of the challenge. Now you have to put the new chip in the board. Here's how to do it.

1. Desolder and replace the suspected malfunctioning chip with one of the same type and specification.

2. Line up the pin-1 end (with the notch or dot) with the pin-1 hole on the board.

3. Place the chip over the holes, lining up one row of pins with its holes.

4. With the chip at a slight angle, press down gently, causing the row of pins in contact with the board holes to bend slightly, which lets the other row of pins slip easily into their holes.

5. Press the top of the chip down firmly to seat the chip completely into the board. Be careful not to flex the board too much. If necessary, support the circuit

boards with the fingers of your other hand as you press the chip into place.

6. Solder the IC chip into the circuit.

7. Check for and remove any solder bridges between IC pins.

Now, that wasn't too bad, was it? Even so, it is pretty easy to make mistakes in chip replacement. Here are two more tips to help you avoid Murphy's law.

Make sure you don't put the chip in backwards. The notch or dot marking the pin-1 end of the chip is intended to help you correctly line up pin 1 on the chip with the pin-1 hole on the board. Don't force the chip down in such a way that one of the pins actually hangs out over the hole or is bent up under the chip.

Test and verify. After the repair action, reassemble the VCR enough to power up and test the repair. This is important. You need to know that all is now well with the VCR.

NOTE

It's a good idea to log the repair action in a record book to develop a maintenance history for the machine.

If the troubleshooting steps in this book don't help you find the failed component, you have two choices: either take the machine to a service repair center; or break out (or borrow) some test equipment, open the schematics, and start hunting for the failed or malfunctioning stage. Try signal tracing with an NTSC signal generator for the recording function. Use a tape with a known good recording on it for playback. Use an oscilloscope and a digital voltmeter (DVM) to test the discrete components such as transistors, capacitors, and resistors. Make voltage and resistance tests to locate the bad part. Test hardware and advanced troubleshooting methods are discussed in Chapter 11.

WHEN THE TIME COMES TO USE A SERVICE CENTER

You may clean the video heads and tape path and replace minor mechanical components, but there are adjustments and alignment required in the electronic and mechanical sections that should only be done by someone who has a good knowledge of VCR theory, the proper training, and the appropriate test equipment.

Although professional repair is costly, there are some things you can do to minimize the expense. So, before you call for help, read through the following list:

1. Identify the function that is affected.

2. Determine if the problem is mechanical or electronic.

3. Determine if the problem was caused by operator error.

4. Determine whether or not the problem is an intermittent failure.

5. Describe the problem in writing.

6. Carefully select a qualified servicer.

7. Record the serial number of your VCR.

8. Request an estimate.

9. Ask for a repair listing.

10. Determine if the repair is covered by warranty.

11. Get a receipt when you leave the unit.

12. Test the VCR before you accept the repair.

13. Request that the failed parts be returned to you when you receive the repaired VCR.

This checklist is a handy guide and should be used both before, during, and after the service center repair. Each step is expanded here for clarity.

Identify the function that is affected. Find out if the problem is catastrophic and affects all operations of the VCR.

Determine if the problem is mechanical or electronic. Be certain the problem isn't in the tape itself. Try to run several tapes you know are good.

Determine if the problem was caused by operator error. Be certain you didn't cause the problem.

Determine whether the problem is an intermittent failure. If the problem is intermittent, and you take it in for repair, it could be quite a while (at quite a fee) before the problem reappears and is fixed. You may just want to live with the problem until the intermittent becomes permanent (a hard failure). At least then you have something concrete to troubleshoot. Intermittents are discussed further in Chapter 3, "Routine Preventive Maintenance."

Describe the problem in writing. Before taking your VCR to the shop, make a list of each of the symptoms it displays describing under what conditions the malfunction occurs. Is it intermittent or does it happen continually? Be sure each of the symptoms and complaints is written on the service order.

Carefully select a qualified servicer. Selecting a service facility that can give your VCR the proper attention is not as easy as picking up the phone book. It is much like seeing a doctor, and is important in prolonging the life of the VCR and video tapes. This is something that should be given careful consideration.

These suggestions will help you to select a servicer who will treat you fairly, professionally solve your problem, and help you avoid the frustration of dealing with incompetence.

- Do not assume that the most expensive is the best. The rates charged for labor do not necessarily correspond to the quality of workmanship.

- Check to see if the service facility has all of the required city and state licenses.

- Check the Better Business Bureau, Chamber of Commerce, and appropriate licensing bureaus (e.g., State Board of Repair) for any registered complaints.

- Ask if the service facility is authorized by the manufacturer to service your brand under warranty. Ask the manufacturer for recommendations of qualified servicers in your area.

- Determine if the servicer has been properly trained for VCR repair. Service centers often display certificates of training.

- Make sure that the service center uses only the VCR manufacturer's original replacement parts and not after-market or generic substitution parts.

- Find out if they will be able to get parts on a timely basis. Most manufacturers have adequate parts available for servicers who deal with them regularly.

- Check to see if the service facility appears neat and orderly, the technicians and employees neat in appearance, the VCRs that have been checked in for service properly stored and not stacked on top of each other.

- Pay attention to the customer service representative. See if he or she seems knowledgeable about VCRs and concerned about your individual needs.

- Ask your local video rental club for recommendations. Ask friends, neighbors, and video sales businesses whom they recommend.

Record the serial number of your VCR. Jot down the serial numbers of all the accessories and the VCR you'll be turning over for repair. The service order should contain the manufacturer, model number, serial number, and a list of any accessories included. Also have the condition of the VCR cabinet noted on the work order when you check it in. Is it in good shape or is it all scratched up?

Request an estimate. Request a written estimate for labor and parts. Ask how long the shop will keep your VCR to complete the ser-

vice. Find out how the labor rates are determined. Some repair facilities charge on an hourly basis; others charge a flat rate.

If possible, ask one of the technicians what may have caused the problem, and what the solution may be. An experienced technician should be able to estimate required repairs but this does not mean the preliminary diagnosis will be 100 percent correct.

Ask for a repair listing. Ask for a detailed listing of repairs or replacements including a complete list of all charges.

Determine if the repair is covered by warranty. Make sure that parts and labor are warranted for at least 30 days. What are the terms of the guarantee?

Get a receipt when you leave the unit. Get a copy of the work order showing the model and serial number of the unit, a list of accessories, the listed defects and the written estimate.

Test the VCR before you accept the unit. Test run the VCR before taking it out of the shop and again as soon as you get it home. Servicers who guarantee their work expect the unit to be operated during the guarantee period. But it is your responsibility to determine that the unit is in good working order before the guarantee expires.

Request that the failed parts be returned to you when you receive the repaired VCR. Some states require that service centers return the defective parts to the customer. If your state does not require this, you might want to request it when you leave the VCR for repair.

CAUTION

As you would with any electrical device, you must observe certain precautions to prevent injury to yourself or damage to the VCR. Be sure to review the safety precautions listed under the heading of "Safety Precautions During Troubleshooting and Repair" found near the beginning of this chapter before opening your VCR. Observing these precautions can save you time, money, frustration, and possible injury.

Special Handling

Always turn the power off, touch a grounded metal object like a desk lamp, and then pull out the power cord before touching anything inside. Many failures are caused by people who don't follow this rule.

Handle your video tapes carefully. Don't lay tapes on a very dusty, dirty surface. Keep cigarette ash away from the tapes and VCR. Don't touch the tape surface. Don't set your tapes on or in front of a TV or color monitor.

Don't cycle the power on and off quickly. Wait 7 to 10 seconds so the capacitors in the power supply discharge fully and the circuits return to a stable (quiescent) condition.

Keep liquids away from the unit. It's amazing how sticky soft drinks become after frying components all over the inside of the unit.

Handle components with care. Don't let chips lie around. The pins will get bent. Watch out for static electricity. Chips may need special handling.

Some logic devices inside the VCR require extra care when you touch or handle them. The metal oxide semiconductor (MOS) chip family (MOS, CMOS, NMOS, etc.) needs some extra care since these chips are susceptible to damage by static electricity, so be sure to ground yourself by touching a metal lamp or grounded object before you reach for a chip inside the VCR chassis. In addition, conductive foam provides protection from static charge during storage or transportation of MOS-type chips.

Additional precautions to take when you use test equipment with your VCR will be covered in Chapter 11, "Advanced Troubleshooting."

REPAIR PARTS

Finding that trouble really exists is only part of the problem. You must locate the specific fail-

ure and then make the repair. This, too, can be challenging. Most of the chips in VCR boards are custom or special purpose parts and are not readily available at the local electronics store. You may be able to substitute the transistors with a generic brand, but the best way to replace any VCR part is with the original manufacturer's part type. The resistors and capacitors can be easily obtained at most electronic supply houses. Be sure to replace any critical safety components with only the manufacturer-recommended replacements. This is important for your safety.

The VCR chips and each of the other parts should be available from your local VCR repair service center.

SUMMARY

So there you have it. In this introductory chapter on troubleshooting, you've learned the troubleshooting steps to success. You've also learned how to recognize the components inside the VCR, how components and mechanisms fail, as well as various methods for localizing failures in your own machine.

CHAPTER REVIEW QUESTIONS

1. How is pin 1 identified on an integrated circuit?

2. If you have an axial lead resistor with color bands reading left-to-right brown, black, red, and silver, what is the value and tolerance of this device?

3. A component marked 1S1519 is what type of device?

4. Leakage describes one type of failure in transistors, diodes, and integrated circuits. Explain leakage.

Routine Preventive
Maintenance

*I*n Chapter 2 you stepped through basic troubleshooting and corrective maintenance of the VCR. Chapter 3 discusses another type of maintenance, one that is intended not to fix a problem but rather to prevent a problem from ever happening. Preventive maintenance is in every way as important as corrective maintenance. In this chapter you'll learn what factors damage your VCR and cause it to fail, and what you can do to prevent these failures.

Often, the price you pay to buy your VCR is actually a small part of the overall system cost. The life-cycle cost of the equipment can be much larger than the initial purchase investment. This total cost increases dramatically as the costs for prerecorded movies, books, magazine subscriptions, those extra connecting cables, blank tapes, and service center repair charges are added in. Service costs alone can grow to 10 to 50 percent of the system cost.

Occasionally we find a repair expense that exceeds the value of the broken equipment. It's when we look at high repair costs that terms like mean time between failure (MTBF) and mean time to repair (MTTR) become important. While your VCR has an excellent reliability track record, the way you operate your machine and the environment in which you place it become important to that MTBF number. Another factor to consider is that those bargain VCRs and blank tapes probably have less than an excellent reliability record. You get what you pay for.

Your VCR is sturdy, fun to use, and under most operating conditions, a very reliable machine. But, like other machines, it can wear out and fail.

As the conveniences of your VCR become increasingly essential in your home and business, your dependence on the machine increases. However, if repair is necessary, you may have to give up the machine for one to three weeks (although many problems can be fixed within a day).

Most large video production companies protect their huge investment in equipment because accidents and unnecessary failures can cost thousands of dollars in lost business. A small business, with a single video camera, two VCRs, and other assorted equipment will experience just as catastrophic a loss by system failure, yet most don't take steps to prevent it.

VCRs don't burn out. They wear out or are forced out by human error or adverse operating conditions. If you misuse your VCR or don't protect it from environmental elements, you can be the cause for its failure.

A few moments of care can yield many more hours of good, consistent performance. We call this care preventive maintenance. Just as you periodically check the oil and water in your car's engine, and lubricate, wash, and wax the body to keep it running right, so you should care for and protect your VCR. You can get good, reliable operation from your VCR for many months, if not years, if you provide timely and proper maintenance to keep the system in peak condition.

Proper preventive maintenance begins with an understanding of what we are fighting. Six factors can influence the performance of your equipment (not including the tape-eating dog or the cable-breaking baby). These factors are:

- Extremes of temperature.

- Moisture.

- Dust and foreign particles.

- Noise interference.

- Power line problems.

- Corrosion.

- Magnetic fields.

Each of these items acts to cause VCR breakdown. This chapter helps you successfully battle these enemies of reliable performance.

EXTREMES OF TEMPERATURE

The chips and other devices in your VCR are sensitive to high temperatures. During normal operation, the machine generates heat that is generally tolerable to the circuitry. Usually, leaving your VCR on for long periods won't hurt it because the slots and air vents let enough of the heat dissipate to the outside of the case.

As long as the components on the VCR boards and chassis are not too hot to touch, the amount of heat being produced should not cause any damage. However, heat can become a problem if you leave the dust cover on or confine the unit to a tightly enclosed cabinet during operation.

Excessive heat within a component causes aging and failure. The heat produced during operation is not uniform across the device, but appears at specific locations on the chip (generally at the input/output connectors where the leads meet the chip itself). The usual effects of heating and cooling are to break down the contacts or junctions in the chip or other device, causing open circuit failure. When they are hot, these devices can produce intermittent operation of the VCR. The continual heating and cooling action during normal operation can also result in poor contact at solder points on the circuit boards.

Heat can contribute to cassette tape failure. Tape cassettes act just like audio cassettes when exposed to heat, especially the heat of the sun. If you leave cassettes in a hot car, some warpage will occur to the case or the tape itself will be damaged. If the tape is damaged too much, you will lose whatever is recorded on it. You could try to copy onto another tape, but the success rate for this "repair" isn't very high.

Heat is seldom a problem for the intermittent VCR user. When use is increased, generated heat can reduce performance and component lifetime. The following suggestions should help you prevent heat-related failure.

- Keep the dust cover off during operation.

- Allow plenty of ventilation around the unit.

- Keep the cooling vents clear.

- Keep the system dust-free inside and outside. Do regular preventive maintenance actions.

- Store cassette tapes in a cool, dry location.

- Stand the tapes up on end when storing them.

The effect of cold on VCRs is interesting. High-technology companies in the United States have developed extremely fast computers that operate supercold. Electronic components operate quite well in cold temperatures, but mechanical components have trouble functioning when the temperature drops. Consider the cassette tape drive mechanism for example. The operating range for a standard cassette tape drive is approximately 40°F to 115°F. At the low end, mechanical parts become sluggish and the possibility of improper tape movement increases. In addition, the tape itself can become brittle as it gets cold.

The rule of thumb for cold temperatures is to let the system warm up to room temperature (stabilize) before turning on the power. If the room temperature is comfortable for you, it's fine for the system.

MOISTURE

It's possible that humidity will build up and condense on the metal chassis parts and video head drum if the internal temperature of the VCR changes drastically. If the dew warning comes on after the unit is brought in from the cold, let it set at room temperature for a few hours. If you can't wait that long, use a hair blow-dryer to dry out the machine.

Condensation on the head drum causes the plastic tape to stick to the high-speed rotating head drum. This can be disastrous to the tape and to the video heads. For this reason the dew sensor was designed to prevent operation of the VCR in the presence of moisture.

DUST AND FOREIGN PARTICLES

Just as flies descend at a picnic, dust seems to descend on VCR equipment. Static electric charges that build up in the VCR and the television monitor attract dust and dirt. That's why professional video tape editing systems are kept in cool, clean rooms. They require air conditioning and dust-free spaces because the large equipment generates more heat, and is just as susceptible to failures caused by dust buildup.

Dust and dirt buildup coats circuit devices. This insulation blanket prevents the release of heat generated during normal operation. If the equipment can't dissipate this heat, the inside temperature rises higher than normal, causing the mechanical and electronic components to wear out faster. Dust is a major contributor to video head failure. Dust seems to be attracted to heat. Have you ever noticed that dust builds up on light bulbs in your lamps or on the tops of stereos and televisions more than it does on cooler objects? The charged dust particles become attracted to the magnetic field around electrical equipment. VCR problems increase in direct proportion to the increase in dust.

Mechanical devices like video cassette tape drives fail more often than solid-state electronic devices because mechanical and electromechanical devices have moving parts that get dirty easily, causing overheating and earlier failure.

Foreign particles such as dirt, smoke ash, and tiny fibers can cause catastrophic problems in the cassette cartridge and in the VCR itself. The air we breathe is full of airborne particles, but most of these are too small even to be seen, let alone become a problem. The larger particles in the air cause VCR problems. Cigarette ash, for example, can settle on the tape path and move from place to place inside the VCR creating improper tape travel and destroying record and playback heads.

Video tape sheds tiny particles as it moves through the machine. These particles prevent the heat generated during normal operation from escaping off the components and into the air. They also combine with the dust and dirt and become lodged in the lubrication ma-

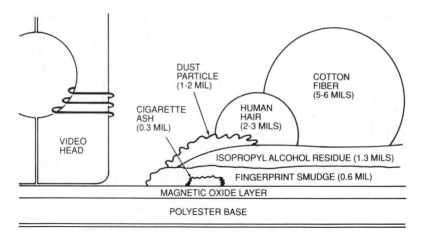

Fig. 3.1. *Any small piece of foreign material can cause problems with a head that rides on the surface of the tape.*

terial for the mechanical parts causing the moving parts to wear out faster.

Video heads and the tape path have more dust-related problems than the other mechanical moving parts. The tape must travel an exact path to record and play back properly. If dust is in the pathway, the tape can be lifted slightly as it passes. During operation, the video heads press slightly against the tape. As the head rides on the cassette tape surface, dust and dirt can cause major problems as suggested in Fig. 3.1.

Some current professional ½-inch video tapes are designed to minimize the attraction of dust and dirt by static electricity. This doesn't mean you can get careless about dust and dirt. Dirt on a tape can be swept off the video head and gouge out a path on the cassette tape surface. It can stick to the head and cause other cassette tapes to be gouged, or cause the head itself to corrode and wear out.

Smoke from cigarettes and cigars can coat the internal surfaces of the cassette tape path with a gummy soot that not only produces tape path errors, but also increases wear by interfering with mechanical operation. Tobacco smoke is also believed to cause rapid oxidation on pins and connectors, increasing the likelihood of intermittent errors. Also, it increases the failure rate of rubber tires and belts in the tape drive mechanism.

Controlling Dust and Dirt

Dust buildup can be controlled. Thoroughly cleaning the VCR area every week will do much to keep your system in top condition. Dirt and dust can be removed from the equipment housing using a damp cloth lightly coated with mild soap. Clean electrical equipment with the power turned off and the unit unplugged from the AC outlet. Be careful you don't moisten the electronic components.

After washing the surface, rewipe the outside of the equipment with a soft cloth dampened with a mixture of one part liquid fabric softener to three parts water. The chemical makeup of some liquid softeners is almost the same as antistatic chemical spray. Wiping the cabinet helps to keep static charges from attracting dust to the top of the VCR. The fabric softener is antimagnetic and prevents the attraction of dust. The chemicals in this inexpensive solution can last longer than some antistatic sprays and help make your plastic VCR cabinet less susceptible to scratching.

Another successful technique involves blowing dust away from the VCR with a pressurized can of antistatic dusting spray. Using this product means you don't have to wipe off the equipment first. In any case, wiping a VCR should be done carefully, because you could scratch the paint, the plastic cabinet, or the plastic windows if there are hard dust or dirt particles on the cabinet or cloth. Here are some manufacturer-recommended cleaning methods.

- Use one part fabric softener to three parts water to clean indicator screens (including the television screen).

- Use mild soap and water, with a soft cloth for drying.

- Use a window cleaner spray. (*Note:* Be careful in selecting the correct solution. Common household aerosol sprays, solvents, polishes, or cleaning agents may damage your cabinets and the TV screen. The safest cleaning solution is mild soap and water.)

- Use an antistatic spray.

CAUTION

Make sure the power is off and the plug(s) pulled out of the power socket(s). Use a damp cloth. Don't let any liquid ruin or get into your equipment.

Use compressed air to blow out dust and dirt from the VCR. Any brushes and vacuum nozzles poked into the unit may bump, scratch, or damage the video heads or other critically aligned components.

You may not have an air-conditioned, air-purified room in which to use your VCR, so dust covers become very important. Plastic covers, made static-free with an antistatic aerosol or by wiping the surface with the fabric softener–water mixture, will provide good dust protection for your unit.

Don't place the VCR on the floor. Dust and dirt rise from the floor into the chassis and insects can crawl into the VCR, all of which will increase the potential for trouble. Use a cabinet or shelf. Here is a summary of ways to counter dust.

- Use dust covers.

- Keep windows closed.

- Don't operate the VCR on the floor.

- No smoking near your VCR.

- No crumb-producing foods near the unit.

- No liquids on any equipment.

- Don't touch the surface of any video tape.

- Vacuum around the unit area of your VCR weekly.

- Clean your VCR cabinet with static-reducing material.

NOISE INTERFERENCE

Your VCR is sensitive to noise interference, which can affect the proper operation or video picture. But what is noise, where does it come from, and how can you get rid of it?

Noise can be described as those unexpected or undesired random changes in voltage, current, picture, or sound. Noise is sometimes called "static." It can be a sudden pulse of energy, a continuous hum in the speaker, or a garbled display of characters.

Three types of noise cause problems: noise that affects the user (acoustic), noise that affects your VCR, and noise that affects other electronic equipment. For example, acoustic noise includes the crying of a baby, the blare of an overpowered stereo, and the loud consistent tap-tapping of a noisy typewriter. Noise that affects the VCR and other equipment can be radiated, conducted, or received. It takes the form of electromagnetic radiation (EMR). EMR noise can be further classified as low- or high-frequency radiation (EMI or RFI), as shown in Fig. 3.2.

If the noise occurs in the 1 Hz to 10 kHz range, it is called electromagnetic interference (EMI). If it occurs at a frequency above 10

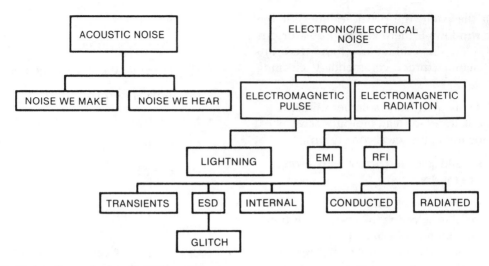

Fig. 3.2. *Various forms of noise affect VCR equipment.*

kHz, it is called radio frequency interference (RFI).

RFI can occur in two forms: conducted RFI and radiated RFI. RFI that is fed back from the VCR through the power cord to the high-voltage AC power line is classified as conducted RFI. In this case, the power line acts as an antenna transmitting the noise interference out. When your VCR and its cabling transmit noise, this noise source is called radiated RFI.

EMI has three primary components: transient EMI, internal EMI, and electrostatic discharge. Transients occur when a large voltage pulse, or spike, smashes through the circuitry. Power line transients and electrostatic discharge from the human body are the two most severe forms of externally generated EMI.

Internal EMI is the noise generated within and by the chips, motors, and other electronic devices. With current microelectronic designs, internal noise levels are so low that other factors such as connections and the length of leads have become the main source of noise in printed circuits. Internal noise does become a problem when the components are excessively heated or when the motors and components begin to fail.

The last form of EMI, the electrostatic discharge (ESD), is the same effect you get from walking across a carpet and then getting shocked by touching metal. ESD can cause the notorious "glitch" in electronic circuits.

All these types of noise interference can produce undesirable or damaging effects in the VCR. They can cause programs to stop in the middle of an operation, garbage to appear on the TV screen, and chips to be destroyed. Noise interference must be prevented by reducing or eliminating its cause. This is not an insurmountable challenge, but it is a substantial practical and analytical task.

Origin of Noise Interference

Noise in the VCR can come from many places including power supplies, fans, the VCR itself, the television, other equipment, connectors, cables, fluorescent lights, lightning, and electrostatic discharge. The lower portion of Fig. 3.3 shows how noise interference appears on a TV screen.

High-powered components used in switching power supplies have led to noise being conducted back into the power lines. Switch-

Fig. 3.3. Noise interference causes the white streaks visible in the lower portion of this TV image.

ing power supplies have been found to generate EMI in the 10–100 kHz frequency range.

Noise can even be passed or coupled to nearby equipment that's on a totally different circuit and not physically connected to our system. If two wires are placed next to each other, one wire can pick up signals coupled across from the other (crosstalk). Just 10 volts of electricity on one wire can cause a measurable (0.25 volt) signal on the other wire. Imagine how much crosstalk there could be if the voltage on one line were increased to 100 volts. The induced voltage on the other wire would be 2.5 volts, which is enough to change information in a stream of data being sent through that section of wire.

Everything has some capacitance associated with it. Unwanted high-frequency noise signals can be introduced into a circuit through capacitance. Some typical capacitance values (in picofarads) are shown in Table 3.1.

Table 3.1.
Typical Capacitance Values

Source	Capacitance (pF)
People	700.0
$\frac{1}{2}$-watt resistor	1.5
Connector	2.0

Engineers have found that even 0.1 pF of capacitance can produce 5-volt spikes in digital circuits such as those found in the VCR.

Power line noise can feed into the VCR circuits whenever it exceeds the blocking limits of the power supply. Nearby high-voltage machinery such as stamping mills, saws, air conditioning units, vacuum cleaners, or clothes dryers can produce strong magnetic fields in the area around them and in their power cords.

Cables that vibrate and move in a magnetic field can also cause problems. Relays and motors can produce high voltage transients when they are turned on or off. And televisions and radios can be a source of noise coming into the VCR.

Finally, EMI can come from industrial, medical and scientific equipment, electric motors, home appliances, drills, saws, lighting dimmer controls, and tool speed controls. It's important to understand noise and how it can be generated. VCRs must be able to operate without causing interference to nearby electronic equipment. They must be able to function without radiating noise, and in an environment that includes noise from outside sources.

Controlling Noise Interference

The most effective approach to noise reduction is prevention. If you can't prevent noise, you can at least take steps to minimize its impact. Here are five methods for dealing with noise.

- Filtering noise.

- Shielding from audible noise.

- Improved circuit board bonding connections.

- Improved wiring.

- Improved component design.

The most common approach is a combination of these methods, although filtering and shielding are most widely used.

Filtering involves the use of capacitors and inductors. There are many kinds of filters that respond to voltage, current, and frequency. For example, one filter prevents high-frequency voltage spikes from leaking out a switching power supply into the circuitry being supported.

While most VCRs don't generate enough audible noise to require acoustic shielding or enclosure it may occasionally be necessary. If some form of audible noise reduction shielding is used, be certain there is adequate air circulation on the top, bottom, and sides of the unit.

Electromagnetic Interference (EMI)

EMI is an unplanned, extraneous electrical signal that affects the performance of the VCR system. It can cause automatic timer errors and unusual functioning of the unit. It can appear as circuit crosstalk, voltage, ripple, or power drift.

While circuit designers try to minimize EMI, it is a natural by-product of aging components, bad solder joints, damaged or corroded connector contacts, and loose connections. It is also produced when a burst of electromagnetic or electrostatic energy is conducted or induced through the circuitry. Externally produced EMI enters the VCR through the cabling or openings in the case.

Some VCR cases are made of metal that is lightweight, durable, and generally rustproof. While these are good qualities, the most important feature of these cases is that metal conducts electricity away from sensitive components so it provides protection against EMI/RFI and even ESD noise. VCR design engineers have tried many techniques to reduce EMI/RFI:

- Use decoupling capacitors (0.01 μF to 0.1 μF).
- Carefully lay out components.
- Keep traces as short as possible.
- Shield sensitive circuits.
- Reduce noise sources.
- Use fewer components.

- Carefully route wires.
- Use a shielded cabinet with minimal openings.

Cables are a source of EMI/RFI. Both the internal cables and those external cables leading to the TV and the antenna or cable system can radiate interference.

You can improve on the VCR's design to counteract EMI. Since you won't be changing the circuit board design, you reduce EMI in two ways: (1) by preventing it from reaching the circuits and (2) by keeping it contained within shielded enclosures. To do this, use shielding, grounded cables, filters, and transient absorbers.

A shield is a conductive coat or envelope placed around a conductor wire or group of wires to provide a barrier to electromagnetic interference. EMI-reduction devices include the ferrite "shield beads" which are placed on power supply leads, connections to ground, or between stages on the circuit board.

Ideally, the shielded connectors should provide a continuous shield from the device, through the connector, and into the shielded cable. Otherwise, the weak shield point becomes a transmission hole for EMI to get out and interfere with other devices or appliances in the area.

One excellent countermeasure to EMI and RFI is the use of fiber-optic cables and connectors. This technology is just starting to become popular with video cable companies, but the cost is still too high for most private video users. The day is coming when fiber optic video transmission will be the norm rather than the exception.

Electrostatic Discharge

It sometimes appears that a wizard with a weird sense of humor secretly loads into every VCR timer a program that intermittently produces random glitches to drive the user wild. Chasing and catching the elusive phantom glitch is a challenge even for experienced repair technicians using expensive and complex troubleshooting equipment. But you can learn how to prevent this intermittent problem from affecting your VCR operation.

Glitches are short-lived electrical disturbances that often exist just long enough to cause problems in digital circuitry. They are usually the result of an electrostatic discharge (ESD), also one of the most severe sources of EMI.

People and objects can accumulate a substantial electrical charge or potential. The human body can accumulate static charges up to 25,000 volts, and it's not unusual to build up and carry charges of 500 to 15,000 volts. Charged objects or people can then discharge (quickly get rid of the voltage) to a grounded surface through another object or person. Remember the times you dragged your feet across the carpet and then shocked someone nearby? This electrical charge is static. It can discharge through your VCR, and when it does, all sorts of undesirable things can occur. If a program is running and a VCR user carrying a large potential of electrical charge touches a function key on the unit, the arc of discharge will find the shortest route to ground, usually through the sensitive microelectronic ICs, and the VCR will stop. If the VCR has a screen display showing the particular mode of operation or the automatic timer program shows on the TV, it can go wild and display weird characters. Sensitive components can be damaged or destroyed. Even a charge of only 3 volts is enough to create erroneous information in most logic circuits.

Electrostatic charges can be of any voltage. Here are some of the sources of ESD glitches:

- People in motion.

- Overhead components.

- Improper grounding.

- Poorly shielded cables.

- Improperly installed shields.

- Missing covers and gaskets.

- Circuit lines too close.

- Poor solder connections.

- Low humidity.

We know that static occurs when two objects are rubbed together. Your movement, walking while wearing wool or polyester slacks, can cause a tremendous charge of electricity to build up on your body. When this charge reaches 10,000 volts, it is likely to discharge on any grounded metal part.

Litton Systems, Inc., has developed the *Triboelectric Series* chart shown in Table 3.2. Cotton is the reference material since it absorbs moisture readily and can easily become conductive. If any material on the list above cotton is rubbed with any material on the list below cotton, the item listed above will give up electrons and become positively charged. The items listed below cotton will absorb electrons causing it to become negatively charged.

Table 3.2. Litton Systems, Inc., Triboelectric Series

Air

Your hand

Asbestos

Rabbit fur

Glass

Your hair

Nylon

Wool

Fur

Lead

Silk

Aluminum

Paper

Cotton (*reference material*)

Steel

Wood

Hard rubber

Nickel, copper

Brass, silver

Gold, platinum

Acetate, rayon

Polyester

Polyurethane

Polyvinyl chloride

Silicon

Teflon

Two oppositely charged materials will tend to cling together. If they are separated, a static charge difference occurs. If Teflon is rubbed in your hands, a large electrostatic charge is built up. The farther apart the materials are listed in Table 3.2, the larger the charge that can build up. Notice that hair is listed above cotton. Hard rubber is below cotton. Paper is listed just above cotton. Have you ever pulled a rubber or plastic comb through your hair and then used the comb to pick up pieces of paper? This is electrostatic charge in action.

A problem occurs when this charge builds and becomes quite large. Just walking across a carpet can generate over a thousand volts of charge. If the humidity is low and the air is dry in the room, the charge can be substantially higher. When the relative humidity is 50 percent or higher, static charges generally don't accumulate. However, in a dry room, built-up static charge will readily arc to any grounded metal, such as a VCR chassis.

An ESD release on the chassis case won't hurt you, but it can be very damaging to the electronics. The discharge pulse drives through the case to the microelectronic circuitry where it can burn out some of the chips. Even if no components are "fried," the damage that is done by this overvoltage spike accumulates and starts to degrade the functioning of some circuit board components. ESD damage costs have been estimated at millions of dollars annually. This figure is even higher when we consider components that are not totally destroyed, but are only degraded. Sooner or later the degraded components fail completely.

In low humidity, walking across a synthetic carpet can charge your body to 35,000 volts. Walking over a vinyl floor can charge you with 12,000 volts. A poly bag picked up off a table can develop 20,000 volts. Even sliding off a urethane foam padded chair can load you with 18,000 volts. Is this a hazard to your VCR? As Table 3.3 describes, some electronic devices are very susceptible to low values of ESD voltage.

Table 3.3. Voltages That Can Damage Electronic Devices

Device	Damaging Voltage (Minimum)
CMOS chips	250–3000 volts
Diodes	300–2500 volts
EPROM memory	100 volts
Operational amplifier	190–2500 volts
Resistors	300–3000 volts
Schottky (S, LS) chips	1000–2500 volts
Transistors	380–7000 volts
VMOS chips	30–1800 volts

If your VCR occasionally gets the "shock treatment" or functions erratically, there are some things you can do. The following list offers some specific solutions to ESD problems:

1. Apply antistatic spray with a soft cloth on rugs, carpets, and VCR equipment to reduce and control static.

2. Install a static-free carpet in your VCR area.

3. Mop hard floors with an antistatic solution. The antistatic floor finish works well for up to six months, but this is an expensive solution more suited to electronic manufacturing facilities.

4. Place your VCR on a conductive table top.

5. Use static-free table mats.

6. Keep chips in conductive foam (that black Styrofoam-looking material).

7. Touch a grounded metal object before touching the VCR.

You can defeat ESD glitches by paying attention to static charge in and about the VCR. By making static charge elimination part of a preventive maintenance program, you help extend the life of your system.

Radio Frequency Interference

Radio frequency interference (RFI) noise is much the same as EMI except it occurs at higher frequencies (greater than 10 kHz). RFI is what causes five other garage doors on your street to open when you operate your new automatic garage door opener. You'd really rather it didn't work like that.

Although RFI isn't a health hazard, it can be a real bother. Many RFI problems are caused by placing the VCR too close to the television set. Sometimes the high-frequency signals generated inside the TV interfere with the VCR's carefully timed signals. This usually occurs when the VCR is placed directly on or under the set. The trouble disappears when the unit is moved to the side or up on a shelf a few feet above the television receiver.

Another common source of RFI is the location of the nearest radio station transmission antenna. The color-under frequencies in your VCR occur around 600 kHz. It is not uncommon to have a VCR cause color rainbow-like patterns on the TV screen when playing back a color program if you live near a transmitter that operates on a frequency close to 600 kHz (600–800 on the AM radio dial) due to some harmonic (doubling or halving) of that frequency. The color signals inside the VCR can mix with the RFI and cause herringbone patterns like those in the test pattern in Fig. 3.4.

The rainbow effect can appear as vertical, horizontal, or diagonal color stripes, or as red–green–blue herringbone or zigzag shapes. This particular problem is normally seen in the playback mode. You can verify the cause by following these suggestions.

First, check to see if you live near a transmitter at or near 629–743 kHz (or some harmonic of these frequencies).

Second, when the problem appears, turn the color control on the set all the way off. If the pattern goes away when the color is turned off, you can assume that RFI from an interfering radio station is the cause.

Fig. 3.4. RFI can cause the tweed or herringbone pattern to be produced on a TV screen.

Third, place the unit in PLAY and completely wrap the VCR in tin foil. If the trouble disappears, the RFI is from an external source and not internally created. (DO NOT LEAVE THE UNIT WRAPPED IN THE FOIL OR IT WILL OVERHEAT.)

Finally, operate the unit in playback when you know the radio transmitter is off the air. (This may be very early in the morning.) If the tape plays without the rainbow interference you have isolated the problem.

A third form of RFI is unique to the audio circuits in an 8mm circuit or a VHS Hi-Fi audio unit. The cause is similar to the color rainbow interference just described, but the frequencies and symptoms are different. VHS Hi-Fi and 8mm sound are recorded and played back on the video tape using FM frequencies ranging from about 1.3 MHz to 1.9 MHz. Sometimes when a unit operates near transmitters that broadcast in this frequency range (e.g., in the upper AM radio band), the VCR may receive and record or play back the station along with the taped program. You'll notice this symptom when you play back the recorded information and hear the interfering radio station signal. You can isolate this problem by following these steps.

First, confirm that all connecting audio cables have a good ground connection. Then, check to see if you live near a transmitter broadcasting in the range of 1.3 to 1.9 MHz (or some harmonic of those frequencies).

Next, place the VCR in the mode in which the symptoms occur and completely wrap the unit in tin foil. If the interference goes away, the RFI is external (not internally created). (DO NOT LEAVE THE UNIT WRAPPED IN TIN FOIL OR THE UNIT WILL OVERHEAT.)

Finally, operate the VCR when the transmitter is off the air. If the symptoms disappear you can assume that the transmitter is the source of the problem.

Other RFI problems find their way into the FM video circuit preamplifiers. The symptoms can vary, but usually they are caused by nearby transmitters that interfere with the video signal in the first stages of amplification.

The only sure way to completely block RFI from getting into the VCR is to completely enclose it in a shield. Some manufacturers have designed special lead covers which provide suitable ventilation for the unit while it's operating and stop most external RFI troubles. In some cases these covers are expensive but more often they are simply not available due to limited quantities. Many manufacturers are now using all metal cabinets instead of the popular plastic cabinets common to earlier VCRs.

There are other ways to minimize or reduce RFI problems. For example, a smaller number of components in a system reduces the number of sources for RFI and improves system operation. Reliability improves in direct proportion to RFI improvements. Here are some more ways you can improve your system's RFI condition.

- Locate your VCR at least a couple of feet away from any television set.
- Use a directional outdoor TV antenna.
- Subscribe to cable or satellite TV.
- Connect traps or line filters to your VCR.

- Replace the antenna twin-lead wire with 75-ohm coaxial cable.
- Use only good quality audio and video cables.

POWER LINE PROBLEMS

One important environmental factor for your VCR is good, clean power. If you depend on the local utility to supply this power in steady, reliable consistency, you may be disappointed.

While room lighting systems can tolerate line voltage problems that momentarily dim the lights when a large power-hungry machine turns on, VCRs cannot. Your unit, like most electronic units today, is more sensitive to power line disturbances than other electrical equipment. Even well-designed machines such as yours are affected by the quality of power provided. Undervoltage or overvoltage puts severe stress on VCR components. The effect is to accelerate the conditions under which a device gradually weakens, becomes marginal, and finally wears out. There are four types of power-line problems that cause concern:

- Brownouts.
- Blackouts.
- Transients.
- Noise (discussed previously).

Brownouts

Brownouts are those planned (and sometimes unplanned) voltage sags, when less voltage is available to drive your power supply, motors, and solenoids. Brownouts are far more common than you may realize.

Voltage dips are common if you operate the unit near some large electrical equipment such as air conditioners or arc welders. Line voltage can be drawn down as much as 20 percent by the heavy momentary drain caused when this equipment is turned on.

The VCR should still work with line voltages that drop and remain 10 to 20 percent below the 110 volt rating. But if the supply voltage gets too low, the regulators in your power supply won't be able to pump adequate power into the unit. During a brownout, VCRs can operate intermittently, overheat, or simply shut down and lock up.

By the way, your power supplies can also handle a voltage "brown-up," or increased line voltage, but the power supply regulators will generate a lot more heat as they handle the extra incoming voltage.

Blackouts

Power line blackout, a total loss of line voltage, can be caused by storms and lightning. It can also be caused by vehicles accidentally knocking down power lines or by improper switching action by a power station operator.

When power is lost, whatever you were recording may be gone. If you are operating in the automatic record mode when power fails, you will have recorded only a partial program. The part of the program being broadcast while the power was out (as well as anything after that) is not recorded. Many VCRs have a battery backup built into the system control that keeps the clock circuitry running and time activations will save anything you have programmed into the automatic timer. These batteries keep the correct time on the clock and last for up to an hour. The unit will not have a display until the AC is restored from the utility company but the clock and timer are still alive.

The VCR will not turn on at the time programmed in the auto timer if no power is available, but will wait for the next timer-scheduled program to turn on the machine. Not all VCRs have this battery backup feature.

If the weather turns bad and thunder is echoing across the sky, don't turn on your unit. If a blackout occurs or if you see lightning, turn off your machine, pull the plug, and disconnect all antenna leads until the storm passes.

And when the power goes out, be careful. While the room lights are out and you're muttering under your breath as you feel around for a flashlight, remember what is sure to happen when power is restored—a tremendous voltage spike will be produced as lights and motors go on all over the neighborhood. This could damage the VCR system. Always unplug your unit when a blackout occurs. Wait until power has been restored for a few minutes, then turn the system back on. Don't subject the power supply filters to these kinds of spikes.

The antenna leads should be disconnected to prevent any damage to the tuner, RF modulator, and electronic TV/VCR switching devices that could be caused by lightning or falling power lines contacting the incoming antenna wire or cable. Even lightning that strikes a community cable system wire several miles away can damage your unit. Many owners now wish they had disconnected the antenna leads before the storm arrived.

Transients

Other than electrostatic discharge, power line and antenna line transients are the most devastating form of noise interference in electronic circuits. Transients are large, potentially damaging spikes of voltage or current that are generated in the power lines that feed electrical power to your community. Spikes can be caused by lightning, by failure of utility company equipment, or by the on/off switching action that occurs when you use any electrical tool or appliance.

Most of these spikes are small and barely noticeable. But some voltage spikes as large as 1700 volts have been measured in home wiring. Residential areas experience more large-spike transients than commercial areas. The line filters in your power supply will protect the unit from some high-voltage transients, but occasionally a spike overcomes the power supply protection and gets to the logic circuitry. The usual result is erased automatic programs or a strange mode of operation, but if the spike is too large, sensitive circuit devices can be destroyed.

Your VCR power supply is normally not affected by the transients generated by on or off switching actions. These actions can produce a short-lived spike that is five times normal line voltage.

Spikes are not all generated outside the unit. When you begin to play a tape, for example, activating the tape drive mechanism produces a voltage spike inside your unit. Design engineers have placed capacitors in strategic locations on the circuit boards and in the tape drive mechanism electronics to carry the spikes harmlessly away to ground and prevent component damage. If any part of the spike reaches the circuit components, the devices are stressed and performance can become marginal.

Preventing Power Line Problems

If you live in an area where power outages or brownouts are common, where electrical storms occur when you aren't ready, or where your VCR occasionally shuts off and the clock display jumps to some strange hour, you need protection. You can prevent some power-line problems by using various forms of power-line conditioners, including isolators, regulators, and filters.

Isolators provide protection from voltage and current surges. They include transient suppressors, surge protectors, and other isolation devices. These devices maintain the line voltage at a power level even when the line supply is 25 percent over normal. Some surge protectors can filter out high-frequency spikes but cannot respond to slow, low-frequency transients. One form of surge protector is called a metal oxide varistor (MOV), a form of diode that will clamp the line voltage at a certain level, preventing overvoltage spikes from getting into your system. These devices are in-

stalled across the power-line wires leading into the unit. Isolators *cannot* provide protection against brownout or complete loss of electrical power.

Regulators act to maintain the line voltage within prescribed limits. They are essential if line voltage varies more than 10 percent at the unit, but they don't provide protection against voltage spikes and blackout.

Filters remove noise from the input power line. They short EMI/RFI signals to ground and remove high-frequency components of the signals from the low-frequency 60 Hz power line. Power-line filters work best when they are located immediately next to or at the front end of the power supply. Filters don't stop spikes, nor are they effective during low- or high-voltage conditions. When you select a power-line conditioner, consider these factors:

- Speed of response in handling voltage spikes.

- Ability to filter out high-frequency noise.

- Ability to handle repeated transients.

- Amount of line power it can handle.

- Range of input voltages it will handle in producing clean power out.

- Multiple outlets to handle several devices.

The amount of power protection you provide is up to you. Many VCR users are able to get along quite well with unprotected systems. Others prefer to operate their systems knowing that the VCR is protected against unseen environmental upheavals.

CORROSION

The metal connector pins on cables, circuit boards, chip pins, and other component leads are subject to corrosion, a chemical change in which the metal plating of the pins and sockets is gradually eaten away. Corrosion can damage not only the electronic connections in the VCR, but those connections in the mechanical section as well. Tape guides, solenoid plungers, levers, and bearings are all subject to corrosion. If the smooth tape guide becomes corroded the oxide will be scratched off the tape surface. This corrosive buildup on the guide can permanently damage each tape operated in the machine. There are three types of corrosion that can affect the VCR.

- Direct oxidation by chemicals.

- Atmospheric corrosion.

- Galvanic electrical corrosion.

Direct Oxidation

Direct oxidation is a form of chemical corrosion. A film of oxide forms on the metal surface and reduces the pin's contact with the socket or connection. High temperatures and high humidity both cause the oxidation process to accelerate. The metal is slowly worn away as the electrical contact surface is converted to an oxide and the oxide crumbles. It is this same oxidation that builds up on the tape guide surfaces and scratches the tape.

Atmospheric Corrosion

Chemicals in the air attack the metals in the VCR circuitry and metal chassis parts, causing pitting and buildup of rust. In the early stages of this corrosion, sulfur compounds in the atmosphere are converted to tiny droplets of sulfuric acid that lie on the surface of the connector pins. This acid eats away the metal causing pits to form.

When atmospheric corrosion is just forming, the plug contacts and metal parts can be wiped clean, restoring the metallic bright-

ness. But if the sulfuric acid is allowed to remain, long exposure converts the acid to a sulfate layer that can no longer be wiped away.

The effect is reduced electrical contact between the pins and their sockets. A layer of discolored rust that prevents any contact between the pins and their sockets causes an open circuit and can be located. It's the in-between stage, when the "almost-open" condition exists, that produces those hard-to-find intermittent failures.

In addition, you should be especially careful if you live near the ocean. The presence of salt spray or increased levels of chlorides can cause severe pitting of some metals.

Galvanic Electrical Corrosion

In galvanic corrosion, an electrolyte such as salt in moisture penetrates between the metal plating and the underlying base metal through a tiny crack in the metal plating of a pin or connector.

A small galvanic battery forms, and a tiny electric current flows between the two dissimilar metals. The plating surface becomes scaly and rough as the plating is slowly eroded. The corrosive action is concentrated on the underlying metal exposed at the breaks in the scale since this is where the galvanic battery exists.

The effect is the same as for the other forms of corrosion. Electrical contact between the pin and socket decreases, causing intermittent problems, until the scale is so complete that the electrical circuit is broken and the signals are blocked entirely.

You can actually start corrosive action by handling the connectors and boards improperly. The wrong way to handle printed circuit boards is demonstrated in Fig. 3.5.

Never touch contacts with your fingers. The oils on your fingers contain enough sodium chloride to begin oxidation action on those pins.

Fig. 3.5. *Handling a printed circuit board the wrong way can cause corrosion to occur.*

Corrosion Prevention

While metal storage sheds and cars can be spray painted to prevent rust (oxidation), this cannot be done on circuit boards, connectors, and metal chassis parts. The best preventive action is cleaning and proper lubrication. By keeping the electrical contacts, tape guides, and chassis parts clean, you can deter oxidation buildup and prevent the occurrence of intermittent operation and tape path trouble.

You can clean the pins on some plugs by reseating them periodically. Turn off all the power and carefully remove and push these devices back down into their sockets. This action will clean the pin surface and restore (or ensure) good electrical contact.

CAUTION

Always turn off the power, remove the AC cord from the outlet, and touch a grounded surface before touching anything inside the VCR.

Oxidation of some contacts can be cleaned with emery cloth, a soft rubber eraser, a solvent wipe, or a contact cleaner spray.

CAUTION

When rubbing to clean contacts, always rub along the pin (lengthwise). Rubbing lengthwise on the pins prevents accidentally pulling a pin contact up off the board.

If you use emery cloth, be careful not to grind away the metal plating itself. If you use a rubber eraser, keep eraser dust away from the unit.

Solvent wipes are found in the cleaning kits sold by many electronic supply companies. These wipes clean and lubricate the contact surface with a film that helps to seal out atmospheric corrosion without interfering with signal flow. Most solvent wipes are individually wrapped in small packages much like hand towelettes.

Spraying the pins with a contact cleaner spray (also available at most electronic parts stores) is an effective corrosion preventative. Contact cleaner wipes and spray are the best methods for removing oxidation.

There is a trade-off between preventing corrosion and preventing electrostatic discharge, because corrosive action is reduced with a reduction in the relative humidity, but ESD increases.

CAUTION

Select a contact spray cleaner that will not harm plastics and that contains no lubricants. Lubricants contained in spray cleaners may get into the chassis and cause belts and tires to slip. If the spray is not properly controlled, the plastics in the chassis or the cabinet may also be damaged.

Electronics manufacturers are aware of the effects of corrosion, and most connectors are made of a combination of metals that resist corrosion but are good conductors of electrical signals. You can choose the type of connectors to use for your cables. You can buy cables and connectors with tin alloy plating on the pins or with a thin gold plating. Although you will pay more for the gold-plated connectors, they can be worth the price because they provide superior contact reliability. While gold-plated contacts don't wear out as tin alloy surfaces do, even the tin surfaces take a long time to wear away. A sound, consistent cleaning program can really help.

There is one final note on the subject of corrosion. High temperatures will increase the corrosive action in the VCR system, so keep your unit clean and running cool.

MAGNETIC FIELDS

The effects of magnetism are especially important in tapes and tape drive mechanisms, since these two parts of the VCR are based on magnetic principles.

Each cassette tape is coated with magnetic oxide that has millions of tiny pole magnets randomly positioned on its surface. As the tape surface passes over the video head, a magnetic force is induced in the head by the signal current flowing through the head coil. This current causes the pole magnets on the tape surface to line up according to the video information being converted to voltage pulses in the head. This is "good" magnetism.

The voltages used in monitors and television receivers produce strong magnetic fields. These can be bad. If you accidentally place one of your tapes in the field, the tiny pole magnets on your tape's surface can change their alignment. Then when the VCR tries to read the tape, the head may misunderstand or misinterpret the information on the tape. You get garbage on the screen and in the speaker.

Magnetic flux can be caused by the presence of a high (115 V) voltage in monitors and televisions. A color television produces the strongest magnetic flux, but high-voltage areas of monitors, telephones, stereo speakers, ballast in fluorescent lights, and even power strips can be sources of offensive flux and can cause a loss of recorded audio and video. The strength of the flux field depends on the strength of the voltage, which can fluctuate depending on the amount of power being required by the equipment.

The moral is keep your tapes and hook-up cables away from power sources and magnetic fields.

PREVENTIVE MAINTENANCE FOR VIDEO TAPES

Two valuable components in any VCR are the video tapes and the VCR itself. All 8mm and VHS VCR systems use video tape as the storage medium. Since tapes and tape drive mechanisms are such critical components in unit systems, it makes sense to do all you can to protect and maintain them.

The video head in the VCR rides on the surface of the most vulnerable part of the storage system, the cassette tape. VHS video tapes are $\frac{1}{2}$-inch wide and are made of a plastic base and coated with a magnetic iron (ferric) oxide. The 8mm video tapes are approximately $\frac{1}{4}$-inch wide (8mm) and use a pure, nonoxidized metallic coating for their magnetic surfaces.

Now tapes are pretty sturdy, but they are sensitive to magnetic and electrical fields, high temperature, low temperature, pressure, stretching, and dust. Dust and little airborne fibers are particularly bad. With the VCR's video head riding on the surface of the tape, any tiny piece of "junk" lying on your tape looks like a huge boulder. A piece of your own hair is about 40 microns (.0015748 inch) thick. As you saw in Fig. 3.1, hair is a huge obstruction to the video head. Even dust and fingerprints on the tape surface cause obstacles to the even movement of the tape under the head. To protect the tape medium and the head, each tape has a door on the front of the cassette housing that shuts out dirt and fingerprints. The video heads inside the VCR rotate at 1800 rpm, and dust and other particles that may have slipped inside the door or settled on the tape are picked up by the tape guides and rotating video heads. A buildup of these particles can cause the tape to travel incorrectly along its intended path or prevent the heads from contacting the tape surface.

Tobacco smoke is harmful to tapes and tape drive mechanisms. The tars and nicotine that filter up into the air from the ends of cigarettes and cigars can settle on your VCR and form a gummy ash on any exposed surface including tapes and tape drive mechanisms. This material gums up the drive, eats into the video heads, and scratches the surface of your tapes. The effect is similar to taking a metal file to your favorite CD album. Avoid smoking or allowing smoking in your VCR area. If you can't do this, clean the system more often.

Tapes are further protected by storage boxes, or jackets. Use them. Don't let tapes lie around outside the sleeve inviting trouble from dust and dirt.

Not all tapes are created equal. Some tapes are manufactured to better standards, with highly refined magnetic coatings, and better binding materials. Naturally, these tapes are more expensive. Less-expensive tapes have less precise magnetic coatings and shed their layers easily. Compare tape specifications before you buy.

Depending on the quality of the tapes, the cleanliness of the VCR area, and the condition of the tape drive mechanism, tape life could be as short as a week or as long as many years. Assuming the quality, cleanliness, and condition factors are favorable, estimated tape life is based on actual passes through the unit while the video heads are in contact with the tape surface, rather than on total time of existence.

As the heads ride on the tape's magnetic surface, they cause tiny bits of the magnetic surface to rub off. Some of these loose magnetic particles are caught and stick to the heads. Gradually a layer of debris builds up. This layer has two effects on system operation. First, it makes the heads less sensitive to recording and playing back signals. Second, it causes an abrasive action on the tape surface.

As this unwanted layer builds up, it becomes ragged. This roughness scratches even more magnetic coating off the tape, until the remaining magnetic layer on the tape surface is too thin to support information storage. When the magnetic layer is missing from the surface, "dropouts," or spots, occur where data can no longer be stored. The tape fails to record or play back properly, and it becomes useless.

Keeping this unwanted layer from building on the video heads will help extend the life of your tapes and machine. The better tapes are less likely to easily shed magnetic particles, so the heads stay cleaner longer, and the tapes and heads last longer.

Toward Longer Tape Life

Here is a summary of what you can do to help extend tape life.

1. Buy name-brand tapes. Avoid "bargain" tapes.

2. Never touch the tape surface.

3. Never open the tape door on the cartridge (except during troubleshooting).

4. Store tape cassettes in their protective jackets.

5. Never force a tape into or out of a VCR. Place the tape in the unit carefully. Never slam the tape cassette holder down into the unit.

6. Store tapes in a cool, clean, and dry place.

7. Don't lay tapes in the sun. They warp just like stereo audio cassettes and computer floppy disks.

8. Never allow smoking near the tapes or VCR. Smoking causes tars to settle on the tape surface (and inside your VCR), gumming up the works.

9. Never set tapes down on or near monitors, televisions, or speakers. The magnetic fields can erase programs.

10. Avoid placing tapes near vacuum cleaners or large motors. Even freezers and refrigerators have compressor motors that can erase programs on your tapes.

11. Store tapes vertically. Storing tapes horizontally can cause the tape edge to bend and to bind in the cartridge. This causes damage to the edge or tape surface.

12. Don't put tapes through airport x-ray machines. Hand them to the security guard for inspection, and have them bypass the x-ray inspection process.

Fig. 3.6. Turning the head Height Adjust or V-Block Position screws could severely affect the alignment of the mechanics of the VCR.

PREVENTIVE MAINTENANCE FOR VCRS

What kind of preventive maintenance is there for VCR mechanisms? How can you test and maintain your own VCR?

Many VCR manufacturers state that there isn't any preventive maintenance that could be done by a novice. Then they describe head cleaning as the only routine maintenance that a customer should do and then only with a head cleaning cassette. Other manufacturers do not recommend cleaning tapes and suggest cleaning be done only be trained service technicians.

Why is this so? The manufacturers have great concern for the critical mechanical and electronic alignments inside the machine. They are concerned that untrained hands will accidentally damage this complex electromechanical device.

Many adjustments should only be made with high precision alignment jigs and electronic test equipment. There are even some VCR alignments that can only be performed by the manufacturer.

An incorrect half turn of alignment screws in the tape path, such as those shown in Fig. 3.6,

could require hours of realignment work for a trained technician. This could be very expensive for you.

> ### CAUTION
>
> *Make no alignments or adjustments on any tape path or other mechanical or electronic components without being absolutely sure of what you are doing. Follow only the prescribed procedures in the machine service literature.*

The repair business is big business. The less preventive maintenance you do or have done, the sooner your VCR will start giving you problems and the more work you will provide for repair companies. Here are some facts.

Cleaning the Heads

Heads need cleaning to remove debris and unwanted coatings that build up on the leading edge of the head tip. Head cleaning is preventive maintenance that you can do using any of the various head cleaning cassettes that are available. The "wet" kind works with a cleaning solvent.

Some head cleaners are abrasive and can damage the head if they are used for too long. If you buy this type of cleaner, you must use it just long enough to remove the unwanted buildup but not long enough to damage the heads. Nonabrasive head cleaners have been marketed for some time now. One head-cleaning kit uses fabric-tape dampened with cleaning solvent. With this kit, you sprinkle cleaning solvent on the fabric and then insert the cassette into the VCR for spinning action head cleaning.

Since any cleaning cassette works by rubbing action and chemical reaction between the tape fabric and the video head, there is potential for abrasion. So be careful not to leave the cassette in the VCR for too long. Carefully follow the instructions that come with the cleaning kit.

Video heads can also be cleaned with certain types of alcohol, or other special purpose head cleaning solutions, and with lint-free materials like the ones shown in Fig. 3.7.

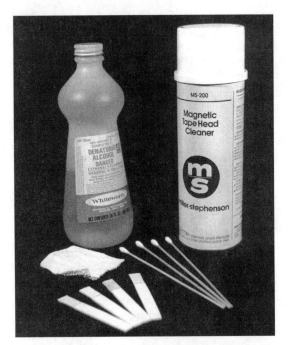

Fig. 3.7. A chamois, some swabs, and the solution or spray used in cleaning a VCR.

Special cleaning materials such as cellular-foam swabs, chamois leather cloth, or a piece of special lint-free cloth are good materials to use for manual head cleaning. Regular cotton swabs are dangerous because the cotton fibers can catch or pull away and lie in the tape path or on the head. They can also catch on the brittle ferrite, or metallic head chip, loosen it from its mounting, or actually break the core and ruin the head.

Special cotton swabs can be used to clean the tape path. These have tightly wound cotton at one end and are suitable only for the tape path—not the heads. Denatured alcohol or methanol can be used as a cleaning solvent as can typewriter cleaner or trichloroethane. The solvent must not leave a residue when it evaporates, so most other alcohol solvents should be avoided. Always have plenty of ventilation and make sure the solvent has evaporated before you operate the machine.

The frequency of cleaning depends on how much the VCR is used and the type of cassettes that are used. A quality tape is good for several hundred passes through the machine. Remember both record and playback time must be considered.

Almost all manufacturers recommend that video heads and the entire tape path be cleaned at least every 500 hours of operation. This includes record and playback hours. If rented movies are played in your unit on a regular basis, more frequent cleaning is a good idea. You never know the condition of the previous user's machine and the debris that machine deposited onto the tape.

Bargain tapes are good for about one-tenth the life of a name-brand tape. This means that instead of 200 hours of record/playback time, you might get 20 hours or less before the head gets caked with oxide or the tape surface becomes too worn to record or play back.

Keeping the cassette door closed on the VCR unless you are inserting or removing a tape will help prevent dust and dirt from getting inside. It also prevents unwelcome visitors (insects and even mice) from climbing into the unit. Believe it or not this does happen!

Preventive maintenance also involves regular cleaning and lubrication of mechanical parts, as well as replacement of worn or stretched rubber tires, belts, and torque limiter assemblies.

Cleaning Cassettes

Before you groom your VCR using a cleaning cassette, there are some things you should know. Generally speaking there are two types of cassette cleaning tapes on the market today. One is known as a "dry" cleaning tape, the other as a "wet" cleaning tape. The wet cleaning tape is a chamois material with drops of cleaning solution placed on it. In some cases the material is much thicker than normal video tape.

Once the wet cleaning tape has been moistened as directed by the instructions, it is placed in the video recorder and the VCR is switched to the play or record mode. The moistened chamois is designed to travel in the tape path and, as it is pulled against the video and audio heads, the heads are wiped clean. It takes less than a minute to wipe off any contaminants that may have accumulated on the video head and in the tape path.

The dry video cassette cleaning tapes are used the same way, but do not require a cleaning solution. Dry tapes come in two basic forms. One of these uses a fibrous material to wipe over the heads with a gentle rubbing action. The other type looks like normal video tape, but is more abrasive. The abrasive material actually scrapes off contamination.

Video cleaning cassette tapes have occasionally damaged video tape recorders. The cleaning cassettes using a chamois material have been known to unscrew or break tape guides. They have also lodged under the video heads and broken the assembly. A contributor to this problem is the thickness of the chamois in the tape path. At least one manufacturer of "wet" tapes has reduced the thickness of the cleaning tape (Fig. 3.8) to decrease the potential for damage.

Fig. 3.8. An improved wet tape cleaning kit with thinner cleaning tape.

If the instructions for the wet cleaners are not carefully followed, it's possible for the cleaning chamois to become saturated with foreign particles and cause machine damage. This can occur if the cleaning cassette is used too many times. If this happens, the cleaning tape can deposit contamination in the tape path.

Occasionally a video head becomes so clogged by excessive use that a wet cleaning tape may not completely remove the contamination, and the video head assembly will seem defective, even after several cleaning attempts. You may have to use a dry tape.

As previously mentioned, there are two types of dry cleaning cassette tapes. One type contains cloth fibers, which are woven into the tape surface. The other looks like normal video tape. The latter cleaning tape is very abrasive. The manufacturers of these tapes state that operating a cleaning cassette in your machine for 30 seconds produces the equivalent head wear of running a normal tape through your machine for 6 to 12 hours. These dry video tape cleaners are not as likely to damage the tape guides and rollers because the tape is the same thickness as normal video tape. Nor are we aware of any of these tapes breaking a video head. However, if used too often, this type of cleaning system will cause excessive wear on the video heads and tape path.

Manufacturers of abrasive cleaning tapes recommend that you only use them when you actually experience symptoms of a clogged head (snow in the screen display or no video). If you use this type of cleaning tape too often, you will wear out the video heads faster than normal.

There have been instances in which a wet cleaner and repeated hand cleaning did not remove buildup from the video heads. Then an abrasive dry head cleaner was used for a few seconds, and it did the job.

The other type of dry cleaning tapes use fibers in the tape that are not as abrasive.

If you choose to use a cleaning tape, carefully follow the instructions provided. This is the key to getting cleaning cassettes to work properly for you. If excessive snow or no video persists after using these tapes, have your unit checked by a trained servicer. The failure may be in the electronic circuitry or in the video heads themselves.

A good-quality, head-cleaning cassette is worth owning. Nothing is more frustrating than to be enjoying a good movie with your friends and to suddenly lose video because of a clogged head. One head-cleaning cassette can make an evening.

If your recorder is still under warranty, check with the manufacturer first before you use a cleaning cassette. Use of a cleaning cassette may void your warranty.

Follow these hints to clean video heads with a wet cleaning cassette:

1. Read and understand directions thoroughly.

2. Turn on the VCR power.

3. Dampen the cleaning cassette with the solvent that comes with the kit.

4. Insert the cleaning tape in the unit.

5. Place the unit in PLAY.

6. After the instructed time, stop the VCR and remove the cassette.

7. Turn off the VCR.

8. Let the tape path dry completely before operating the system.

To clean video heads using a dry cleaning cassette, follow steps 1, 2, 4, 5, 6, and 7 in the preceding list.

Metal Tape Format VCRs and Brown Stain

All VHS format VCRs use tape with an oxide formulation for the tape's magnetic coating. VCRs based on the 8mm format use tape with a metallic surface for magnetic signal retention. These metal-tape VCRs require the use of metal heads to sufficiently magnetize the tape surface used in these machines. (Oxide format machines such as VHS and SVHS typically have ferrite heads for recording information on the oxide tapes that these machines use.)

A phenomenon called *brown stain* can occur when metal heads are used to record onto metal tape (such as 8mm). This phenomenon is an unwanted brown deposit on the head surface where it contacts a metal tape. Brown stain buildup is believed to be caused by a chemical reaction between the binder agent that "glues" the metallic surface onto the tape base and the video head. Heat generated by the frictional contact between the rapidly moving video head and the tape surface is thought to be the catalyst for the chemical reaction.

Like grass stains on your favorite slacks, brown stain cannot be easily removed. Hand cleaning the video heads and using a wet cleaning tape in the machine normally do not remove this buildup. Abrasive head cleaners must be applied to the video head with the brown stain deposit. The problem is that the stain is heat-bonded to the surface of the video head. Therefore, it must be "scraped" off by the scrubbing action of an abrasive tape.

The challenge is to use the abrasive cleaning tape just long enough to remove the brown stain deposit, but not so long that it wears down the video head itself.

Some VCR repair centers try to clean a clogged metal video head by hand first. If successive hand cleanings don't restore proper video head operation, they then examine the surface of the head with a special microscope. They want to detect if brown stain buildup has occurred. If brown stain is present, they run an abrasive head cleaner tape through the machine for a few seconds. Stopping the tape, they recheck the surface of the head using the microscope. If the contamination is still present, they run the abrasive tape again and recheck the head. This process is repeated until the stain is no longer present. By running the abrasive tape in short bursts, they remove the brown stain deposit without wearing down the surface of the video head. The principal disadvantage of this repair method is that it is quite time consuming. In addition, the repair center must invest in an expensive special microscope. Yet, removing brown stain without wearing down the head surface itself is a distinct advantage over head replacement.

Currently, many service technicians are unaware of the brown stain phenomenon. These technicians will declare a video head defective after a few attempts at cleaning by hand. As a result, many salvageable heads are needlessly scrapped, increasing the cost to the customer.

Cleaning the Video Heads Manually

To clean the video heads and tape path manually, first be sure you have these supplies:

- Flat head screwdriver.
- Phillips head screwdriver.
- Adequate lighting.
- Tray to hold loose screws.
- Cleaning material.
- Cleaning solution.

When you have assembled the necessary equipment, turn off the power and unplug the VCR from the AC socket. Remove the top cover

of the VCR using the procedures found in the service manual. Also remove all internal shields from around the video heads and tape path to provide more working room inside the unit. Be sure that ground wires removed from the shields are not touching the unit.

Now, reconnect the power and turn on the VCR. Place a tape in the unit, press PLAY, and observe where the tape comes in contact with the tape guides, capstan shaft, audio and erase heads, and the rotating video head drum as shown in Fig. 3.9. Also note how the tape is guided along a ledge on the lower stationary portion of the head drum.

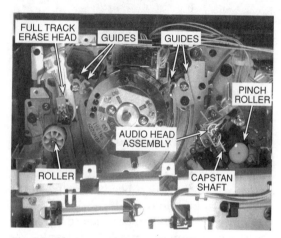

Fig. 3.9. The tape path in a typical VCR.

Stop the unit, eject the tape, turn off the VCR, and pull the power cord plug out of the electrical socket. Moisten the cleaning swab material with cleaning solution and clean all of the guides where the tape makes contact. Before you clean the video head, read this.

CAUTION

Apply only the slightest pressure possible to the video heads. Move the cleaning instrument in the same plane as tape travel. Heads are fragile, and cleaning in an up-and-down motion can break the heads.

Figure 3.10 shows the upper drum (upper cylinder), a video head, the lower drum, and the lower tape guide ledge.

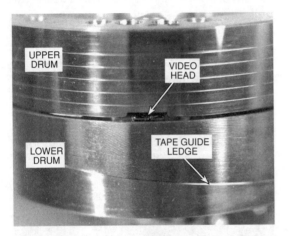

Fig. 3.10. The upper drum and the lower drum with its tape guide ledge are clearly visible in this photograph. A video head can be seen between the upper and lower drum assemblies.

Hold the rotating portion of the video head at the very top of the drum to stop it from moving, and with a gentle side-to-side stroking action, clean the video heads and the upper drum assembly. Hold the head drum only at the very top. Carefully rotate the drum to clean the rest of the upper drum. While cleaning the lower (stationary) part of the drum, carefully rotate the upper video head drum keeping the heads away from the cleaning instrument. This prevents accidental contact with the fragile heads. Pay special attention to the lower tape guide ledge around the stationary portion of the drum. Any buildup here will cause tape path error.

Next, clean the audio/control track and erase heads. When you do, move the cleaning swabs in the same plane as the direction of tape travel. A forward and back movement is good.

Clean the capstan shaft and rubber pinch roller. Do this last because the capstan frequently has the most debris on it and the pinch roller will get your cleaning swabs black. If you did this first, you might add contaminants to the heads or guides.

Now, throw away the cleaning instrument. If it were used again, it could cause future contamination of the heads. Don't be afraid to use more than one cleaning swab on the tape path each time you clean; this is no time to be cheap. If a dirty cleaning stick is dipped into a bottle of cleaning solution, the dirt from the stick will come off in the solution and eventually find its way back into the system again.

Let surfaces dry completely before you reassemble the unit. When the heads and tape path are dry, reinstall the shields, attach any grounds, and replace the cover. Attach the AC cord, make a test recording, and play back to check the audio and picture quality. (*Note:* If you have any problems refer to the specific troubleshooting charts found in Chapter 4 of this book.)

Video Head Cleaning Interval

Cleaning your tape path is like changing the oil in your car. You do it when you feel you've driven enough miles or when the oil looks dirty. Most VCR manufacturers recommend cleaning heads by hand at least every 500 hours. Some repair centers suggest cleaning every six months; others suggest you don't clean the heads until the TV picture has a lot of snow in it or the audio sounds bad. If you live in an area that gets a lot of smog, you may want to clean the heads more often. In any case, it won't hurt to clean them at least annually.

Preventive Maintenance for Tape Drive Mechanism

Mechanical preventive maintenance is not needed as frequently as head cleanings. Belts, rubber tires, and torque-limiting clutch assemblies are changed or cleaned every 1500 hours of operation. Replacement recommendations vary among manufacturers but most are in the 1000 to 2000 hour range. Rubber will stretch and crack with time, so even if you don't use the VCR a lot, you should consider replacing the rubber parts every 18 to 24 months. Unfortunately, failures always seem to occur at the wrong times.

One customer stated that she had 20 children over for her nine-year-old's birthday party. She rented movies as the main entertainment because it was raining heavily. Guess what happened? A two-year-old belt in the VCR broke in the middle of the video movie. Preventive maintenance is designed to prevent such experiences, but it only helps if you do it regularly. The Appendix has a suggested preventive maintenance schedule for VCRs.

This procedure is not recommended for the person who is not mechanically inclined or who doesn't like getting a little dirty. If this is you, refer the job to a trained servicer. If you decide to try anyway, you'll need a service manual which you can buy from the manufacturer. In addition, there are some important things to know before you start.

Don't get grease or lubricant on the belts, pulleys, or rubber tires because oil or grease can cause these parts to slip and lose traction. Loss of traction prevents proper operation. And use only small amounts of lubricant. Half a drop of oil goes a long way in a VCR. Too much lubrication is bad because oil can migrate to and damage other parts of the machine.

Use lightweight high-grade machine oil or an equivalent. Transmission fluid is an excellent substitute because it is light, clean, and does not gum up with time like many oils. But it can also migrate easily, so use it sparingly.

Use only the greases that are recommended by the manufacturer. Some units require different types of grease at different points. The wrong grease (such as standard petroleum grease) will react with the nylon, plastics, and metals and cause the moving parts to stick. Petroleum greases gum up faster than most synthetic VCR greases, but even some synthetic greases cause the machine to malfunction much sooner than expected. Fortunately, new silicon- and lithium-based greases have recently been developed that don't thicken.

Never disassemble anything in the tape path including the blocks and supports that the guides travel and seat in. Alignment of these areas is critical and in some cases cannot be done except at the factory (and most of these factories are overseas).

Be very careful if you remove any gears, linkages, or levers in the mechanical section. You may mistime the crucial operation of the drive mechanism. Many current VCR models use a single motor to perform several functions including loading and unloading tape, fast forward and rewind, play, and tape cassette tray elevator operation. If any precisely phased gears are incorrectly positioned during reassembly after cleaning and lubricating, the unit will not function correctly. Before removing any of these mechanical parts, check the mechanical alignment section of the service manual for correct gear phasing and linkage alignment procedures.

CAUTION

If you do not feel comfortable performing the manufacturer's alignment methods as described, or you do not understand the complete procedure, do not attempt to remove these parts.

You may be able to remove most of the old grease and replace lubricant without removing these critically aligned parts in the drive mechanism. You won't be able to remove all contaminated lubrication from the gear shafts and opposite side of gears without removing the parts. Therefore, if you are not proficient at this task, it's best to take this job to an experienced service technician.

If you decide to remove, clean, and lubricate the entire drive mechanism, thoroughly remove all the old grease or oil before applying the new. And don't forget the shaft and collars in which the pulleys are mounted. Remove the pulley shaft and wipe the shaft and collars with a cleaning solvent on a surgical cotton swab, like the one you used on the tape path. Remember to clean and lubricate the pulley shaft before reinstalling it.

Never force anything. If it won't move, it's either incorrectly positioned or it isn't supposed to move. Never use spray cleaners or spray lubricants in the chassis. Chances are good that the spray will cause more problems than it will prevent.

Check all belts, rubber tires, and the rubber pinch roller for wear or age before removing them. Stretch the belts as shown in Fig. 3.11, and pull the rubber slightly away from the tire rim to check for cracks as is being done in Fig. 3.12.

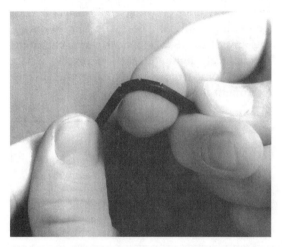

Fig. 3.11. Stretching and bending the belt out can reveal hidden cracks.

Fig. 3.12. Rotate the tire away from the rim and look carefully for signs of cracking.

If the belts or tires appear dry, cracked, worn, or stretched, they must be replaced. Worn pinch rollers don't crack like belts and tires when they wear or age. Defective pinch rollers look shiny or glossy. A new pinch roller has no sheen. Wear makes the roller shiny, and the glazed surface won't clean away. Some pinch rollers can be replaced without mechanical realignment. Others require mechanical disassembly and realignment after part replacement. If you must replace the pinch roller, note if any linkages or levers require removal and realignment after roller replacement. It's important to have replacement parts at your side *before* you remove the old ones. This way you'll remember which belt went where. Also, if the parts are not readily available, your unit is not out of commission for an extended period of time.

Read the service disassembly instructions and carefully note any warnings. Do only those things that you are confident you can handle. If you are unsure about a procedure, don't do it. You may end up paying a servicer to untangle a mess you created.

Take notes or use some other method to remember where each of the screws, washers, etcetera, goes in the unit. The sizes, lengths, and threads are often different for each area in the chassis. An ice cube tray or a muffin tin are both good small parts holders. The screws from each dismantling step can be organized in these holders to help you during reassembly.

As you work, watch for tiny washers beneath the tires and pulleys which often stick to the item being removed and disappear onto the floor. These washers, often made out of graphite, are essential to the smooth operation of the unit.

With these observations in mind, you're now ready to disassemble your own unit. Remember to work patiently and carefully.

First, disconnect all power and antenna connections. Place the unit on a flat, clean working surface. Now, carefully reread the service literature, paying special attention to those sections on which you are going to be working. Remove the top and bottom covers of the VCR as directed in the service data. Remove the cassette tray assembly, and open the necessary circuit boards on the top and bottom. Instructions for these procedures are in most service manuals.

Each VCR has a "service position" for opening the boards and working on the bottom part of the chassis. This may require standing the unit on one of its sides. On most units the circuit boards are on hinges that swing open for easy access as shown in Fig. 3.13. These boards don't require complete removal.

Fig. 3.13. *Most printed circuit boards swing out for easy access during troubleshooting and repair.*

Occasionally, manufacturers identify the disassembly screws in the circuit boards and cassette tray assembly with special colors—red is common. If your machine is one of these units, your job will be much easier.

Remove all of the belts and clean the pulleys the belts ride in with a special cotton swab. These swabs are ideal because the cotton is tightly wound on the stick and doesn't fray easily. Figure 3.14 shows the belts and pulleys to be cleaned on one VCR.

Fig. 3.14. *Check and clean the belts and pulleys on both sides of the chassis. Shown here is the bottom side of one VHS VCR chassis.*

Carefully remove the tires on the top side of the unit. Be sure to collect those tiny washers! Reel tables, a torque limiter assembly, rubber tires, and belts are shown in Fig. 3.15. The reel table, lower left, has a tire that is shedding rubber dust in the slide lever to its left.

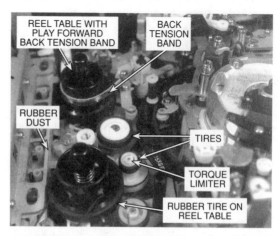

Fig. 3.15. *Rubber dust is quite visible on the slide lever to the left.*

Pull the underside pulleys (attached to the shafts associated with the tires you just removed from the top of the unit) out of their collars. Any more washers?

With a swab soaked in solvent, carefully clean the metal collar (often brass colored) in which the pulley shaft spins. Figure 3.16 shows a pulley shaft after removal. The finger points to old grease buildup which should be wiped off.

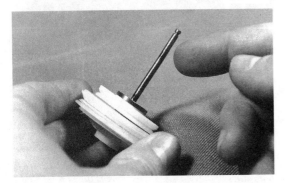

Fig. 3.16. Carefully remove the grease buildup and lubricate the pulley shaft before reassembly.

After the solvent has dried, apply a very small amount of lubrication to the pulley shaft; replace the shaft and new tire. When you replace the tire, make sure it spins freely. If the tire is placed on the shaft without a little clearance, it will friction bind after the unit warms up and the collar swells. If the tire snaps on, as many do, there is no clearance adjustment. Don't forget the washers!

Replace the worn torque limiter assemblies. If these are on shafts, clean and lubricate the new shaft just as you did the shafts of the rubber tire pulley.

Install new belts on the unit. Remove any twists in the belts. A twist can cause a belt to work out of the pulley groove. Replace the bottom circuit boards and return the unit to its normal operating position.

Now you have a decision to make. Part of the mechanical preventive maintenance involves checking and adjusting tensions and torques in the cassette drive mechanism. These adjustments require expensive gauges like those shown in Fig. 3.17, so to perform

Fig. 3.17. Typical gauges used in tension and torque adjustment.

these checks you will need to borrow or purchase the gauges. If this is inconvenient or not desirable, have a service shop check and align the tension and torques for you.

The following section on adjusting torques and tension is essential to extending the life of video tapes and video heads. If you aren't going to perform this adjustment, you can skip to the checkout and reassembly instructions. On the other hand, you might want to read this section so that you understand why tension and torques are so important.

Adjustment Tension and Torques

The video tape must wrap around the video head drum under precise tension. If the wrap is too tight, the heads and tape surface will wear out from excessive friction. If the wrap is too loose, the heads do not contact the tape properly and the picture is lost. Correctly setting this tension is called *play or forward holdback tension adjustment.*

Conversely, too much play tension will cause the tape surface to wear faster than normal from the increased friction between the tape and the parts it contacts in its path. Increased friction from high tape tension will also cause video heads, tape guides, and audio heads to wear out faster.

Torques are similar to tension but they serve a different purpose. The torques are the amount of power the reel tables have as they rotate. For example, the tape is pulled through the VCR in fast forward or rewind by the torque of the engaged reel table. There is a play torque too. The tape is pulled back into the cassette in play or record by the take-up reel table that's on the right side facing from the front in a VHS unit. The amount of pull this reel table has is known as *take-up* or *play torque*. If this pull is too high, the tape may be stretched. If it is too low, the tape won't be pulled evenly into the cassette and may not be pulled at all. Tape will spill into the machines and be damaged. By the way, the VCR senses the take-up action, and if it senses no take-up torque and the tape is spilling, the unit will stop functioning.

When the VCR is in the STOP mode, brakes are applied to the reel tables to halt the tape movement. The amount of braking force is called *brake torque*. If the brakes aren't doing their job, the tape comes crashing to the end of the leader each time you FAST FORWARD or REWIND, increasing the possibility of the tape snapping. If the brake torques are too high, tapes will stretch each time the VCR is stopped. Proper brake torques are very important.

There are several ways to check and adjust tensions and torques. A typical device to measure tape tension is shown in Fig. 3.18. Figure 3.19 shows how to measure rewind torque and supply reel table brake torque.

Fig. 3.19. Measuring rewind torque using a special test instrument.

Consult your individual service manual for the method and types of gauges to use. Basically, place the unit in the mode to be tested, and measure the tension or torques with the appropriate gauge. Some units use direct drive reel tables in which the motor and reel tables are one part and there are no connecting belts or tires. These must be adjusted electronically. If tires and belts are used to drive the reel tables, play holdback tension is controlled by a felt brake band and a spring, while other tension and torques are derived from the belts, tires, and torque limiter assemblies (felt clutches). Follow this series of steps for checkout and reassembly:

1. Replace the cassette tray assembly.

2. Connect the VCR to the antenna and TV.

3. Plug in the power cord and energize the machine.

4. Make a test recording and play it back. Use an expendable tape so you don't lose valuable material.

5. Observe the tape travel. Watch for damage to the tape as it moves through the machine. Watch the tape in both load and unload operations as well as fast forward and rewind functions.

Fig. 3.18. A device measuring tape tension adjustment.

6. If you notice incorrect operation, stop the unit and refer to the specific troubleshooting section in Chapter 4.

7. Turn off the VCR, disconnect the power cord, and install the top and bottom covers.

SUMMARY

This chapter has covered aspects of routine preventive maintenance necessary to keep your VCR in peak operating condition. It discussed six major contributors to VCR failures: excessive temperature, dust buildup, noise interference, power line problems, corrosion, and magnetic fields. For each of these factors, this chapter presented one or more preventive countermeasures. You learned that video cassettes are to be protected from dust and dirt, discovered how tapes and video heads can be damaged, and most important, learned how to extend the life of tapes and tape drive systems.

CHAPTER REVIEW QUESTIONS

1. What are two reasons for preventing dust from building up in a VCR?

2. Why should video heads be cleaned on a regular basis?

3. Why should video tapes be stored vertically?

4. When cleaning video heads by hand, why must you move the cleaning instrument along the same plane in which the tape travels?

5. Why is it important to use only manufacturer-recommended grease to lubricate the mechanism?

6. What are two negative effects of increased tape play tension in a VCR?

Specific Troubleshooting and Repair

Chapter 4 is a specific troubleshooting and repair guide focusing on a large variety of VCR failures. The section is divided into six parts.

- *Audio problems.*
- *Video problems.*
- *Color problems.*
- *Power supply problems.*
- *Improper operation or no functions.*
- *Clock and timer problems.*

Each fault can be associated with one of these areas. By letting your "fingers do the walking" through the Troubleshooting Index, you can quickly locate the page where your particular problem is addressed.

Part I of the Troubleshooting Index covers audio-related problems including no sound at all, varying levels of loudness, unwanted audio still on the tape, weird-sounding audio, and static-filled sound.

Part II discusses all the symptoms related to video problems including no video, snowy pictures, intermittent video, strange video shapes, and bands of interference on the screen.

Color problems are addressed in Part III. This section covers no color and strange color.

Part IV addresses all symptoms that can occur after you turn your machine on and discover power partially out, or totally gone. This includes the problems of no working functions, only eject working, and operational malfunctions.

In Part V you'll find the symptoms related to improper or inoperable functions such as sluggish rewind, mode functions that change after initiation, no play, fast forward, rewind, or record capability, and tape spilling into the machine.

Clock and timer problems are covered in Part VI. This section covers the situations where the timer loses 10 minutes every hour and where the machine won't auto-record using the timer.

All troubleshooting charts apply to both VHS and 8mm format VCRs unless otherwise indicated.

TROUBLESHOOTING

VCR failures can be categorized as either electrical or mechanical. Each part of this chapter is subdivided into specific failures and provides symptom, troubleshooting steps, and possible cause. The step-by-step troubleshooting instructions quickly guide you to the electrical or mechanical cause for most VCR problems.

Many malfunctions are caused by contamination or foreign objects blocking the tape path or mechanical moving parts.

Isolating a failure to a mechanical or electrical breakdown can sometimes be challenging. Mechanical problems may cause the electronic circuits to act strangely and appear at fault. There is a close relationship between the electronic and mechanical functions in a VCR.

This chapter will help you identify and localize a fault to not only an electronic or mechanical failure but, in most cases, to the circuit or function in which the problem has occurred. Possible causes are listed to assist in determining the exact malfunction. The possible causes listed under the "Advanced Troubleshooting Required" headings are not listed in a specific order. The faults, however, are categorized into functional areas. By matching the symptom that you discover with the closest description in the Troubleshooting Index, you will quickly be able to determine if the problem is one that you can (or want to) correct yourself.

As you progress through the troubleshooting charts, you can decide if you have the ability to service the unit yourself. Once you step past a certain point, most problems require service center action, or at least advanced troubleshooting techniques and test equipment. This book does not assume you have these skills. If you'd like to try the advanced methods, refer to Chapter 11 for guidance. Always observe good troubleshooting procedures. However, if you are not experienced in electronic servicing or don't have the right test equipment, you should refer the advanced troubleshooting to a qualified servicer. Nearly all component replacements require soldering skills. Be sure that you understand the information contained in Chapters 5–11 before attempting any advanced troubleshooting.

WHEN YOU SHOULD CALL THE REPAIR SHOP

The message is clear: if, during your repairs, you reach a point where you're unsure about your techniques or skills, STOP and refer the repair to a qualified technician. It's less expensive to pay up front for a repair that requires skills beyond those that you have, than to pay to undo a "mistake."

Several years ago, an article in *TV Guide* magazine described an Electronics Industries Association (EIA) study showing that 32 percent of all VCR failures were corrected by cleaning. Cleaning, belt and rubber tire replacement, and mechanical adjustment are the most common cures for VCR malfunctions. The next most common repair action (15 percent) was to the rewind mechanism; this was closely followed by video head replacement (11 percent). We've found that video head problems account for less than 6 percent of all failures in our business. Nevertheless, most VCR failures can be solved by a good cleaning, belt and tire replacement, and mechanical adjustment.

Cleaning and simple mechanical adjustments/replacements are types of service you can do. This chapter is designed to help you diagnose and correct most failures, or decide whether to attempt the repair or refer it to a trained technician.

SPECIAL TERMS

In the troubleshooting charts there are some VCR terms you should be familiar with. These terms are described here.

E-to-E (electronic-to-electronic): the VCR video and audio output signals generated when the TV/VCR button on the unit is in the "VCR" position. In E-to-E, the VCR output signal to the TV is the same signal received by the VCR's tuner or line input jacks and processed by the unit.

Linear track audio: the audio signal recorded on the tape by the stationary audio head of all VHS VCRs. Current 8mm models don't record audio on a linear track.

AFM (audio frequency modulation): an option available in VHS VCRs for recording audio in high fidelity stereo. In this method, heads mounted on the rotating video head drum assembly produce high-quality audio playback. AFM is the method used by all 8mm format VCRs to record audio.

Hi-Fi audio: a common term describing AFM record/playback.

PCM (pulse code modulation) audio: a technique for recording audio digitally on 8mm tape. PCM is available as an option on some 8mm format VCR models.

ATF (auto track find): a method used to perform tracking control in an 8mm VCR. This process uses four pilot frequencies recorded on the tape with the video signal to automatically perform playback tracking.

USING SERVICE LITERATURE

For those problems that require service manual support, we recommend you obtain the manufacturer's service manual covering your machine. These manuals are available from most manufacturers. The service manual is different from the operator's manual. It is a detailed troubleshooting manual that includes circuit voltages as measured, oscilloscope waveforms at key points in the circuitry, and circuit resistance measurements. Since many VCRs are designed and built in Japan, a mixture of Japanese-to-English writing style sometimes called "Japanenglish" or "Jenglish" is often found in much documentation. This makes some of the expressions difficult to understand.

Service manuals frequently contain block diagrams, a theory of operation, expanded troubleshooting charts, and a lot of additional service information. Combined with this VCR troubleshooting and repair guide and a service manual on your own machine, you are in optimum position to conduct your own advanced troubleshooting and repair.

NOTE

The following troubleshooting techniques may require soldering. If you are uncomfortable with this, go as far as you can without soldering and then consult your local VCR service center. Desoldering or soldering on your VCR may void your warranty.

I. AUDIO PROBLEMS

SYMPTOM: Won't record audio. Video O.K.
TV has audio on normal TV. (VHS only)

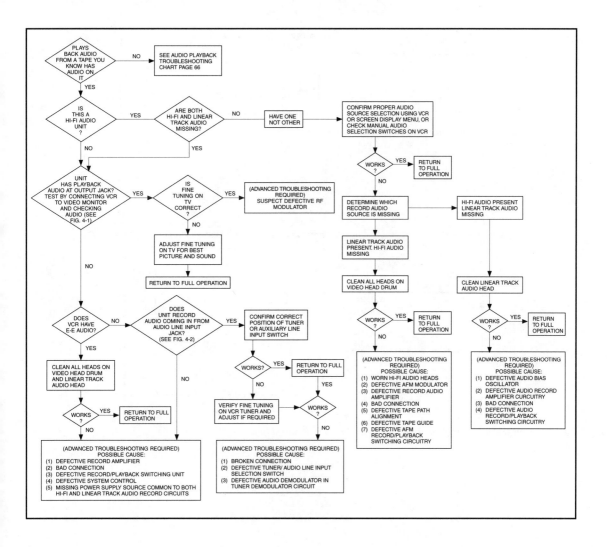

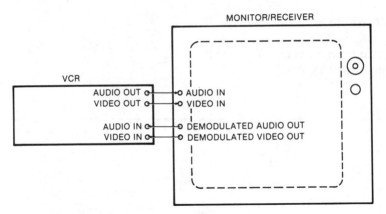

Fig. 4.1. *Connect the VCR Audio Out to the monitor Audio In.*

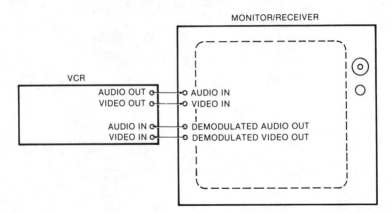

Fig. 4.2. *Connect monitor Demodulated Audio Out to VCR Audio In.*

SYMPTOM: Won't record audio. Video O.K.
TV has audio on normal TV. (8mm only)

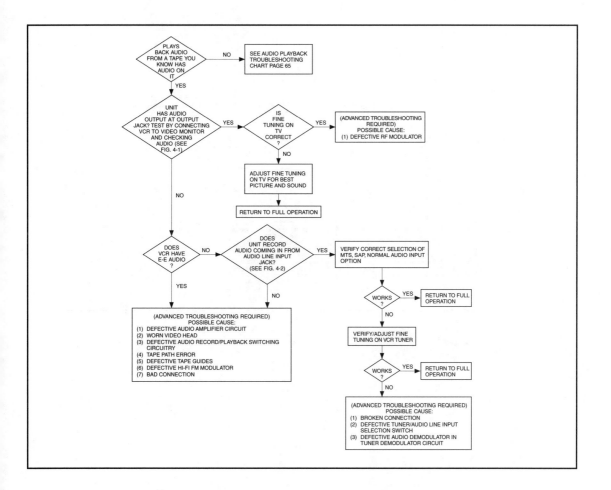

SYMPTOM: Playback volume level fluctuates.
(VHS only)

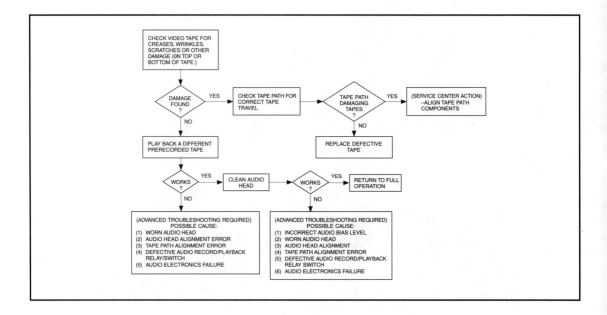

SYMPTOM: No audio during playback. Video
O.K. (VHS only)

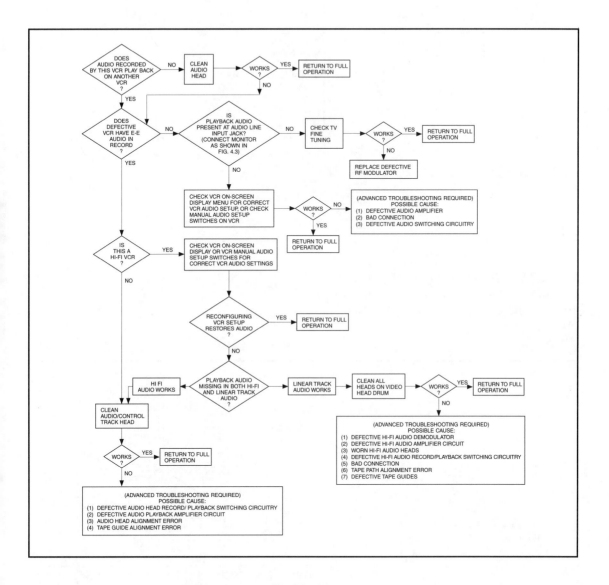

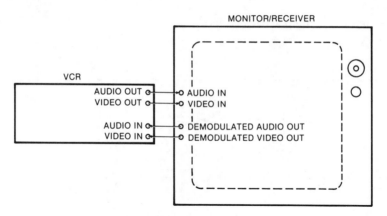

Fig. 4.3. *Connect VCR Audio Out to monitor Audio In.*

SYMPTOM: No audio during playback. Video O.K. (8mm only) (*Note:* May have horizontal noise bands in top one-third of screen display.)

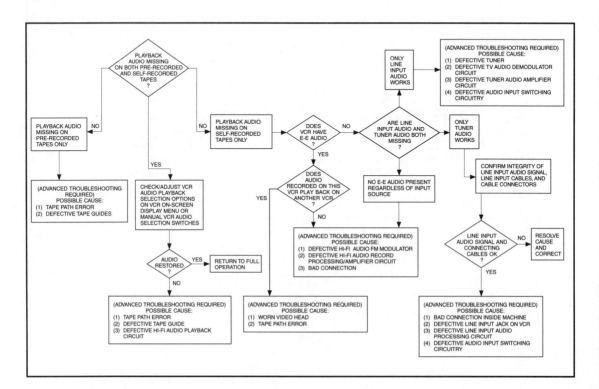

SYMPTOM: Audio has wow and flutter. Sound varies in pitch (not volume) and is especially noticeable on sustained notes in music. (VHS only)

Scenario

You just made a recording of one of your favorite musicals. When it is played back the sound seems to vary in pitch at regular intervals. The problem is especially noticeable in the music portions of the recording. The pitch of the sustained tones seems to vary. The volume level doesn't change but the pitch changes. One description of the symptom might be to say the VCR has a bad case of vibrato.

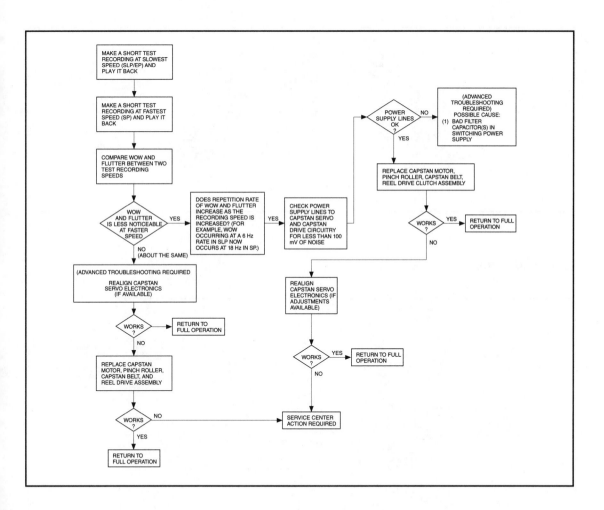

SYMPTOM: Audio from previous program still heard. Records video O.K. (VHS only)

Scenario 1

You are anxious to record a program on TV. You select a video tape you have recorded on before and quickly start recording. When the program is over, the tape is rewound and you start to play back the new recording. You can hear the sound of the new recording OK, but mixed with the new sound is some sound from the program you previously recorded on that tape.

Scenario 2

Same as Scenario 1 except that during playback, you hear only the previous audio. (VHS only)

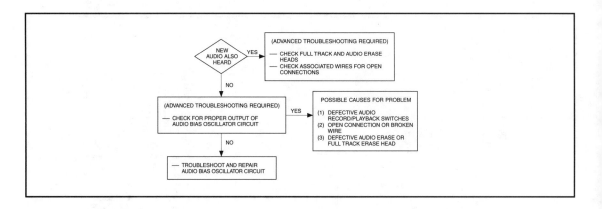

SYMPTOM: Buzz or hum in audio playback.

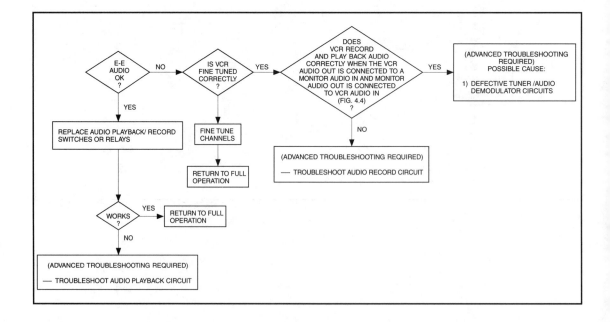

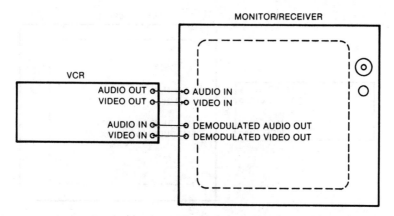

Fig. 4.4. *Connect VCR Audio Out to monitor Audio In and monitor Demodulated Audio Out to VCR Audio In.*

SYMPTOM: Linear track audio produces low audio level in playback of self-recorded tapes. Prerecorded tapes play back with normal sound level. Hi-Fi audio plays back O.K. (VHS only)

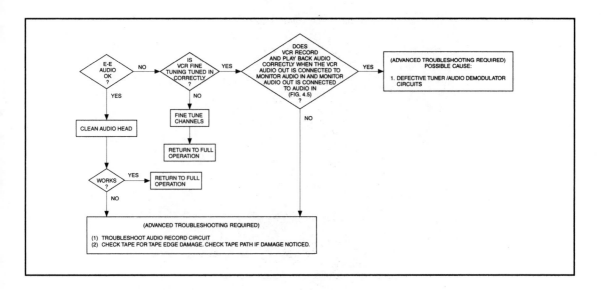

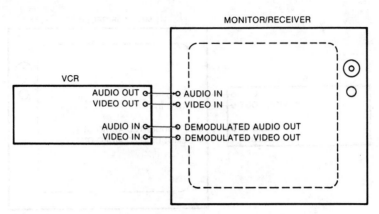

Fig. 4.5. *Connect VCR Audio Out to monitor Audio In and monitor Demodulated Audio Out to VCR Audio In.*

SYMPTOM: Linear track audio has low volume level during playback of both self-recorded and prerecorded tapes. Hi-Fi audio O.K. (VHS only)

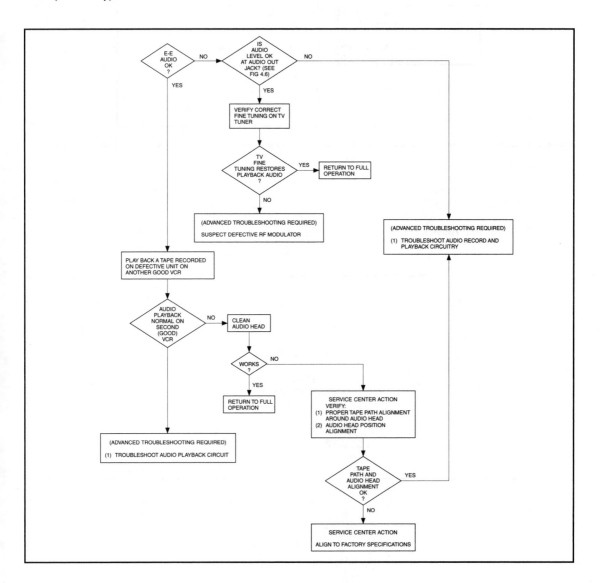

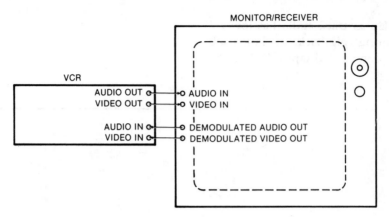

Fig. 4.6. *Connect VCR Audio Out to monitor Audio In and test for proper audio levels from the monitor speaker.*

SYMPTOM: Audio plays back at wrong speed (VHS only).

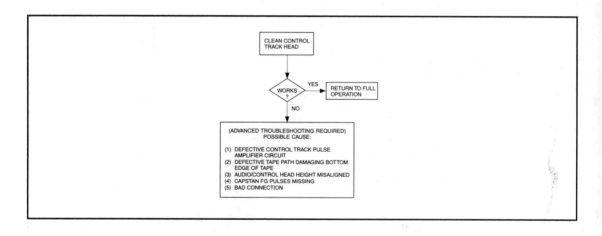

SYMPTOM: No audio or video in playback.
Screen display all blue or black (no picture).
Tape moves E-to-E audio/video O.K. (VHS only)

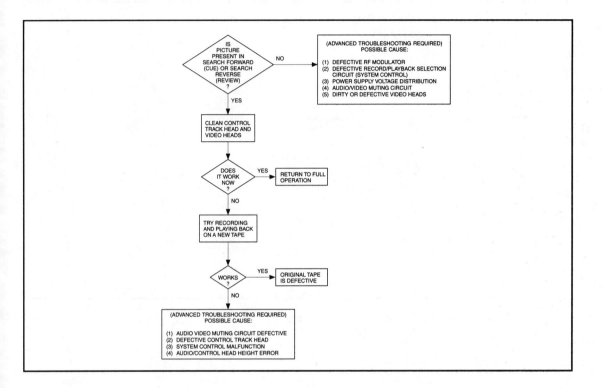

SYMPTOM: Static or popping sound in Hi-Fi audio.

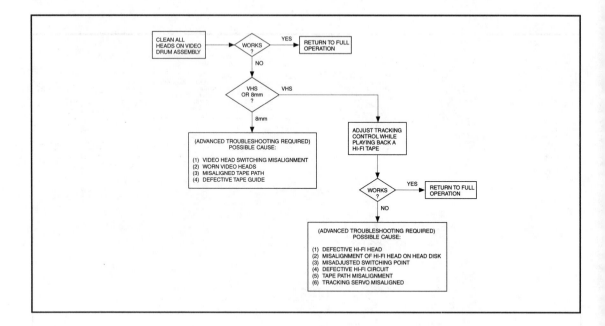

II. VIDEO PROBLEMS

SYMPTOM: Snow on screen during play-back like that shown in Fig. 4.7. Tracking con-trol won't remove. May or may not have some viewable picture in background. If this symp-tom occurs on a VHS VCR, the linear track audio may be present but will probably have lots of static. There will be no Hi-Fi audio play-back. For 8mm machines, the sound will be noticeably absent or overridden by static.

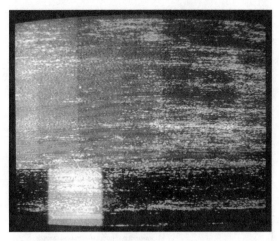

Fig. 4.7. *Snow on a display screen.*

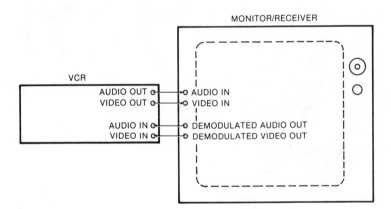

Fig. 4.8. *Apply a video signal from the monitor Demodulated Video Out to the VCR Video In and make a recording.*

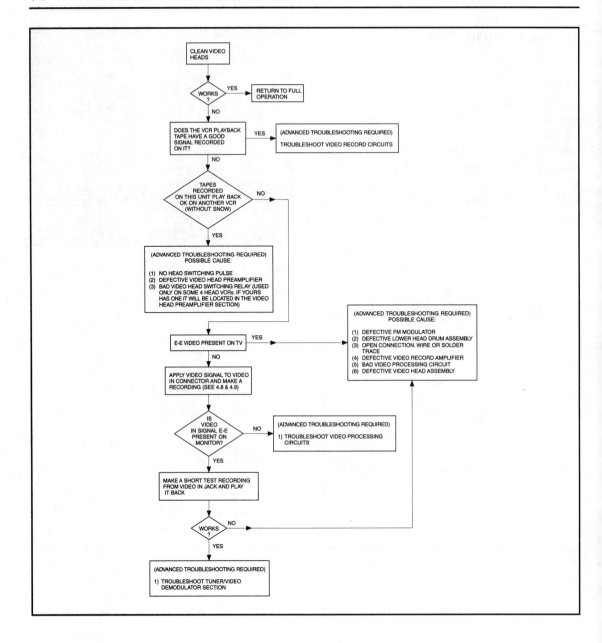

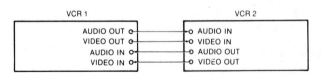

Fig. 4.9. *Apply a video signal from VCR 2 Video Out to VCR 1 Video In and make a recording.*

SYMPTOM: Fine horizontal line floats through picture (Fig. 4.10).

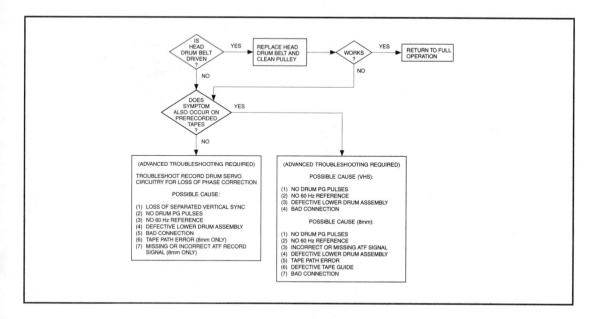

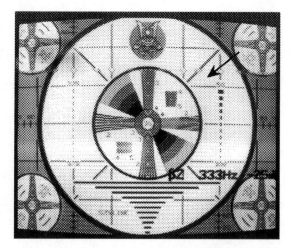

Fig. 4.10. *A fine horizontal line can be seen in the top portion of this photograph.*

SYMPTOM: Video picture alternates between good video and snow. May occur at regular intervals.

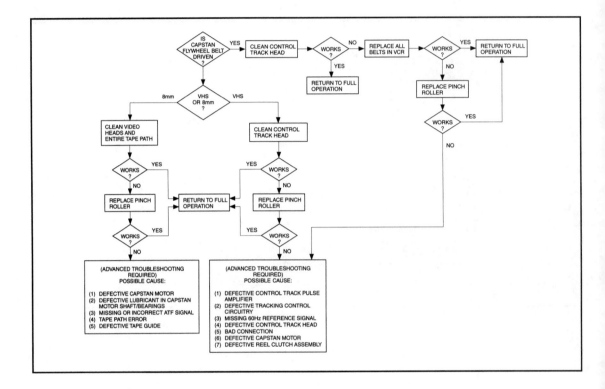

SYMPTOM: No video or audio in playback.
May display a blue screen in playback.

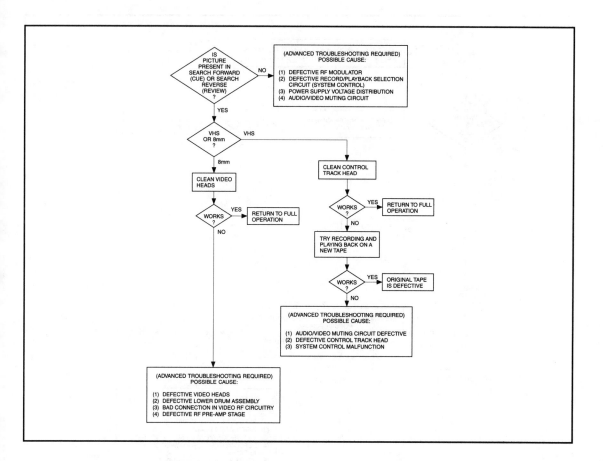

SYMPTOM: Vertical images bend or tear near top of the screen (Fig. 4.11).

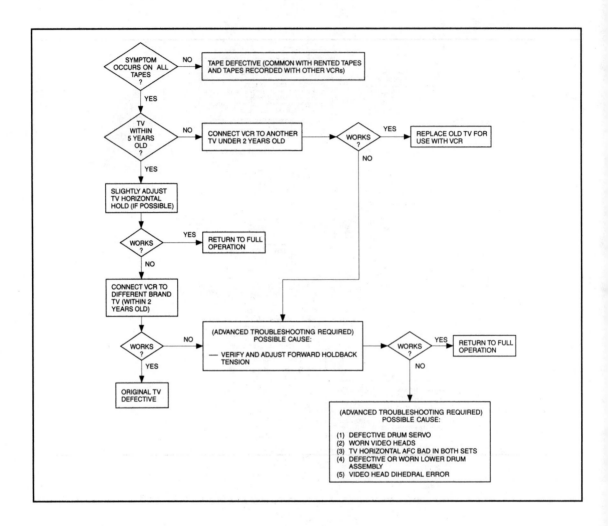

Fig. 4.11. *A vertical image bending and tearing at the top of the screen.*

Fig. 4.12. *Snow bands at the bottom of a screen display.*

SYMPTOM: Snow bands or lines of interference at top or bottom of screen (Fig. 4.12).

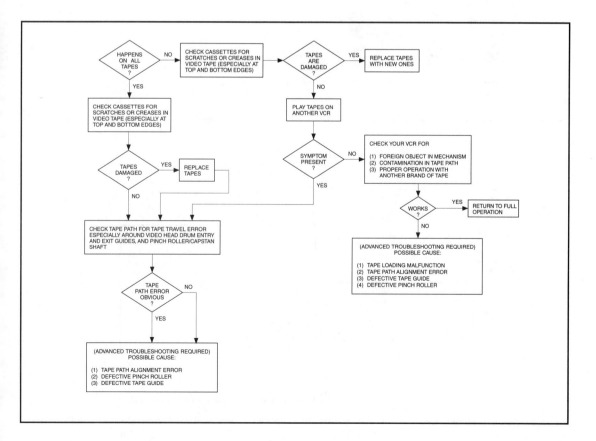

III. COLOR PROBLEMS

SYMPTOM: No color on newly recorded tapes. Get color on playback of prerecorded tapes.

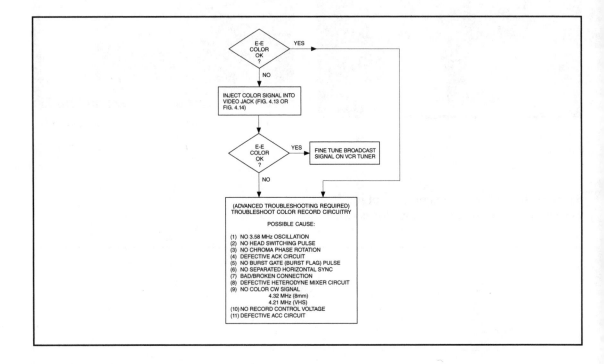

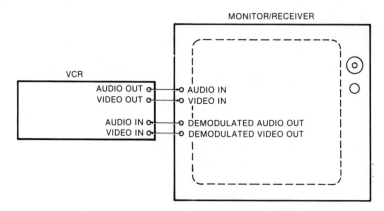

Fig. 4.13. *Inject a color signal into the VCR by connecting the monitor Demodulated Video Out to the VCR Video In.*

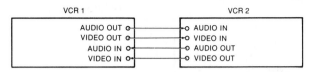

VCR 1   VCR 2

AUDIO OUT   AUDIO IN
VIDEO OUT   VIDEO IN
AUDIO IN   AUDIO OUT
VIDEO IN   VIDEO OUT

Fig. 4.14. Inject a color signal into VCR 1 by connecting VCR 2 Video Out to VCR 1 Video In.

SYMPTOM: No color in record or playback. Prerecorded tapes have no color.

Scenario

The TV has color when you are viewing normal broadcast signals but no color coming from the VCR when recording or playing back. You are not sure if the VCR is recording color or not. The defect might be in the color playback. In this case the VCR would record color all right but you would never know because it won't play it back.

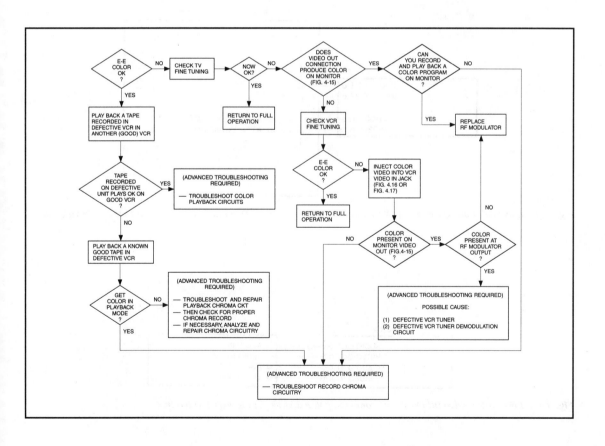

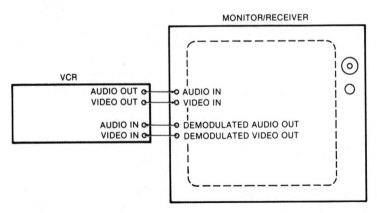

Fig. 4.15. *Check for color by connecting VCR Video Out to monitor Video In.*

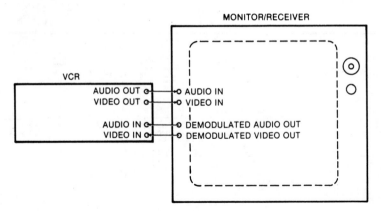

Fig. 4.16. *Inject color video into the VCR by connecting the monitor's Demodulated Video Out to the VCRs Video In.*

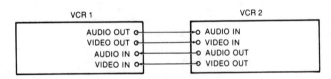

Fig. 4.17. *Inject color video into VCR 1 by connecting VCR 2 Video Out to VCR 1 Video In.*

SYMPTOM: Bands of color on screen during color playback.

Scenario

When you play back a color program on the VCR, bands of color are present on the TV screen. These bands are diagonal and may have the appearance of red, green, blue, pur- ple, or a combination of these colors. The bands may show up at unpredictable times, but once they are present they are fairly con- sistent. This symptom is sometimes known as the "barber pole" effect because of its similar- ity to the diagonal stripes on a rotating bar- ber's pole. The problem may look like hundreds of bands that are close together on the screen, or there may be only a few spaced farther apart.

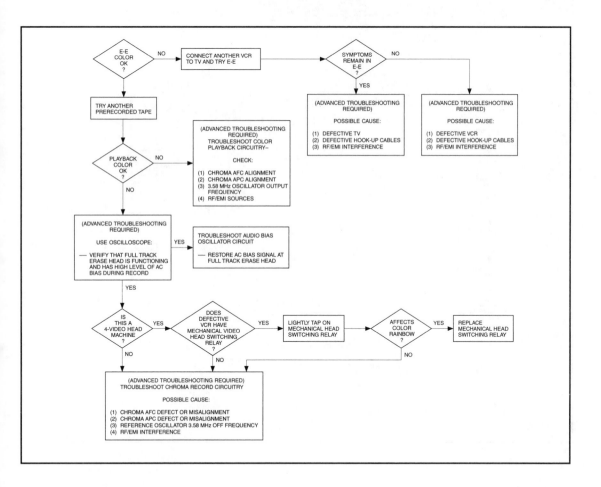

IV. POWER SUPPLY PROBLEMS

SYMPTOM: No functions. Clock may or may
not have a display.

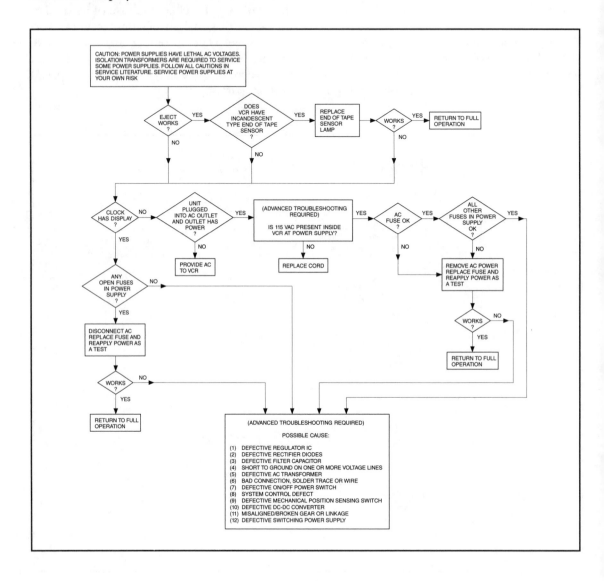

SYMPTOM: No power (dead VCR).

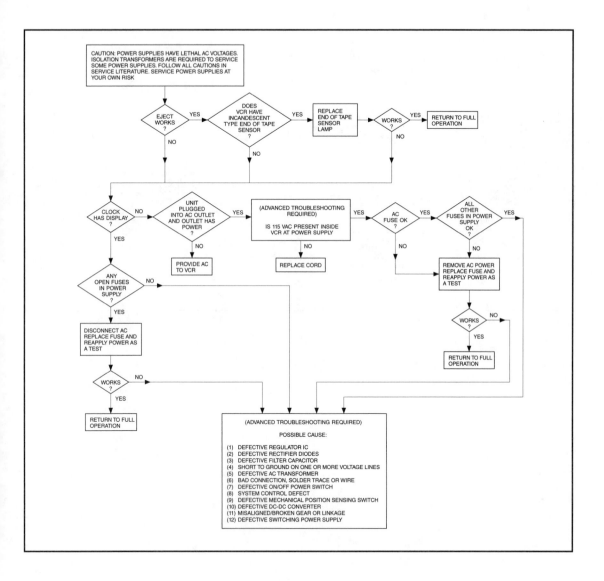

CAUTION: POWER SUPPLIES HAVE LETHAL AC VOLTAGES. ISOLATION TRANSFORMERS ARE REQUIRED TO SERVICE SOME POWER SUPPLIES. FOLLOW ALL CAUTIONS IN SERVICE LITERATURE. SERVICE POWER SUPPLIES AT YOUR OWN RISK

EJECT WORKS ?
YES
DOES VCR HAVE INCANDESCENT TYPE END OF TAPE SENSOR ?
YES
REPLACE END OF TAPE SENSOR LAMP
WORKS ?
YES
RETURN TO FULL OPERATION
NO
NO
NO

CLOCK HAS DISPLAY ?
NO
UNIT PLUGGED INTO AC OUTLET AND OUTLET HAS POWER ?
YES
(ADVANCED TROUBLESHOOTING REQUIRED) IS 115 VAC PRESENT INSIDE VCR AT POWER SUPPLY
YES
AC FUSE OK ?
YES
ALL OTHER FUSES IN POWER SUPPLY OK ?
YES
YES
NO
PROVIDE AC TO VCR
NO
REPLACE CORD
NO
NO
REMOVE AC POWER REPLACE FUSE AND REAPPLY POWER AS A TEST

ANY OPEN FUSES IN POWER SUPPLY ?
NO
YES
WORKS ?
NO
YES
DISCONNECT AC REPLACE FUSE AND REAPPLY POWER AS A TEST
RETURN TO FULL OPERATION

WORKS ?
NO
YES
RETURN TO FULL OPERATION

(ADVANCED TROUBLESHOOTING REQUIRED)

POSSIBLE CAUSE:

(1) DEFECTIVE REGULATOR IC
(2) DEFECTIVE RECTIFIER DIODES
(3) DEFECTIVE FILTER CAPACITOR
(4) SHORT TO GROUND ON ONE OR MORE VOLTAGE LINES
(5) DEFECTIVE AC TRANSFORMER
(6) BAD CONNECTION, SOLDER TRACE OR WIRE
(7) DEFECTIVE ON/OFF POWER SWITCH
(8) SYSTEM CONTROL DEFECT
(9) DEFECTIVE MECHANICAL POSITION SENSING SWITCH
(10) DEFECTIVE DC-DC CONVERTER
(11) MISALIGNED/BROKEN GEAR OR LINKAGE
(12) DEFECTIVE SWITCHING POWER SUPPLY

SYMPTOM: No modes functional. May or may not eject a tape. May appear as if power is missing to half the machine.

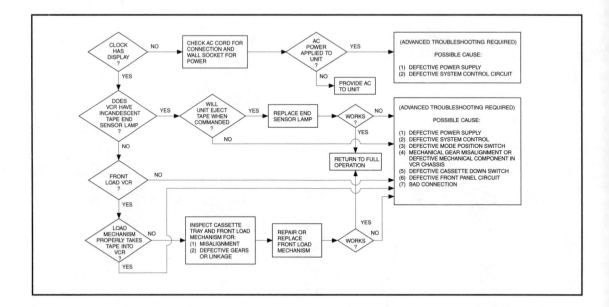

V. IMPROPER OR NO FUNCTIONS

SYMPTOM: Shuts down or returns to STOP after a few seconds. Tape may spill into unit. Power may turn off a few seconds after pushing PLAY.

Scenario

You have just placed a tape into the machine and pushed PLAY. The VCR makes some of its normal noises for the first few seconds as the tape loads up and then the unit returns to STOP. You push the PLAY button again and the same thing happens. On some models, the power may turn off a few seconds after pressing PLAY.

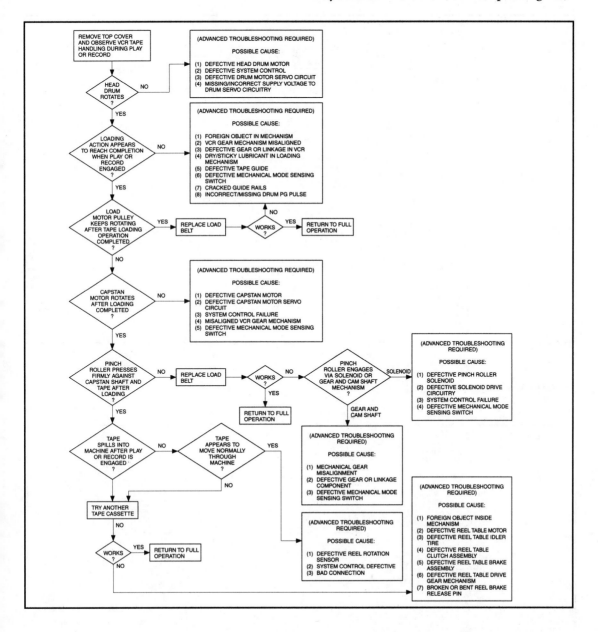

SYMPTOM: Won't properly rewind tape.

Scenario

You place a tape in the unit and attempt to rewind it. It seems as though the time required to rewind is a lot longer than when the unit was new, or perhaps the VCR doesn't quite have the power to rewind the tape. Maybe it tries to rewind but stops before the tape is fully rewound. Or it starts rewinding but as it gets to the end of the rewind cycle the tape slows down and stops, even though the motor is still trying to finish rewinding. It is as though the unit is slipping.

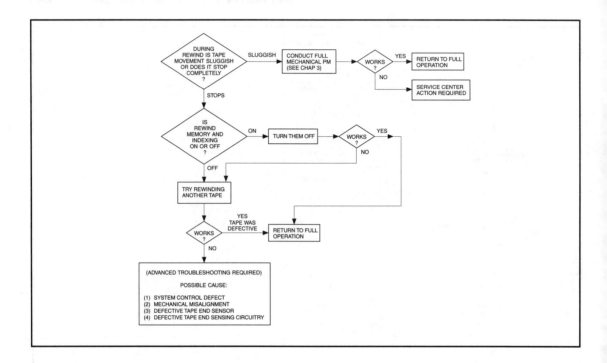

SYMPTOM: Won't rewind at all.

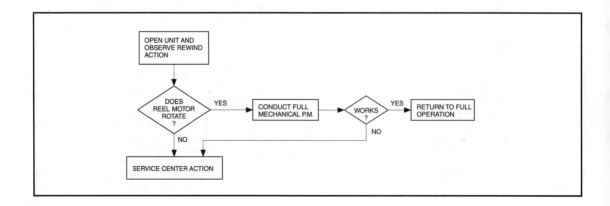

SYMPTOM: No play, fast forward, rewind, or record. Here are the possible causes.

1. No power.

2. End-tape sensor lamp out.

3. Mechanical problem:

   a. Broken belt, slipping tire, etc.

   b. Broken or misaligned gears or linkages.

4. Electrical problem (head drum, capstan doesn't rotate, pinch roller doesn't engage, tape thread time-out, etc.).

5. Defective or misaligned mechanical position sensing switch.

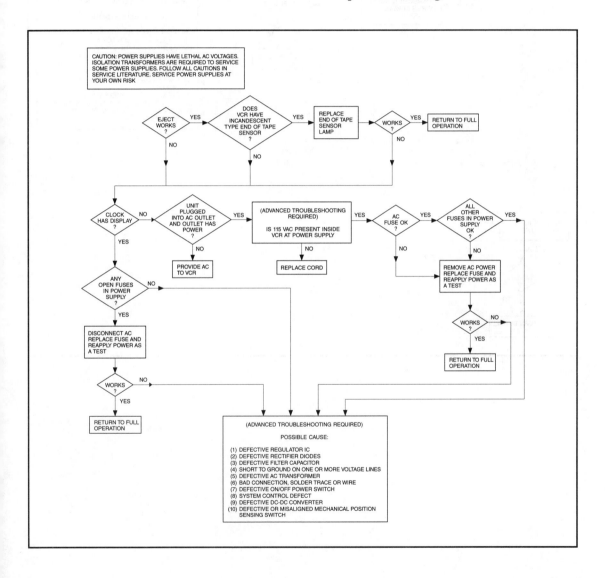

SYMPTOM: Unit acts like it doesn't have power.

Scenario

The VCR acts like power is not applied to some or all circuits. Few, if any, functions work. Clock may not display.

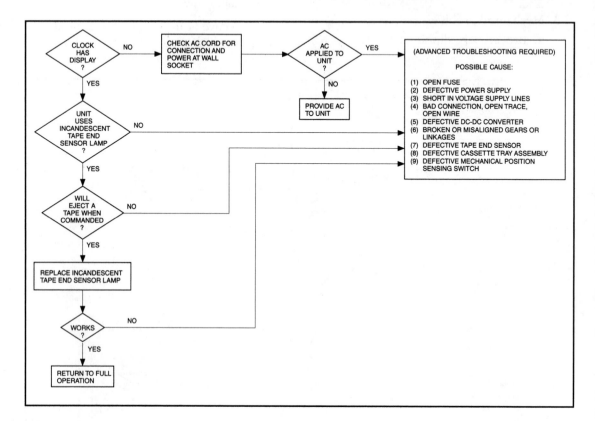

SYMPTOM: One or more of these symptoms occur:

1. No functions work.

2. Unit may return to STOP a few seconds after PLAY or RECORD are pressed.

3. Tape may spill into unit.

4. Power may also turn off a few seconds after pushing PLAY or RECORD.

Scenario

You have just placed a tape into the machine and pushed PLAY. The VCR makes some of its normal noises for the first few seconds as the tape loads up, and then the unit returns to STOP. You push the PLAY button again and the same thing happens. On some models, the power may turn off a few seconds after pressing PLAY.

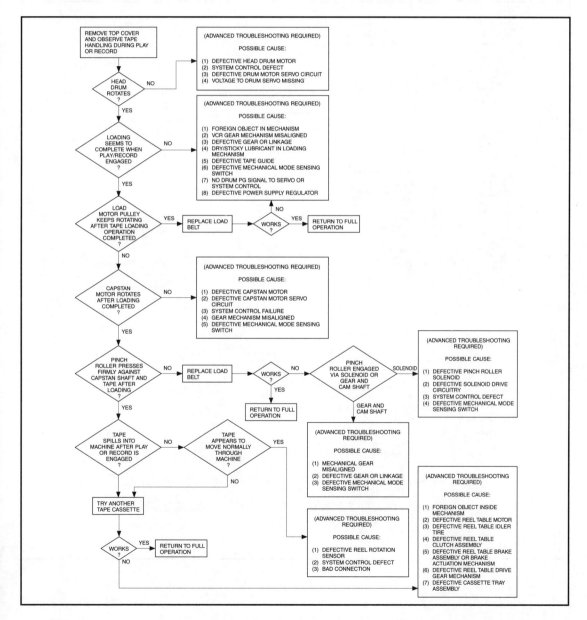

SYMPTOM: VCR damages tapes.

Scenario

You eject and remove the cassette and discover tape hanging out of the cassette. Some tape could still be caught in the VCR.

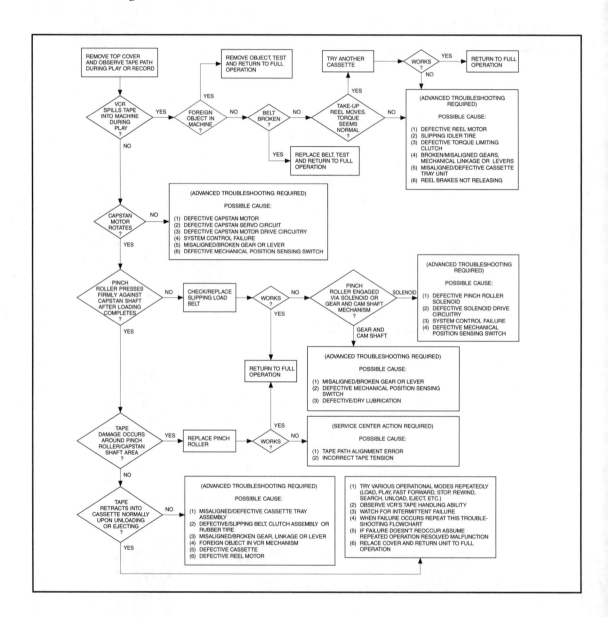

SYMPTOM: Slow or no fast forward.

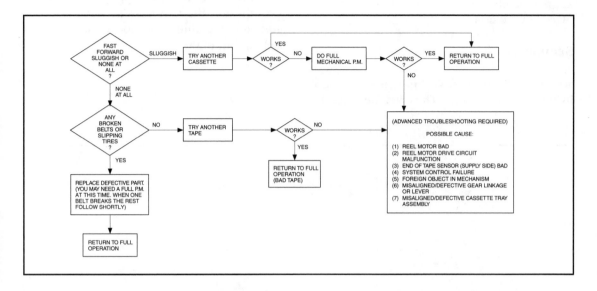

SYMPTOM: Mode button won't engage on
first try.

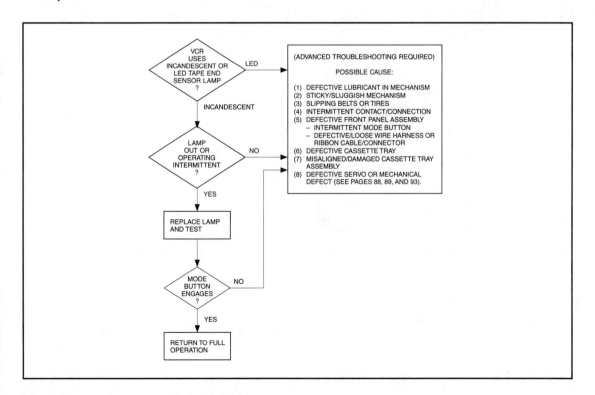

SYMPTOM: Front load VCR ejects the tape approximately three seconds after cassette is placed in the unit.

Scenario

You have a front load VCR in which you have inserted a tape cassette. The machine takes the tape from out of your fingers and lowers it into the mechanism normally. Approximately three seconds after the tape cassette has been lowered into the unit, the tape cassette comes back up out of the VCR.

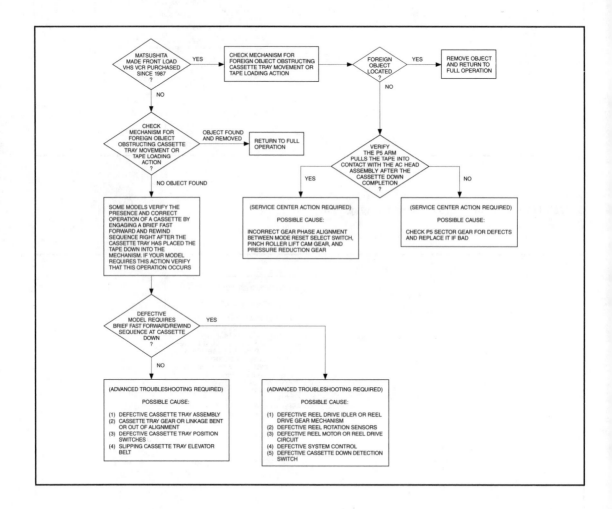

SYMPTOM: Power turns off a few seconds after play or record is pressed. Will not eject tape cassette after eject button is pressed. Power goes off after unit fails to eject tape cas- sette. Tape is unloaded when review mode selected. (*Note:* These symptoms may occur individually or in conjunction with one another.)

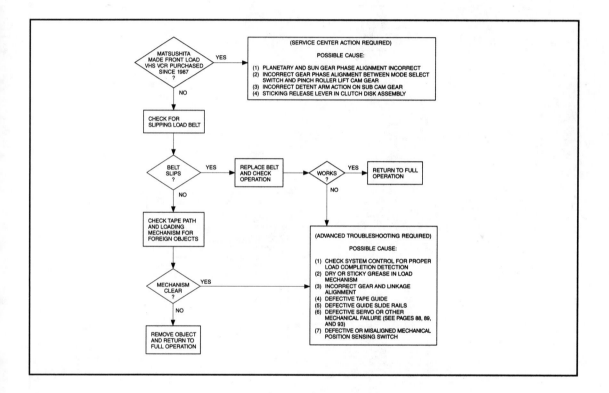

SYMPTOM: Front load unit starts to accept tape cassette, then ejects cassette before taking it all the way in. Tape may go in crooked, even after careful insertion.

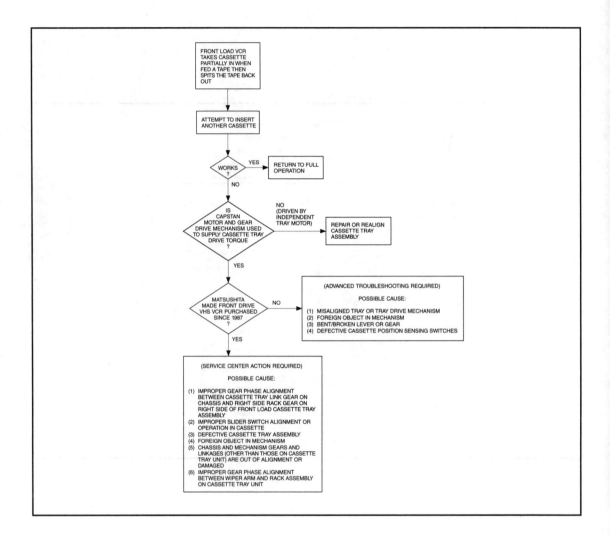

SYMPTOM: Eject problems.

1. Power shuts off a few seconds after eject is pressed.

2. Will not eject tape. Unit operated normally until eject was pressed.

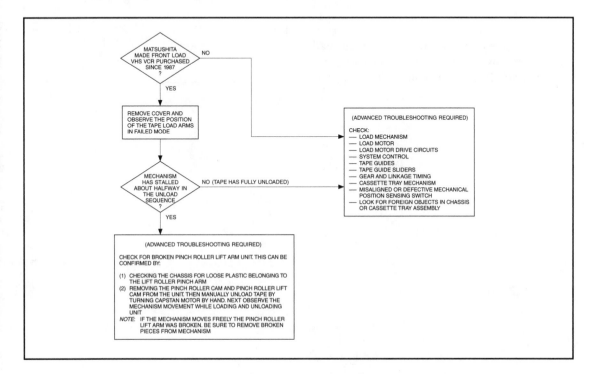

SYMPTOM: During play or record, the unit stops and goes into rewind by itself. (May occur intermittently.)

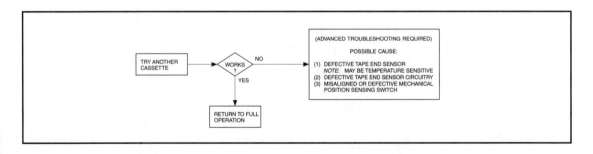

SYMPTOM: At power-on, the front cassette tray assembly lowers into the VCR without a cassette installed in the unit. The tape "in" indicator also comes on when no tape is in place.

Scenario

As soon as power is turned on, the front load mechanism lowers into the VCR as if a cassette were inserted into the machine. This happens even when no cassette is in the machine. When the defect occurs the tape in indicator also lights. The defect may be intermittent.

(ADVANCED TROUBLESHOOTING REQUIRED)

POSSIBLE CAUSE:

(1) DEFECTIVE OR MISALIGNED CASSETTE-IN DETECTOR SWITCH
(2) DEFECTIVE TAPE END SENSORS ON CASSETTE TRAY MECHANISM
     NOTE: THEY MAY BE TEMPERATURE SENSITIVE
(3) DEFECTIVE TAPE END SENSOR CIRCUITRY

VI. CLOCK OR TIMER PROBLEMS

SYMPTOM: Clock loses 10 minutes each hour.

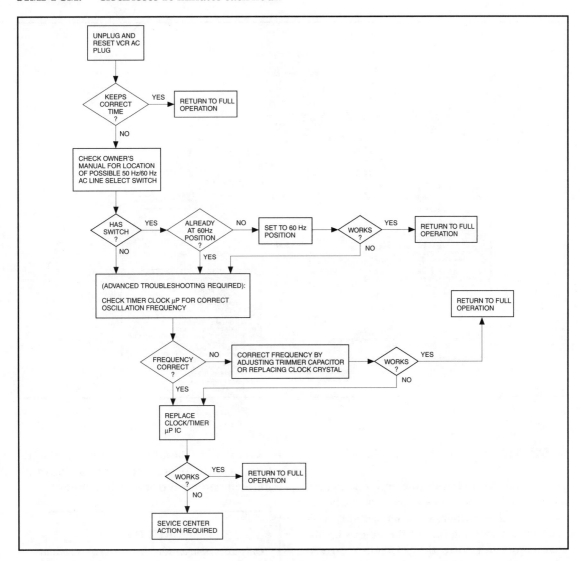

SYMPTOM: No or intermittent record using timer.

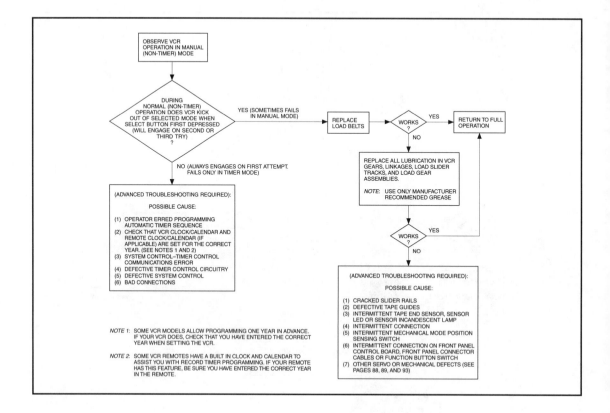

SUMMARY

This detailed troubleshooting and repair chapter has covered most of the general problems experienced by owners (and servicers) of VCR systems. If following one of the guides in this chapter doesn't solve the problem, you can take the final step ("Service Center Action" or "Advanced Troubleshooting Required") yourself if you feel qualified. Chapter 11 will provide assistance if you decide to really dig into your machine. Remember that the information in Chapter 3, "Routine Preventive Maintenance," can help prevent many of the problems analyzed for repair in this chapter.

> **CAUTION**
>
> ***Only experienced service technicians should work on power supply and CRT (TV or monitor) displays.***

Magnetic
Recording Theory

Chapter 4 was a step-by-step guide for the troubleshooting and corrective maintenance of video cassette recorders. Chapter 5 explores the fascinating subject of magnetic recording theory. This chapter provides technical detail on how magnetism is used to record and play back audio and video information.

As you learned in Chapter 1, the technology used to store audio and video signals on a magnetic tape is quite straightforward. Magnetism and the phenomenon of pole attraction have been understood for many years. However, the techniques used to optimize the magnetization of the oxides on a long thin strip of mylar tape vary greatly. The fact that magnetic principles are used remains constant with all forms of recorders.

Most of us have experimented with a simple electromagnet in school science courses. These magnets were made by taking a piece of iron core material and wrapping a copper wire around it. By applying a voltage from a battery

or some other power source, this iron core could be made to radiate magnetic lines of force called flux which attract and magnetize other metallic objects. When the electrical current flowing through the copper windings increases, so does the strength of the generated magnetic field. VCRs use this same principle to record video information on a magnetic tape. In a VCR, a head becomes the iron core and a thin film made of either ferrous oxide or metallic particles is pasted on a flexible mylar or plastic tape to become the object to be magnetized.

MAGNETIC TAPE RECORDING PRINCIPLES

The process of storing and retrieving magnetic video tape information is not as easy as it may first appear. Several phenomena have major effects on the magnetic recording process. These include the composition and thickness of the metallic coating on the tape, and the strength and frequency of the magnetic signal applied through the recording head.

Limitations of Video Recording

Some of the limitations of video recording are shown in the flux-density or BH curve shown in Fig. 5.1. The BH curve is a graph showing magnetic tape characteristics during recording and the ability of the tape to retain the information once the external magnetic influence has been removed. The remaining flux density on recorded tape is designated "B" and is measured in nanowebers per square meter. The horizontal line represents magnetic field intensity, H. This intensity is measured in ampere turns per meter. It's the amount of current (in amps) applied directly to the coil windings on the recording head. As shown on the graph, flux density, B, increases in direct proportion to the field intensity generated by an increase in head coil current. At very low record current levels, little flux density remains after the coil current is reduced to zero. This is because residual magnetism attempts to keep the atomic structures polarized in their previous orientation. The atomic structures naturally resist any force to realign in the same direction as an externally applied field. This is a form of reluctance. Reluctance is defined as the resistance of a material to align magnetically with the influence of an externally applied field. An analogy is the resistance of some materials to the flow of electrons in electronic theory. The graph shows that the low record current is enough to cause some residual magnetic influence to remain on the tape.

As the coil current increases, the flux density begins to increase linearly. A point is reached where all the magnetic particles on the tape are oriented to the same polarity as the externally applied field, and current increase no longer produces a change in flux density. This is called the magnetic saturation point and is represented by a flat horizontal line on the graph.

Figure 5.1 also shows the effect of record current on field intensity once the recording signal level has been reduced to zero. As indicated by the dotted line on the graph, as the external field is reduced, the flux density level does not return to zero but drops to a level slightly less than it was when the external field was being applied through the video heads.

To obtain zero flux density, the current passing through the coil of a magnetic head core must generate a magnetic field of opposite polarity in the head gap area. The field intensity necessary to produce zero flux is called coercive force. As coercive force increases, magnetic polarization is driven negative until it eventually reaches the magnetic saturation point of opposite polarity, as shown in Fig. 5.2. If the current in the record head coil is made to alternate at equal amplitudes in both the positive and negative directions, a closed hysteresis loop is formed, as shown on the graph. The magnetic coating on the video tape must have high magnetic retentivity and should reach high flux densities before saturation. These characteristics are important in evaluating the differences between normal and high-grade video tapes.

The magnetic coating on the video tape must have high magnetic retention and should reach high flux densities before saturation. These characteristics are important in evaluating the differences between normal and high-grade video tapes. Metal powder and metal

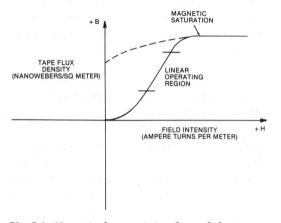

Fig. 5.1. Magnetic characteristics of recorded tape.

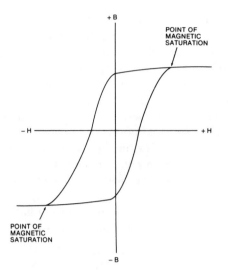

Fig. 5.2. *Hysteresis loop formed by alternating currents of equal amplitude.*

evaporated tapes have significantly higher magnetic saturation points than oxide tapes. Figure 5.3 compares the magnetic characteristics of oxide and metal formulation tapes. Notice that metal tape not only has a higher saturation level, but also requires increased coercive force to magnetize the tape surface.

Magnetic heads that record audio on the longitudinal audio track are caused to operate

in the linear region of the BH curve. In this performance region, there is a uniform relationship between the flux density impressed on the tape and the flux density retained by the tape's magnetic surface. The record head must operate in this linear region so that all input audio recording levels will properly magnetize the tape for accurate reproduction during playback. If the head operated outside this linear region, distortion would occur during playback. This is because low levels of input audio would cause the record head to function near the intersection of the BH axis where little retentivity occurs on the tape. During playback, reproduction of these low levels would not be possible because the signal retained on the tape would be too low for recovery. Conversely, high record audio input levels would drive the tape into magnetic saturation resulting in loud unintelligible sounds during playback. In this case, only those input record levels operating in the middle of the volume spectrum would reproduce in recognizable form. Thus the dynamic volume range would be severely limited.

To force the longitudinal audio track record head to operate in the linear region of the BH curve, a high-frequency AC signal known as bias is added to the record signal just before

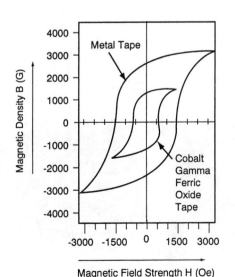

Comparison Between Metal Tape and Oxide Tape (0 dB)

Item		Tape	Metal Tape	Cobalt Gamma Ferric Oxide Tape
Magnetic Material			Fe	γ-Fe_2O_3
Magnetic Characteristics		HC	1,500 (Oe)	650 ~700 (Oe)
		Br	2,470 Gauss (G)	1,000 – 1,300 Gauss (G)
		Br/Bm	0.80	0.80
Electro-magnetic Con-version	Video Charac-teristics	0.5 MHz	+3.3 dB	0 dB
		1.0 MHz	+4.9 dB	0 dB
		3.0 MHz	+5.7 dB	0 dB
		5.0 MHz	+6.6 dB	0 dB
S/N			+6 dB	0 dB

Fig. 5.3. *BH curves for metal and oxide tapes compared.*

the signal is sent to the recording head. This mixed-in audio bias forces the recording audio head to operate within the linear region of the BH curve in Fig. 5.4.

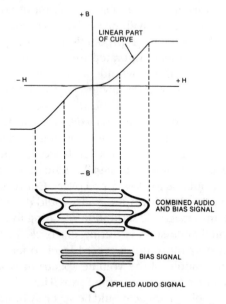

Fig. 5.4. *Record bias causes signal recording only in the linear region of the BH curve.*

If an audio signal were applied directly to the head without bias, and the signal strength of the input signal varied between zero (no volume) and the amount required for saturation, the playback signal recovered off the tape would be severely distorted. If, however, the incoming signal varied from no volume to peak loudness within the linear operating range on the BH graph, the playback signal would represent what was recorded regardless of the input volume.

Typical VCR bias frequencies range from 60 kHz to almost 100 kHz. The VCR, however, doesn't use bias to record video information. The video signal record operation is accomplished using another method. This will be described later.

Video Head Construction

The video head is the device that develops the magnetic force, which magnetizes the video tape. It is a very tiny electromagnet made of a material that can be easily magnetized. Such a material is called permeable and is symbolized by the Greek letter μ. Permeable material has magnetic domains, which align with the flux of an external magnetic field. Shaped loosely like a horseshoe with a tiny gap at one end, the head is mounted so it barely makes contact with a flexible tape at the point of the gap as shown in Fig. 5.5. The size of a video head is compared with a one-cent coin in Fig. 5.6.

Magnetic fields or flux lines radiate from this head gap into a nearby tape surface. Flux

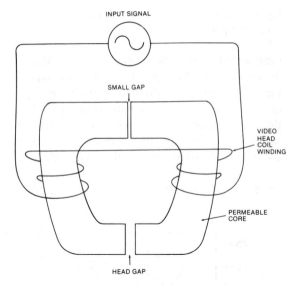

Fig. 5.5. *Video head construction.*

Fig. 5.6. *Photograph comparing a video head and a coin.*

extends across the gap between the two poles on the video head core. Thus, this core material adds its own magnetic influence to the external signal and increases the intensity of the field. The shape of the head and the core material itself affect the permeability of the head. See Fig. 5.7.

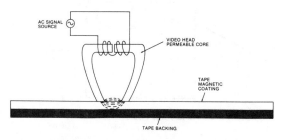

Fig. 5.7. Magnetic lines of flux radiate from the high reluctance head gap through the magnetic coating of the tape.

With electrical current flowing through the wire coil, the head gap is moved with respect to the video tape. This movement causes magnetic particles on the surface of the tape to align so that the tape coating becomes magnetized in an arrangement determined by the current of the electrical coil.

The polyester tape is thinly coated with a binder material containing very small metallic particles. The tape's magnetic surface is comprised of either iron oxide for VHS and SVHS machines or a pure, nonoxidized, metallic powder for 8mm format units. Both 8mm and Hi 8 equipment are designed to use two types of nonoxidized metal tape—metal powder and metal evaporated tape. The most common 8mm tape has a metal powder magnetic surface.

The magnetic particles on the tape's surface material tend to retain their newly aligned state after an externally applied magnetic field action has taken place. This ability to retain magnetic flux after the externally applied field has been removed is called retentivity. Retentivity values specify the ability of the tiny magnetic particles on the tape surface to retain

their magnetic alignment polarities after the tape has passed the recording head.

The flux of the video head is concentrated and focused at the gap by the design of the head itself to produce a maximum field intensity at the point where the head gap contacts the tape. If an alternating signal is applied across the coil of the video head, the flux field produced will magnetize the tape at the same alternating rate producing columns of particles aligned north-to-south, north-to-south, etcetera. The recording current passing through the coil of wire wrapped around the video head material produces a magnetic flux field that alternates proportionally to the signal. This alternating current is essential for recording intelligent information on the tape.

The tape is then moved relative to the head as an electric current in the coil modifies the magnetic field in the core and head gap area. If the tape and head were to remain stationary with respect to each other, the magnetic polarization would vary in only one location on the tape and a single spot would be constantly magnetized in alternating directions (polarities).

The material for the video head permeable core is carefully selected to minimize reluctance (resistance) to magnetic flux. The head core must have minimal reluctance, but it must also be highly permeable with minimum retentivity. A video tape must have good retentivity to retain an externally applied field, but a video head must not become permanently magnetized after the applied current is removed from the head windings.

There are losses in a magnetic head that result from circular flows of electrons, called eddy currents, which occur in the core material itself as it is exposed to rapidly changing magnetic fields. Eddy currents behave like short circuits and dissipate heat energy in the core. Eddy currents increase with the increase in signal frequency applied to the head coil. As these currents increase, they resist the magnetic field being applied to the surface of the tape, producing heat in the core material and weakening the strength of the flux field focused at the head gap.

To minimize these losses, the video head core is constructed of very poor conductive material. Many VHS VCR heads are constructed of ferric oxide mixed with oxides of other metals such as zinc, magnesium, and nickel. Collectively called ferrite, this crystalline material is characterized by a high resistance to electrical conduction. It resists current flow. Ferrite heads are efficient at high frequencies so their characteristics are ideal for the frequencies associated with video tape recording.

The core is typically constructed from two pieces of material. At one end, the core pieces are tightly connected to minimize reluctance. The other end of the core has a gap filled with a nonmagnetic material which maintains critical tolerances in the gap area. It's also important to control the thickness of the core material itself (in the area of the gap). The head gap material on many video heads is silicon dioxide (glass).

Because the video head core is highly permeable with a gap characterized by high reluctance, the magnetic lines of flux actually flow into and through the tape surface, which is also highly permeable.

The video heads of 8mm machines are constructed with different materials than the typical VHS head. The 8mm heads cannot have ferrite in the head gap region because of the higher flux density required by metal tape during record. Ferrite heads reach a point of magnetic saturation at the head gap long before the magnetic record signal is strong enough to completely impress the record signal onto a metal tape. Instead of ferrite, 8mm heads use a metal alloy in the gap region that permits a much stronger magnetic signal to be developed by the head before the pole tips reach their saturation point.

Some 8mm heads have a hybrid metal and ferrite construction while other 8mm heads are manufactured using only metal alloy. The hybrid heads have a ferrite core with a metal alloy head tip. This metal tip is the section of the head that presses against the tape surface and includes the region of the head that contains the head gap as shown in Fig. 5.8.

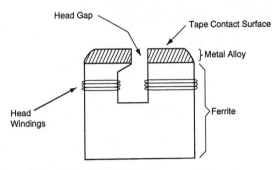

Fig. 5.8. The hybrid metal head.

The noncomposite metal alloy head is shown in Fig. 5.9. In this design the entire head structure consists of metal alloy. There is no ferrite in the core.

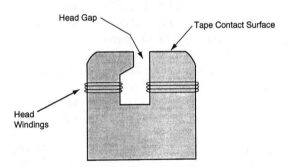

Fig. 5.9. The metal alloy video head has a metal core and metal tip.

There are generally two basic types of noncomposite metal alloy heads—the single metal block head and the laminated/multilayered head. The single metal block head is a continuous sheet of metal alloy. Its metallic composition forms a permeable head with a high saturation rating that is capable of saturating metal tape. The Sendust head is an example of this type of head. The stationary audio head stacks found in VCRs are frequently made of Sendust.

The laminated/multilayered metal head is constructed of layers of metal alloy separated by laminations. The advantages of using laminated metal layers over the single sheet metal Sendust construction include increased head strength and a reduction in eddy currents. Eddy currents are of more concern to a designer of metal heads than they are to a designer of ferrite cores. An example of the laminated/multilayer metal video head is the amorphous head shown in Fig. 5.10.

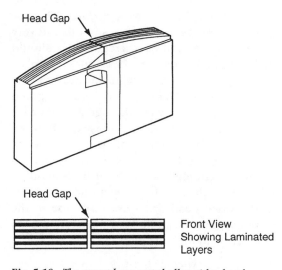

Head Gap

Head Gap

Front View Showing Laminated Layers

Fig. 5.10. The amorphous metal alloy video head.

While some professional and high-end consumer SVHS VCRs use metal alloy video heads, most VHS equipment uses the ferrite type head. All 8mm equipment use some form of metal head.

Head strength is a concern in all types of heads, whether they are constructed of metal, ferrite, or a combination of the two. Of special concern is the inherent weakness existing between the poles in the head gap. A nonmagnetic glass substance is placed in the gap of all head types to provide strength to the head tip.

The video head manufacturer must select a glass material that will properly bond to the head core material and that also has similar wear characteristics. The supporting glass must wear at the same rate as the head tip or the effective head life will be affected.

When amorphous heads are designed for a VCR, the lamination material between the metal layers is formulated with the ability to bond well to the metal alloy and have a wear rate matching that of the alloy.

NORMAL TRACK AUDIO HEADS

The head for a VCR's normal audio record/playback track is designed similar to that of the video head. The magnetic characteristics of the record/playback functions are modeled after those of the video head. There are two major variations between the video and normal audio heads. These variations relate to the size differences between the heads and the mounting configurations. Video heads are significantly smaller than the normal audio heads and are mounted on a rotating cylinder. The normal audio head is mounted on a stationary platform. For reasons that will be explained later, the video head scans across the tape at a slight angle. Because the normal track audio head is stationary, the tape must be pulled across its leading surface for the head to operate properly. Figure 5.11 shows the differences between a video and normal audio head.

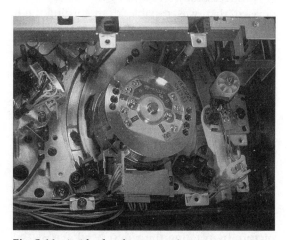

Fig. 5.11. A video head is mounted on a rotating disk while the normal audio head is positioned on a stationary platform.

All consumer VCRs use a single audio head to perform both playback and record operations. The video head, which records the picture information, is also used to reproduce the signal during playback. Some professional and broadcast grade machines contain separate record and playback heads. These added heads provide optimal performance of both record and playback functions and allow the operator some additional record/playback options.

VIDEO TAPE

A critical part of the video tape record and playback system is the video tape itself. Magnetic tape was first developed for recording audio signals several years ago. In the early development stages of video tape recording, standard audio tape was used as the recording media. Important advances in the video tape recording industry have succeeded only because the video tape medium has improved substantially. Recent improvement in the signal-to-noise ratio and the development of higher resolution recording formats have been primarily due to the improvement of the tape media.

For video recording technology to develop successfully, a recording media had to be devised that was capable of recording high video frequencies and that had a smooth tape surface permitting precise contact between the video head and the tape surface. Good head-to-tape contact is important in allowing the video recorder to place a high-frequency video signal on very narrow tracks on the tape. It's also important that the video tape flow smoothly around a cylindrical video head drum assembly without causing undue video head wear.

One example of the rigid requirements placed upon video tape is found in the still playback function of a VCR. When placed in the still/pause mode, the VCR displays a still picture by rotating the video head across the same track on the tape over and over again. During this time, the tape is held stationary in the machine. Video heads mounted on a drum

pass over the same track on the tape at least 1800 times each minute.

Characteristics of Video Tape

The tape used in consumer and professional environments is a result of many compromises that achieved a balanced solution to conflicting commands of the technology. The characteristics of an ideal video tape are summarized as follows:

1. *Smoothness of tape surface.* The surface of the tape must be very smooth and very flat. Surface smoothness irregularities should be less than 0.1 micrometer to minimize playback signal losses resulting from head-to-tape contact errors. This type of error is called dropout. It occurs when the signal level drops below a recoverable level. A noise glitch shows up on the screen when this problem occurs. But the tape surface cannot be overly smooth because extremely smooth surfaces tend to make the tape stick to the stationary portions of the video head drum assembly and to the stationary guides, audio heads, and erase head assemblies in the tape path.

2. *Tape surface hardness.* Both the magnetic coating and the plastic backing to the tape should be durable enough to resist scratches and prevent damage to the tape's magnetic surface. However, if the video tape is too hard, excessive wear will occur to the video head and other components in the tape path.

3. *Optimum head wear.* The video tape must provide just enough friction to gently scrub the head clean and yet not wear down the projecting head tip too fast. Therefore the video tape is designed to cause a certain amount of video head wear to keep the surface of the video head in an optimum condition both mechanically and magnetically.

4. *Coercivity, magnetic flux density, and retentivity.* Coercivity is a rating that defines how much external force is required to magnetize material to a new polarity. In video tape,

the coercivity rating defines how much flux is required to magnetize the tape. High coercivity and maximum flux density are necessary to produce a high magnetic field output at the very high frequencies used in the video signal. The magnetic surface must be capable of retaining the magnetic information over a long period of time and over a variety of ambient conditions.

5. *Tape thickness.* The video tape should be as thin as possible to allow maximum record time by permitting as much tape to be packed on the spools as possible. However, the tape backing must be thick enough to provide stability at high temperatures. Since the video tape backing is plastic, it is desirable to maintain stable physical characteristics throughout a relatively wide temperature range. Tape that is too thin is also very fragile mechanically. Thin tape is especially susceptible to damage along the top and bottom edges.

6. *Record/playback signal flexibility.* The video tape must be able to record and retain energy stored on it by several different means including longitudinal audio with mixed AC bias and application of a high-frequency-modulated video signal. The tape must be capable of retaining signal frequencies ranging from 30 Hz to beyond 5 MHz.

7. *Compatibility.* There must be very close mechanical tolerances and magnetic compatibility between tape manufacturers so a tape machine can perform consistently when different tape brands are used. For example, tape width tolerances must be maintained within a ± 10 micrometer width so a $\frac{1}{2}$-inch VHS tape will properly play and record.

Tape Manufacturing

Tape backing

Video tape consists of a very smooth, highly flexible polyester (plastic) substrate with a magnetic coating on one surface. To prepare the tape's plastic backing for the magnetic coating, it is subjected to a process of stretching,

heating, and cooling called annealing. The annealing process produces a tape backing with the desired thickness and strength and that is smooth and highly flexible. The tape's physical dimensions must remain stable throughout its life and while it is used in the various modes of VCR operation. All of a tape's physical properties must be consistent over a long period of time and throughout a variety of ambient conditions.

In a consumer VCR, changes in environmental temperature or humidity that cause video tape shrinkage or swelling of only 0.1% will cause distortions in the playback video. The distortions originate from two causes: (1) mismatched servo control pulses (control track) present on the tape; and (2) video horizontal line synchronization pulses that are no longer standard.

Surface smoothness and resistance to wear are two other important qualities of a tape's polyester substrate. Both of these attributes are needed to provide good head-to-tape contact in the record and playback processes. If the surface of the substrate isn't smooth, the applied magnetic coating is usually uneven as well, creating poor head-to-tape contact when the video head scans across that portion of the tape. These tape surface irregularities cause signal dropout.

A compound of silicic acid and carbon black is sometimes applied to the back side of the substrate to make the surface rough so the tape has good winding characteristics on the reels and doesn't slip between layers on the hub during storage. The black coating also precludes the tape end sensors from false triggering by preventing light from penetrating through the tape.

The magnetic coating

The magnetic coating is typically applied to the plastic tape backing through a process of calendering. Three methods of calendering are shown in Fig. 5.12.

There are three basic types of magnetic coatings on video tape—ferric oxide, chromium dioxide, and metal. Ferric oxide is the most

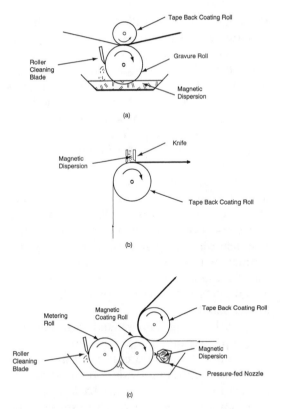

Fig. 5.12. *Three methods of applying the magnetic coating using calendering.*

common type of backing found in consumer quality tape. In the early days of video tape recording, a gamma ferric oxide tape was used. This tape material was originally designed to be used as the magnetic coating on the audio recording tape of that day. It was soon discovered, however, that a different tape formulation was needed to meet the high-frequency demands of the video tape recorder and to reduce the video signal-to-noise ratio. The use of ferric oxide tapes became possible when improved magnetic compounds were developed to meet the demands associated with recording a video signal. The improvements came by adding compounds that modified the oxide to provide additional coercivity and by improving the ferrite crystal's size and shape.

Chromium dioxide tape has magnetic characteristics that fall between those of oxide tape and the higher performance metal tape. This is because chromium dioxide tape is more metallic in content than oxide tape. Chromium dioxide tape first gained popularity in the professional VCR market in the 1970s and continues to be produced by some video tape manufacturers. Chromium dioxide has the advantage of providing a higher coercivity and better picture quality than oxide tape. However, it is more abrasive and more expensive to produce than oxide-based tapes. Currently, the most common VHS and SVHS tapes have oxide-based coatings. VHS and SVHS units are not capable of using metal tape.

Metal powder and metal evaporated coatings

Tapes with a nonoxidized, pure metal coating were first available back in the early 1960s, but these tapes were not used because the video market had no demand for such a high coercivity tape. Although metal tape has superior playback characteristics over the oxide tape media, manufacturing was difficult and costly. The mechanical long-term stability of the tape coating itself was also a concern.

Metal tapes are highly susceptible to deterioration in the form of oxidation and rust caused by humidity and gases in the atmosphere. Once this deterioration begins, the metal particles lose their magnetic retention ability. An oxide coating was significantly more stable in varying ambient environments than a metal coating, so oxide tapes gained popularity rapidly and became the only video recording media for many years. Since an oxide tape is oxidized during the manufacturing process, the magnetic particles tend to be much more stable after exposure to the environment. When video tape was first produced, oxide tapes were also far less expensive to manufacture. Technological advances in tape manufacturing and design have improved the stability of metal magnetic media and have also reduced the manufacturing costs making metal tape a viable option for video recording today.

Metal tape is not compatible with machines designed for oxide tape. This limits the metal tape format to the 8mm and Hi 8mm consumer markets. However, new consumer formats are coming that will also use metal tape.

The binder

The material used to glue the magnetic particles to the tape surface is called the binder. It consists of solvents, lubricants, nonmagnetic fillers, and antistatic ingredients. Mixed into the binder are particles that are magnetized to represent the video signal.

The binder is highly elastic and interfaces well with the plastic backing. It's also designed to keep the particles glued to the tape for a long period (years). The lubricant in the binder performs during the life of the tape.

During the application and calendering process, the magnetic particles tend to congregate and settle in the mixture. To prevent this, the binder solution contains a dispersion agent to keep the particles evenly spread throughout the mixture. Even dispersion of the particles is an important objective of the application process. Uneven particle displacement will cause erratic retention of magnetic information resulting in interference in the playback picture.

Ideally, the lubricant in the binder will provide a single molecular layer on the surface of the magnetic coating to minimize wear on the heads and tape surface. As the tape is used, the top layer of lubricant wears off, and is replenished by migration and diffusion from down within the magnetic layer. This continual replenishment must continue throughout the life of the tape. By proper design, a delicate balance is maintained between too much lubricant, which will cause deposits to occur on the video heads, and too little lubricant, which will cause both the heads and tape surface to wear rapidly.

Antistatic agents are added to the binder of some professional and broadcast quality video tapes to reduce static electricity buildup on the tape surface. The antistatic additives prevent the collection of unwanted debris on the tape's surface. This unwanted debris can damage the tape surface and cause signal dropout.

Nonmagnetic fillers may also be added in small quantities to improve wear resistance and the frictional properties of the magnetic coating. These additives increase the hardness of the tape surface. They also serve to clean the video heads by providing a controlled amount of friction between the tape and the heads. As the tape moves, a gentle scrubbing action is produced, which keeps the front face of the video head clean. By design, these fillers do cause an acceptable amount of video head wear as they keep the surface of the video head in an optimum condition both mechanically and magnetically.

The final manufacturing process and its importance to good tape performance

During manufacturing, about 5000 meters of annealed tape is pulled from a wide spool and passed through calendering rollers at about five meters per second where the tape is coated with a magnetic mixture. The calendering process that presses the magnetic particles and binder against the tape occurs without interruption while the large spools of tape pass through the rollers.

After the tape has been coated, it is dried at 60°C to 100°C. After initial drying, the magnetic coating is not completely solid but is compressible like a sponge. At this point, the tape goes through a second calendering process. It's passed between a highly polished steel roller and an opposing roller that has a paper fabric or plastic surface. These rollers put the tape under extremely high pressure and temperature. The steel roller is often heated to 90°C. This calendering operation produces a smooth magnetic coating with tape surface peak-to-valley roughness of less than 0.05 micrometers. A smooth surface is especially important in the new higher resolution VCRs such as the SVHS and Hi 8mm equipment where head-to-tape contact spacing is critical.

It's during this final calendering process that the interactive relationship between the tape surface smoothness and the additives in the binder must be carefully controlled. A surface that is too smooth will tend to adhere to the stationary portion of a video head drum assembly, to the stationary tape guides, or to other stationary heads in the tape path. This tendency to stick is called slip/stick, or stiction. In this situation, the tape sticks then slips repeatedly over the stationary surfaces as the mechanism attempts to pull the tape smoothly through the tape path. Since the high-speed rotating video heads require a smooth tape surface for proper head-to-tape contact, the prevention of stiction depends upon the proper amount of lubricant and nonmagnetic particle additives in the tape binder. There is a compromise between the smooth surface required by the video heads and the surface roughness required to keep the heads polished and prevent stiction.

Most rotating video head drums contain air bearings to help minimize stiction. Air bearings are small gaps in the rotating disk that produce a fine film of air in the tape path around the head drum. This air film reduces stiction in the video head drum area of the tape path. In addition, the stationary portion of the head drum has a series of fine grooves carved into its surface or has a finely pitted surface to aid in smooth tape travel.

Once the coating process is completed, the tape is slit into the appropriate width and cut into the length specified by each individual format. The abrasive nature of the magnetic particles and the strong polyester tape backing make precise slitting a challenging procedure. Therefore hardened rotary sheer blades are used, and the slitting process is continually monitored to ensure that the tape is slit to the precise width required. Tape movement is precisely guided under carefully controlled tension to maintain uniform tape width.

All VHS machines record the normal audio and control track information on longitudinal tracks located on the extreme top and bottom edges of the tape. If the tape slitting process is imprecise, the information recorded at the tape edges could be disturbed due to poor head-to-tape contact. Occasionally, new batches of tape are found to not record one channel of normal stereo audio properly or will produce control track disturbances. (Control track disturbances appear as sudden speed changes in tape travel motion and cause the picture to break up intermittently.) If the improperly recorded audio channel is located on the outermost edge of the tape, a defective video tape may be the cause, and the trouble may have occurred in the splitting process. Defective tape edges produced during the splitting process may not be visible to the eye but may have enough deformity to create problems in a machine's playback mode.

Fortunately, tape-edge deformity caused by the manufacturing process is rare. Damage to the tape edge usually occurs as a result of a defect in the video tape machine itself, a defective video cassette shell, or improper tape storage. Video tapes should be stored vertically (standing up), not horizontally, and should be kept in an ambient environment of $70 \pm 10°F$ with a relative humidity of 60 ± 10 percent. Adverse ambient conditions can cause the tape to either stretch or shrink, making correct playback impossible. Storing the tapes horizontally can cause the tape to slip from its proper packing position on the reels and form a conical position. When this happens the tape edges can be damaged as they press against the top or bottom disks on the reel hubs. The information recorded on the tape edges can also be distorted, causing improper playback of that information.

High-Resolution Recording and Tape Formulation

Recent improvements in VCR reproduction quality are directly related to an ability to resolve fine detail. Fine detail can only be produced on a display screen if the system is capable of recording and playing back high-frequency video signals. Thus, high-resolution VCRs must place more information on a tape within the same fixed surface area as a conventional video recorder. This high-density recording involves writing higher frequencies onto a video track within the same length as a lower frequency (lower resolution) machine. This means that more information (more changes in magnetic polarity) exist within the same linear distance on the tape surface. Higher frequencies equate to shorter wavelengths. By definition, a recorded wavelength is the length of tape required to record one complete cycle of a given frequency. Wavelength is determined by the frequency of the signal being placed on the tape and the relative speed of the tape as it travels across the recording head. More detail on recorded wavelength is provided later in this chapter. Since high-density recording dictates that more changes in magnetic polarity will be present within the same length of tape, the size and geometry of the magnetic particles in the coating on a tape's surface are important characteristics for high-density recording.

High-density oxide tapes

The importance of particle density and shape on tape performance can be illustrated by the following example. Much of the area shown in Fig. 5.13A and B contains no magnetic particles. This occurs because nonuniform size reduces the packing density of the particles on the tape. In part A a relatively high-frequency sine wave is recorded onto tape containing the uneven magnetic particles. The energy in the sine wave is distributed relatively even throughout the tape. The missing oxide particles have negligible effect on the output signal because there is sufficient continuity in the playback signal for the waveform to be reconstructed. Part B shows the same tape after a much higher frequency is recorded on it. Notice that higher frequency recording requires increased particle density to completely recover the entire signal. The tape with the lower particle density is not able to ad-

equately store information at a higher density because of the higher frequency used in recording. Places on the tape where oxide particles are missing will not produce an output and much information (changes in polarity) will be lost. The uneven distribution and nonuniform size of magnetic particles will also cause more noise in the playback signal. Figure 5.13C shows why smaller particles with dimensional uniformity are needed when recording shorter wavelengths (higher densities).

A conventional gamma ferric oxide video tape does not perform well when recording high-density information because the oxide crystals dispersed in the binder are too large and are not uniform in size. The particles are therefore not able to be as densely packed as a tape whose particles are smaller and more uniform.

Within the last few years technological improvements using highly refined oxide-based formulations in oxide video tape and advanced recording machine design have increased the resolution in the record and playback signal. The improvements in video tape include a smaller size and better shape of magnetic particles. Cobalt has also been added to oxide formulations to produce results close to those of chromium dioxide formulations.

Cobalt-modified magnetic coating has enabled higher density recording on oxide tapes. This was important in the development of Super VHS (SVHS). The higher resolution SVHS equipment is downward compatible, which means that it will not only use the improved high-density oxide tape for higher resolution recordings, but it will also accept conventional VHS tape for normal record and playback. Using the improved cobalt-modified tape design, SVHS units do not need metal tape to achieve higher resolution.

The key features of SVHS tape are:

1. The ferric oxide particle size has been reduced by 20% to an average size of 5 micrometers.

2. The particulate has greater dimensional uniformity.

(A)

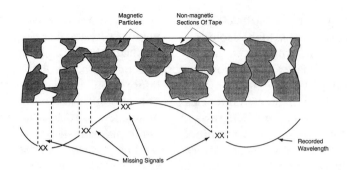

Small portions of missing recorded wavelength.

These sections are small enough that they don't effect the playback recovery process.

(B)

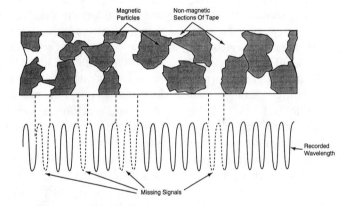

Missing recorded wavelengths will cause a distorted reproduction of the original signal.

(C)

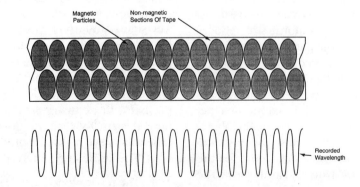

No missing portions of the recorded wavelength are present during playback.

***Fig. 5.13 A,B,C.** Particle size and uniformity are important qualities of high density recording. (A) Conventional oxide tape works well with longer recorded wavelengths. (B) Much of the information is lost when shorter recorded wavelengths (high-density recordings) are attempted on conventional video tape. (C) Small particle size and geometric uniformity are required for high-density storage and retrieval.*

3. The tape has a new undercoating of only 0.3 micrometers between the magnetic coating and backing substrate.

4. The undercoating and substrate are "ultra" smooth to reduce surface asperities.

5. The substrate is 14 micrometers thick.

6. The new back coating is 0.7 micrometers thick.

Figure 5.14 compares SVHS and conventional VHS tape performance.

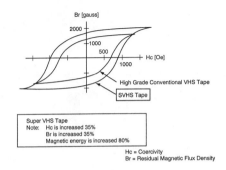

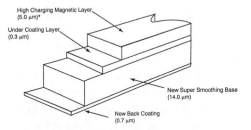

Fig. 5.14. *SVHS tape provides many improvements in oxide tape performance.*

Coercivity

Another important factor in higher density recording is the tendency of the magnetic coating on the tape to self-demagnetize due to relatively low coercivity. Conventional ferric oxide tape has a coercivity value of 400 to 600 oersteds (OE). Metal tape exhibits coercivity between 1300 and 1600 OE. The closer that magnetic information is packed on a tape, the greater potential for nearby and opposite domains to affect each other, creating a form of self-erasure. Higher coercivity tape prevents this from happening because it requires a greater magnetic moment to reverse domain polarity and return a particle's magnetized state to zero. Modified cobalt ferric oxide tape for higher density recorders has a coercivity value of around 850 OE.

SVHS machines use this higher coercivity oxide tape to record the higher resolution signals. These machines increase the record current in the head coils during the record mode to properly write a signal onto a tape. When a conventional tape is recorded in an SVHS unit, the recording signal strength is reduced to accommodate the lower OE rating of the tape.

During playback, an SVHS machine adjusts the amplification of the playback signal for the type of tape being used. SVHS tape has a higher retentivity rating, which equates to a stronger playback signal level than that of conventional VHS tape. The SVHS VCR recognizes the presence of an SVHS tape by sensing a special hole in the bottom of the SVHS cassette and tripping a switch in the VCR mechanism.

The high coercivity rating of metal tape is one reason that it has become popular for use in high-resolution machines. The coercivity value of MP metal tape is nominally 1500 OE. The significantly higher coercivity rating makes high-density recording much less susceptible to the self-erasure phenomenon. This is a major reason why engineers are designing future high-resolution machines to use metal tape.

One drawback to metal tape is that ferrite heads cannot be used to record a signal on the magnetic surface. Ferrite heads currently don't develop enough magnetic field intensity to properly saturate the metal surface during record. This is because ferrite head gap materials reach magnetic saturation long before the surface of a metal tape is completely magnetized. Presently, only metal heads can be used to record on metal tape. Ferrite heads can play back a prerecorded signal from a metal tape but cannot record on the metal tape.

Metal tape and high-density recording

The first video cassette containing metal particle tape was introduced in the mid-1980s for use in the 8mm format. There are two basic types of metal tape: metal powder (MP) and metal evaporated (ME). The magnetic coating on both types of metal tape is pure metal rather than oxide crystals. MP tape is coated with an alloy such as iron, nickel, and cobalt (Fe-Ni-Co).

One advantage of metal tape over oxide and chromium dioxide tape is that the size and shape of magnetic particles are more uniform. This enables dense packing of particles in the coating—thus higher resolution recording and playback. This feature makes metal tape a favorite for engineers who design high-resolution VCRs. Another advantage to metal tape is that metal coatings have a significantly higher coercivity rating than that of any other form of magnetic coating.

A third advantage of metal tape relates to retentivity. In a ferric oxide tape, only one-third of the particles in the coating are magnetic. Most of the particles are used as special purpose fillers as previously described. In metal tape, all of the particles are magnetic. Therefore, metal particle (MP) tape has three times the retentivity of ferric oxide tape. Higher retentivity means that more magnetic flux remains in the tape coating after the head has left the tape.

The design of MP tape magnetic coatings from various manufacturers is very similar. Slight differences involve the quality of the polyester substrate, the uniformity of the binder applied to the substrate and the precision of the tape slitting which determines the width and length of the tape used for the various formats.

During the MP tape manufacturing process, the outside layer of each of the needle-shaped metallic particles in the tape coating is oxidized slightly in a very carefully controlled environment. Slight oxidation provides better flux retention stability over long periods after the magnetic coating has been exposed to atmospheric gasses and humidity. Following oxidation, the magnetic particles are coated with resin, then mixed with lubricants, abrasives, dispersants, and binder agents for application onto the plastic substrate. The resin aids in insulating the magnetic particles from further oxidation by elements in the operational environment. Figure 5.15 is an expanded view of typical metal tape.

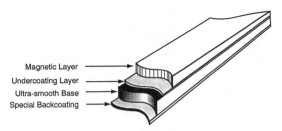

Magnetic Layer
Undercoating Layer
Ultra-smooth Base
Special Backcoating

Fig. 5.15. *Metal tape layers enlarged.*

There is a limited availability of metal evaporated tape due to problems in the manufacturing process. ME tape differs from MP tape in that ME tape uses a coating process based on vacuum evaporation of the particle medium to place the metallic surface directly onto the tape substrate. No binder solutions are used as adhesives between the magnetic coating and tape backing as required with MP and oxide tape. The evaporation process for layering the coating onto the backing results in a magnetic surface with a better response at shorter wavelengths (higher density recording) than that found in MP tape. This is because the binder compound in MP tape containing adhesive and other agents effectively reduces the MP particle coating density as compared with the particle coating density on ME tape. ME tape also has very high magnetic saturation—the result of directly depositing the magnetic layer without a binder.

The disadvantages of ME tape include the lengthy time required for evaporating of the magnetic layer. This results in relatively low productivity. The evaporated layer is mechanically brittle and susceptible to cracking. This makes it a fragile form of flexible media. ME tape has a thin (less than 0.02 micrometer) protective oxidized layer to protect it from corrosion and oxidation. ME tape presently suffers from lubrication problems similar to those common to all hard materials. Special lubricants are required to treat the surface of the ME metal alloy tape.

Figure 5.16 compares the relative output level and recording currents required by gamma ferric oxide, chromium dioxide, cobalt ferric oxide, metal particle, and metal evaporated tapes.

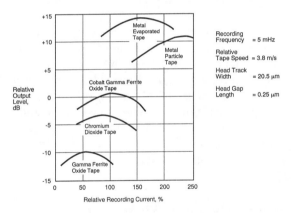

Fig. 5.16. Relative signal strength output versus relative recording current for various formulations of video tape.

MP tape doesn't experience the same mechanical stress-related cracking of ME tape and is easier to manufacture, but MP tape does not provide the high coercivity, retentivity, or high-density recording characteristics of ME tape.

Extending Tape Life

The ability of video tape to store information for long periods has been a concern in the video industry. The length of time that video tape can archive information varies and is influenced by several factors. The operational life of video tape is influenced by its ability to yield to stress generated by the video tape recorder during thread, unthread, play, fast forward, rewind, and pause conditions, and by the tape's sensitivity to the environment including temperature and humidity. The tape's ability to withstand friction is also important and is largely influenced by the lubricant used in the binder.

Tape Wear and Failure

In a video tape recorder, the stationary audio and moving video heads wear gradually as a result of friction caused by moving tape. On the other hand, wearing of the tape's magnetic coating generally starts with the loss of lubricant on the tape surface. This is normally followed by a failure in the cohesiveness of the binder. Mechanical stress and heat caused by friction from the rotating video head as it presses against the tape are major factors in the gradual erosion of the magnetic coating and the cohesive failure of the binder. Foreign materials such as dust, fingerprints, and cigarette ashes are also detrimental to tape life.

Video tape wear is directly proportional to the hardness of the video head. A compromise is made between the hardness of the head (that determines the longevity of head life) and the wear the head has on a tape coating. VCR heads must be hard enough to provide a reasonably long operating life, but soft enough to prevent excessive wear of a tape surface.

Tape wear will accelerate rapidly if the tape is used in adverse ambient conditions. Dust and debris that accumulate on the tape surface will affect head-to-tape contact and in some cases may become permanently imbedded into the magnetic surface, damaging the tape. Using the tape at temperatures above 80°F or at relative humidity levels above 70%, or a combination of the two, has a distinct and measurable effect on the rate of tape wear.

Relative wear rates for oxide and metal video tapes strongly depend on the quality of the binder agent used. The long-term adhesive action between the polyester backing and the coating mixture, the sizes of the magnetic particles in the coating, and the long-term replenishing ability of the lubricant are important aspects that determine tape life.

VCR mechanical condition

One factor that accelerates tape wear is the condition of the tape path within the video recorder. The surface of a tape will wear and sometimes become scratched as it is drawn across contaminated components in the tape path. The tape surface damage may occur all at once, or may happen gradually as the same surface area of the tape is repeatedly pulled through the mechanism.

If the surface area of the video head drum assembly is too smooth, stiction occurs. Normally, this problem is the result of a worn head drum assembly. Initially, the head drum has a combination of a rough surface and tiny air bearing passages on its face to reduce the friction between the tape and the drum surface. As the hours of VCR use accumulate, the drum surface becomes polished by the abrasive action of the tape increasing the tape-to-drum friction coefficient. This causes the video tape to stick to the assembly as the tape is pulled around the drum.

Stiction can also be the result of failing lubricant in the video tape binder. Whether stiction occurs because of a failing binder lubricant or a worn head drum, it causes the tape's magnetic coating to be scraped off and deposited on the drum's surface. This increased deposit multiplies the wear factor between the drum and the tape surface and decreases the operational life of the tape. Stiction will distort the playback video and audio signals coming from the tape.

The following are typical symptoms that a tape is failing. One or any combination of these symptoms may be noticed.

1. High levels of dropout observed using a particular tape cassette.

2. Stiction.

3. Video and/or audio heads require frequent cleaning during each use of a particular tape cassette.

4. Poor normal audio reproduction from a particular tape.

5. Poor tracking each time a particular tape is used.

6. Poor picture quality reproduction.

7. Poor Hi-Fi audio from a particular tape.

SYMPTOMS OF VIDEO HEAD FAILURE AND WEAR

So far in this chapter you've learned basic principles of magnetic theory and how they apply to the tape and heads used in the VCR. Before continuing with more theory, let's explore a few points of practical service diagnosis. It's easy to misdiagnose a defect caused by a faulty video head or by the electronic signal processing section in a VCR because many of the symptoms appear similar.

Video Head Clogging

Head clogs are caused by an unwanted accumulation of foreign debris or the scraped off magnetic coating on the surface of the head. When enough contamination has built up on the head contact surface, an unacceptable separation develops between the head and the magnetic layer on the tape causing lost signals. This spacing loss contamination interferes with both the recording and playback processes. Good head-to-tape contact is a serious concern with shorter wavelengths (higher frequency recording). The impact of spacing loss thus makes higher resolution machines more sensitive to head clogs than conventional VCRs.

Head clogs occur in different levels of severity, and as such they affect picture quality in varying degrees. Most consumers generally don't notice a head clog until the spacing loss is large enough to create complete loss of playback or recorded signals.

Many things can cause a head clog, but the most common clogs are caused by contamination of the tape surface from fingerprints, dust, or any other form of debris. Heavy friction between the head and tape can also cause a clog. A worn video head can scrape the magnetic coating off a tape and build up contamination on the head surface. A loss of lubricant or a defective binder adhesive can also cause contamination to build up on the head. Friction is particularly serious when both metal tapes and metal heads are used. And extremities of low and high ambient humidity and temperatures also increase the potential for head clogging. As previously described, the parameters for video tape manufacturing are very important in the prevention of head clogging.

As video heads wear, they become more prone to head clogging. One sign of a wearing video head is its tendency to continually clog despite the quality of video tape used. As the video heads rub over the tape surface, the head tip that sticks out slightly from the rotating drum is worn down by friction. A reduction in head tip protrusion reduces critical head-to-tape contact and decreases the natural head-to-tape self-cleaning action. This makes frequent cleanings necessary to maintain a good video signal. Excessively worn heads occasionally develop jagged edges along the head surface. These jagged edges scrape even more oxide off a tape causing head clogs to occur frequently.

If frequent head cleaning is required to clear a snowy video picture, and the tape being used is not the cause, suspect defective heads. Before replacing the video head assembly, verify that a high-quality tape has been used, that the cassette is not defective, and that the head clogging is not occurring because of contamination in the tape path (dust,

shedding oxide, etc.). Clogged or dirty video heads cause heavy snow (noise) in the picture on the television screen. Typically the snow is seen all over the screen. Some video may be present in the background but often the snow is so heavy that no picture can be seen.

Be very careful when cleaning video heads and the tape path because permanent damage can occur to the heads if they aren't cleaned correctly. Head and tape path cleaning was described in Chapter 3.

Brown Stain

Another form of head tip contamination is called brown stain. It occurs when metal video head tips are used to record and playback signals on a metal tape. Brown stain contamination is an extremely thin, nonmagnetic coating on a head. This stain is caused by a chemical reaction between the metal tape and the metal head core material. Heat generated from the head-to-tape friction acts as the catalyst for the impurity. Heat also acts as the agent that binds the stain to the head surface. Moisture in the air serves as a natural lubricant for head-to-tape interaction. When the relative humidity is low (below 10%) the potential for brown stain contamination is increased. If allowed to build on the head, brown stain will eventually cause a substantial spacing loss.

A satisfactory way to prevent brown stain has yet to be found. Currently, a special abrasive cleaning tape is used to remove brown stain. Here is one situation where an abrasive head cleaning tape is recommended because hand cleaning just won't remove the stubborn stain. The abrasive properties of the special cleaning tape work to polish and lap the heads as it removes the brown stain.

Brown stain only occurs when metal heads interact with metal tape. It does not develop when metal heads interface with oxide tapes. As the use of metal tape and the metal alloy head increases with our demand for higher resolution machines, the special treatment of tape surfaces to prevent brown stain will become an important part of technology research in video recording.

Dropout

A high level of dropout is another symptom of worn video heads. Dropout can occur because of imperfections on the tape surface, missing oxide on the video tape, or worn or partially clogged video heads. Poor head tip protrusion substantially increases the dropout potential and typically increases the amount of dropout seen on the screen. Experienced technicians can often diagnose worn video heads by observing the dropout count on a tape whose normal dropout is known to be low.

AUDIO HEAD WEAR

VHS machines record normal audio information on a longitudinal track on the tape. Frictional wear causes spacing loss to occur for a longitudinal track audio head just as it does for a video head. High-frequency audio losses are symptomatic of normal audio head wear in the early stages, while an overall (all-frequency) volume loss occurs with advanced wear.

Audio head cleaning and position realignments may help overcome the audible wear symptoms, but these are considered by professionals to be "quick fixes" which only mask a more severe problem.

Careful visual inspection of the stationary audio head assembly often reveals friction wear around the gap area that prevents proper head-to-tape contact. Looking for distortions in a reflected light beam shining off the normal audio head gap area can reveal signs of audio head assembly wear. The wear is observed as uneven surfaces around the audio head gap revealed by disturbances in the reflected light. The distorted light produces an uneven reflection because the surface of the head is deformed and not flat.

Low audio record and playback and poor high-frequency response can be caused by something as simple as tape debris or contamination lodged on the head surface. When evaluating the normal audio section for head wear, don't discount contamination preventing good contact with the tape. Typical longitudinal track audio head life expectancy is between 3000 and 5000 hours of operation time.

Audio problems in 8mm format machines are not caused by normal longitudinal heads. In this format, audio is recorded using a different method that writes the sound information on the video track along with the video signal. Chapter 9 discusses this method in detail.

THE IMPORTANCE OF WAVELENGTH

Having described some of the challenges of recording information onto a tape, it's appropriate to understand wavelength and the term *recorded wavelength*. Wavelength is the inverse of frequency and is measured in meters. It represents the distance that one cycle of a signal travels in one second. Recorded wavelength is the length of tape required to record one full cycle of alternating current signal.

Wavelength, lambda (Λ or λ), is determined by the speed of the tape and the frequency of the signal being recorded. For example, if a frequency of one hertz (cycle per second) is recorded on a tape traveling at a speed of 10 inches per second, the recorded wavelength would be 10 inches, as shown in Fig. 5.17.

As this 1 Hz signal is recorded, the polarity of the magnetic field across the head gap is initially polarized in neither the north or south direction. As the tape passes across the head and the applied signal becomes more positive, the poles on the head become oriented toward the north. After 2.5 inches of tape travel beneath the head, the applied AC signal has peaked and begins to decrease. The magnetic lines of flux across the head also decrease. At

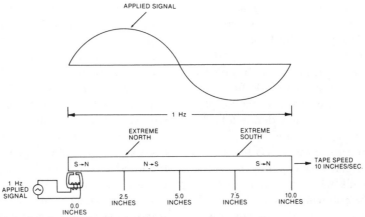

Fig. 5.17. *Recorded wavelength (calculated using frequency, and tape speed).*

the 5-inch mark on the tape, the AC signal reaches the zero flux point and starts in the negative direction. The head gap polarities begin to reverse and the recorded field starts shifting to the south. The applied AC current finally reaches the zero point again at the 10-inch mark. Playback of the recorded signal is accomplished by rewinding the tape and then pulling it across the heads again which causes the magnetic field on the tape to generate a tiny alternating current in the head.

As described in Fig. 5.18, maximum playback output occurs when the gap length is half

that of the recorded wavelength. Once the recorded wavelength signal reaches the one-half point, the output from the playback head suddenly drops to zero within the next octave. This is easy to understand from the following illustration. If both the positive and negative portions of the recorded wavelength signal are present in equal amounts within the head gap boundaries, the north and south poles cancel each other and the resultant output signal from the video head is zero.

During playback, the recording circuit is turned off and a high-gain amplifier is con-

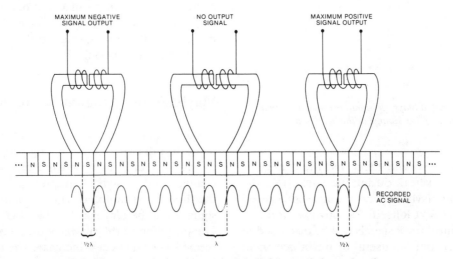

Fig. 5.18. *Maximum signal occurs when the head gap equals half the recorded wavelength.*

nected to the windings of the video head coils through a rotary transformer. As the magnetized tape is pulled across the head gap, the magnetic flux of the tape seeks the path of least reluctance. This path is through the core of the video head. As the magnetic fields expand and contract in the core, a voltage is produced in the windings on the video head. The magnetic field on the tape must be changing to cause an output from the video head coils as shown in Fig. 5.19. Voltage is only produced in video head coils when magnetic lines of force intersect the coil windings, and this only occurs when a magnetic field is expanding or contracting.

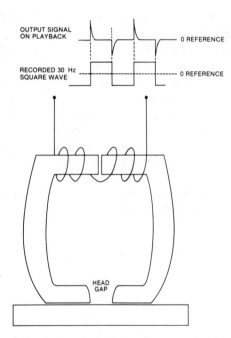

Fig. 5.19. Output playback signal occurs only when magnetic lines of flux change at the head gap.

To observe the effect that the change in flux has on a head during playback, note the record and playback control track signals used within a VHS VCR itself. The purpose and theory of control track signals will be discussed in detail later, but it's useful to introduce them here to illustrate an important point.

During the recording process, a 30 Hz square wave signal is fed into a stationary control track head that writes the signal on the tape. When this signal is retrieved off the tape, the 30 Hz square wave signal is distorted. The playback signal coming off the control head appears to be a differentiated square wave with a series of positive and negative going spikes, as shown in Fig. 5.19. This waveform occurs because the square wave signal is recorded directly on the tape via the control track head using no AC bias.

In playback, the positive cycle of the square wave aligns the magnetic flux of the tape uniformly in one direction. During the next half cycle, the recorded flux is equal in strength but points in the opposite (negative) direction.

The recorded square wave being played back is distorted because the head produces no output when no change is detected in the flux alignment. When the DC portions (the positive and negative plateaus) of the recorded square wave are pulled past the head gap, the detected lines of force don't change in the head windings, so no output is produced. The playback head only produces an output signal when these lines of force are alternating within the coil. The differentiated-looking output spikes are produced because the playback head reacts to both positive and negative transitions of the detected square wave. The polarity of the spike corresponds to the polarity of square wave transition.

Playback Head Signal Strength

The strength of the signal coming off of the head during playback is directly proportional to the frequency at which the magnetic flux changes. If the playback frequency is doubled, the flux rate is doubled. The resultant output signal from the playback head is also doubled. The doubling of the playback frequency is described as a one octave increase. The doubling of the output voltage is a rise of 6 decibels (dB). A decibel is a measurement of signal strength.

In playback, the ability of a pick-up head to detect magnetic lines of flux on the tape is not equal over all frequencies. Typical playback head output voltages increase by 6 dB per octave. This means that the head output voltage doubles with every doubling of the frequency, as shown in Fig. 5.20. The relationship of the playback head's output voltage to the signal frequency present on the tape is characteristic of all magnetic pick-up heads. The head's output voltage increases by 6 dB per octave until the signal frequency nears its highest playback frequency.

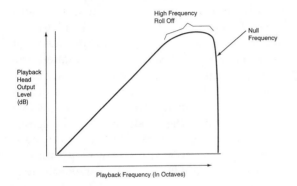

Fig. 5.20. Playback signal increases 6 dB per octave.

As previously mentioned, the upper frequency limit of a magnetic pick-up head is the frequency at which the recorded wavelength equals the length of the head gap. It is at this point that the voltage output from the playback head suddenly drops to zero. Note in Fig. 5.20 that the output from the head doesn't continue to increase at a 6 dB per octave rate until it suddenly plunges to its null (zero) point. Instead, just prior to this point the rate of the voltage increase begins to taper off. The frequency response begins to round off prior to the null point because hysteresis and eddy current losses increase substantially within the core material as the head approaches its upper frequency limit. The roll-off in playback efficiency is also caused in part by thermal losses occur-

ring in the head resulting from shock waves at the magnetic domain boundaries. These shock waves originate from changes in atomic dimensions caused by electron spin. This is a form of friction occurring in the head. It's just originating at the atomic level. The closer the recorded frequency comes to the upper limit of the playback head, the greater the hysteresis, eddy current, and thermal losses become. As the playback frequency advances toward the null frequency, playback efficiency continues to decrease and the playback response begins to flatten out until suddenly the recorded wavelength equals the head gap length and the head output voltage plunges rapidly to zero. The frequency at which this efficiency roll-off begins varies according to the type of head core material used and the head construction. The frequency at which the head output reaches the null point is determined by the recorded wavelength on the tape and the head gap length as previously described. Head gap dimensions, recorded wavelength, and the linear traveling speed of the tape are major factors in limiting high-frequency response.

Equalization

Conventional audio recorders overcome nonlinear playback frequency response through electronic equalization and by head design—the head gap lengths and record wavelengths are well under the high-frequency roll-off point of the head. VHS machines also use this same technology to reproduce sound on the normal audio track.

Electronic equalization decreases the playback head amplifier gain as the playback signal frequency increases. This process compensates for the increased output signal from the playback head by reducing the electronic amplification at high frequencies. The response curve for the equalization circuit resembles a mirror image of the playback frequency response characteristic curve. Low frequencies are amplified or boosted while high frequencies are reduced.

The resultant output is a flat frequency response.

Equalization is not a difficult problem for design engineers. However, there are limitations, which restrict its success. One of these is associated with the dynamic range of the tape itself. Dynamic range is the ratio of (1) the loudest signal that can be recorded on the tape before it reaches magnetic saturation to (2) the lowest level signal that can be detected above the noise threshold. Tape noise is also called hiss.

Current tapes have a dynamic range of approximately 70 dB. This permits slightly more than 10 octaves of bandwidth for recording signals directly onto the tape before the signal completely saturates the tape or becomes lost in the tape hiss. Good quality audio recorders can process signals from about 20 Hz to 20,000 Hz (20 kHz). The frequency response of these recorders spans approximately 10 octaves using a direct recording process with audio record bias.

The frequency range for video signals is from just under 30 Hz to over four million hertz (4 MHz). Since the head's playback frequency response is 6 dB per octave, the video signal frequency span of more than 16 octaves would require a magnetic tape with more than a 70 dB dynamic range. Equalizing such a signal is essentially not possible.

However, recording this wide range of video frequencies can be accomplished by reducing the head gap length, increasing the speed of recording (writing speed), and recording the signal with a process called frequency modulation.

Head Gap

The head gap on a stationary head of a conventional audio recorder or on the normal track audio head in a VHS machine is typically several microns (millionths of a meter) wide. Since video tape recorders deal with much higher frequencies, the gap on a video head is

measured in tenths of microns. Video head gaps in consumer ½-inch VCRs range between 0.3 and 0.6 microns. Reducing the thickness of this gap results in a loss of sensitivity because the reluctance of the gap drops as the two parts of the core (poles) approach each other. Other factors cause losses in video record and playback heads. The resistance of the coil winding, frequency-dependent capacitance effects, and head construction also cause signal level losses in video heads.

A video head coil with only a few turns will have low resistive and capacitive losses, but it will also have lower playback output voltage. If many turns of wire are used, the playback voltage is increased, but the coil can have so much stray capacitance that the head's own inductance will cause the head to resonate (vibrate). It's possible for the resonant frequency to be lower than the highest frequency of the signal to be received and played back. If this happens, severe signal loss occurs at higher frequencies. This situation requires a compromise in the production and design of the video head. Typical video head coils have from 1000 to 2000 turns of wire with a coil wrapped around both halves of the head core.

Writing Speed

VHS machines and conventional audio recorders use stationary heads to record and play back audio program signals. This can be accomplished using relatively low tape speed and a relatively wide head gap because audio frequencies are limited to a small (50 Hz–10 kHz) frequency range.

Video signals require complex design solutions. The luminance or brightness portion of a video picture ranges (in theory) from zero to about 5 MHz. Some of the first video tape recorders (e.g., RCA's 1954 product line) were designed using stationary video heads.

In a stationary video head system, reproducing a signal with a recorded wavelength covering the video frequency spectrum requires a tape speed of approximately 30 feet per second. As shown by the equation below, you would need over 20 miles of tape to produce an hour-long video program on this machine.

$$\text{Length} = \frac{\text{Speed of Tape} \times \text{Time}}{\text{Distance}}$$

Thus for 1 hour of recording:

$$\text{Length} = \frac{(30 \times 3600)}{5280} = 20.45 \text{ miles}$$

where

30 is the speed in feet per second,
3600 is the number of seconds in an hour,
5280 is the number of feet in a mile.

The tape must pass the stationary video head at a speed of 20.5 miles per hour to play a single hour of program.

The minimum tape speed required to write (record) a video signal onto a tape can be approximated by multiplying the frequency of the record signal (in hertz) by the wavelength (in meters) as shown below.

Writing Speed: $V = F \times l$

where

V = velocity or writing speed,
F = the highest frequency of the signal
    to be recorded in hertz (Hz),
l = the shortest wavelength (in meters)
    to be reproduced.

Using this equation, we can estimate the minimum tape speed required for any VCR format. For example, assume that you have a VCR designed with a video head gap of 0.6 microns and an upper recording frequency limit of 5.1 MHz. Earlier you read that the shortest recorded wavelength (highest frequency) to be written on the tape must not be shorter than twice the length of the head gap. Using the values in the previous equation yields a writing velocity (V) of 6.12 meters per second.

$$V = (\text{frequency}) \times (\text{wavelength})$$
$$= 6.12 \text{ meters per second}$$
$$(240 \text{ inches/second}).$$

The head gap length and upper recording frequency limit were used in the Betamax format developed by Sony and popular several years ago. In the Betamax format, the writing speed was actually 6.9 meters per second instead of the lower calculated speed of 6.12 meters per second. This is because the calculation approximates the lowest writing speed possible and does not take into account video head losses, manufacturing tolerances, the five degree helical angle used in consumer equipment, the lower sideband losses that occur in the record and playback process, and the diameter of the rotating head drum used to write and read the video information. In actual practice a higher writing speed is sometimes selected to help overcome these particular losses.

If you use the same formula to determine the VHS and 8mm format writing speeds, you'll find that the design engineers elected to use a writing speed that is notably higher than the calculated value from this formula. The VHS units have a head gap design that is only 0.3 micrometers in length instead of the 0.6 micrometers used in the older Betamax units. The VHS equipment still has an upper frequency recording limit of around 5 MHz. Plugging these numbers into the writing speed formula reveals a minimum writing speed of 3 meters per second. The specified writing speed for the VHS standard play (SP) mode is 5.767 meters per second or nearly twice the calculated value.

At SP tape speed, the 8mm format lists a writing speed of 3.75 meters per second. The head gap for this format is 0.25 micrometers with an upper recording frequency approaching 5.8 MHz. Entering these numbers into the writing speed formula results in 1.45 meters per second as the minimum requirement.

The writing speeds used in both the VHS and 8mm are higher than the minimum calculated values for several reasons. First, a rotating video head is used to write and read the modulated video signal on the tape. The speed and diameter of the rotating head disk both play an important part in the final writing speed used in both of these formats. Design engineers try to establish a VCR format writing speed close to a rotating video head velocity of nearly 1800 RPM (precisely 1798.2 RPM). This is because the video frame rate for an NTSC color signal is 29.97 Hz and the head must rotate fast enough to record that many frames per second (29.97 frames per second × 60 seconds = 1798.2 RPM). Using a head drum that rotates at 1800 RPM requires a writing speed in the 4 to 6 meters per second range.

Since the drums in both the 8mm and VHS format machines rotate at 1800 RPM, the variables in the formula are the upper video frequency record limits, the diameter of the drum, and the head gap length. Notice, however, that the upper frequency limits and the head gap lengths are relatively close when comparing the two formats. Why then are the writing speeds for the VHS (5.767 m/sec) and the 8mm (3.75 m/sec) so different? Why doesn't the 8mm equipment use a writing velocity closer to 5.767 meters per second?

It's because the diameter of the rotating head drum in the 8mm machine is smaller than the drum in the VHS equipment. As you'll soon discover, the diameter of the head cylinder plays a major role in determining the actual writing speed used. Recall that the basic writing velocity formula only approximates the minimum writing speed and that this formula does not incorporate several factors. One of these factors is the diameter of the rotating drum.

Let's step back now and consider another reason that conventional VHS and 8mm models are designed to use a higher writing speed than that actually calculated using our minimum writing speed formula. If you compare the 0.6 micrometer head gap used in the writing velocity formula example with an actual VHS head gap length of 0.3 micrometers, you will discover that the required VHS writing speed could have been reduced from the 5.767 m/sec in the example to nearly 3 m/sec.

The VHS design engineers elected not to reduce the writing velocity proportional to the reduction in the head gap length. This enabled future higher resolution (higher density) recordings such as those used in SVHS machines without facing a requirement to modify the existing rotating head speed, head gap lengths, or any of the mechanical specifications of the equipment. To be downward compatible, the mechanical parameters of the SVHS and Hi 8mm machines must remain consistent with conventional VHS and 8mm equipment. Since it was the desire of both the 8mm and VHS designers to maintain downward compatibility and because higher resolution recordings require shorter recorded wavelengths, only the recorded wavelength could be modified in machines built to reproduce both conventional and higher resolution pictures. Designing conventional VHS and 8mm machines with this future application in mind made downward compatible, higher density recorders possible.

For example, the SVHS format has an upper recording frequency limit approaching 7.5 MHz. Since the head gap length has not changed from the original 0.3 micrometers, the recorded wavelength still remains at 0.6 micrometers. Plugging these numbers into the basic writing speed formula, we can calculate a minimum required writing speed of 4.5 m/sec. The 5.767 m/sec writing speed is comfortably above minimum velocity needed for SVHS recordings.

Then plugging the Hi 8mm parameters into the formula we find that the Hi 8mm minimum writing velocity equals the 3.75 m/sec writing speed selected for this equipment. The Hi 8mm machine has an upper frequency recording limit of just over 7.7 MHz, which equates to a minimum writing speed of 3.75 m/sec. The use of metal tape and metal heads allows an 8mm machine to safely operate close to the minimum calculated writing speed, because the combined use of metal heads and metal tape exhibit improved high-frequency response compared to the combination of oxide tape and ferrite head designs used in VHS equipment.

Rotating the Video Head Cylinder

Let's now expand what you just learned about calculating writing speed and consider how much tape would be used by a VCR if the signal is recorded using a stationary video head. You already learned that the VHS format requires a writing speed of 5.767 m/sec (228 inches/sec). This means that a machine using a stationary head needs over 12 miles of tape to reproduce a one hour program.

$$L = \frac{5.767 \times 3600 \times 3.28}{5280} = 12.9 \text{ miles}$$

where

L is the length of the tape,
5.767 is the writing speed in meters per second,
3600 is the number of seconds in an hour,
3.28 is the number of feet per meter,
5280 is the number of feet in a mile.

Even 8mm equipment with a slower writing speed still requires 8.5 miles of tape. Both of these tape lengths are obviously unacceptable.

One way to reduce the amount of tape required to record a video program is to reduce the length of the video head gap. However, there are practical limits to the size of the video head gap. A head gap size must be large enough to provide adequate sensitivity.

An alternate solution is to design a high-speed relationship between the video heads and the tape while reducing the actual speed of the tape traveling through the machine. This head-to-tape relationship is called writing speed. It's achieved by moving the tape through the machine at a relatively slow speed (slower than an audio cassette player) while rotating the drum assembly containing

the video heads at high speed (about 1800 RPM).

The electrical energy for record and playback is transmitted to and from the video heads by a transformer that rotates with the drum. The tape is drawn around the drum assembly at a slight angle using precisely positioned tape guides in what is called helical-scan recording. In a VHS machine, the tape wraps slightly more than 180 degrees around the drum. An 8mm unit requires about 221 degrees of tape wrap. Just like VHS machines, 8mm equipment only needs 180 degrees of wrap to write and read the video information on the tape. The extra wrap is required for recording an ATF (auto track find) signal for tracking control and an optional PCM (pulse-code-modulated) audio signal. PCM audio is a form of digital audio that is recorded on a different portion of the tape than that used to record the video signal. The additional wrap length produces a longer recording track than a 180 degree wrap produces and thus provides space for the added audio information.

As the video head assembly (also known as scanner, upper cylinder, or upper drum) rotates, the heads on the drum contact the relatively slow moving video tape producing an equivalent writing speed close to 250 inches per second (ips) or about 6.3 m/sec. The result is a stretched wavelength that enables high-frequency recording without high-frequency roll-off.

Precise calculations of writing speed in consumer helical-scan recording recognize that the heads are passing over the tape in the same direction that the tape is traveling and consider the diameter of the scanner. The fact that the heads strike the tape at a slight angle (about five degrees) has little effect on the speed. Other considerations include manufacturing tolerances, the amount of lower FM sideband reclaimed, and miscellaneous high-frequency loss.

If the tape didn't move, the length of the video head track would be half the circumference of the head drum. (For simplification, assume the tape is wrapped exactly 180 degrees around the scanner). The scanner (video head assembly) in a VHS system has a 62.0 mm diameter. Using the following equation for circumference, C:

$$C = 2 \ (P)r$$
$$= 2 \times 3.1416 \times (62.0/2)$$
$$= 194.779$$

Since tape travels around one-half of the drum surface only, the calculated circumference value is divided by 2 to give 97.389 mm.

For simplicity, assume that the five degree track angle has an insignificant effect on the equation so it can be ignored and that the video head track is parallel to the tape run. Since the heads record one field in half the circumference, multiply one-half the circumference by the vertical field rate and subtract the linear speed of the tape because the tape travels in the same direction as the head rotates.

Thus, writing speed is equal to the diameter times P divided by two times the vertical field rate less the linear tape speed.

Writing speed
$$= 97.389 \text{ mm} \times 59.94 - 33.35 \text{ mm/sec}$$
$$= 5804 \text{ mm/sec (5.804 m/sec)}$$

The published specification for the VHS SP (standard play) writing speed is 5.767 m.sec. The writing speeds for VHS LP (long play) and SLP (super long play—also known as EP for extended play) recording are slightly faster because the linear speed of the tape is changed. The head drum continues to rotate at the same velocity but the linear tape speed is slowed down to provide a longer recording time. Since the tape moves in the same direction as the head rotates, slowing the tape down relative to the consistent head motion creates a slightly faster writing speed. The linear tape speed for the VHS LP mode is 16.68 mm/sec (0.66 ips)—11.116 mm/sec (0.4 ips) for SLP/EP. The writing speed for the VHS SLP/EP mode is 5.826 mm/sec.

The 8mm machines have a smaller head drum diameter (40 mm) and operate at slower

linear tape speeds—14.345 mm/sec (0.56 ips) for SP, and 7.173 mm/sec (0.28 ips) for LP. The 8mm machine's writing speed for SP is 3.75 m/sec (147 ips).

FM RECORD/PLAYBACK

An obstacle to achieving suitable video recording is the dynamic range limitation of the tape itself. Current tape media has a dynamic range of about 70 dB. This allows a machine to directly record just over 10 octaves of information on the tape. The obstacles to successful video recording are the tape's limited 70 dB dynamic range and the 6 dB per octave response of the head during playback. The frequencies in 10 octaves uses up the bulk of the tape's dynamic range (60 dB) before the magnetic coating on the tape becomes saturated. To record signals ranging from 30 Hz to 4 MHz requires over 16 octaves. This prevents recording a video signal directly onto tape without some form of modulation.

The tape's dynamic range problem is overcome by modulating the video signal with a narrow bandwidth signal before the video is recorded onto the tape. A frequency modulation (FM) scheme is used to accomplish this task. Frequency modulation means that a certain frequency corresponds directly to the amplitude of the original information coming in from the video input source.

A frequency modulator shown in Fig. 5.21 is a form of voltage controlled oscillator (VCO). The VCO converts an incoming voltage to a frequency. As a frequency modulator, the VCO operates at a free run frequency when no signal is being input to the VCO circuitry. In a VCR, the free run frequency is called the carrier frequency or sync tip frequency. The frequency output from the VCO circuitry increases linearly and in direct proportion to the increase of the input voltage. The change in frequency caused by the change in input voltage is called the deviation frequency. In a VCR, the maximum deviation frequency corre-

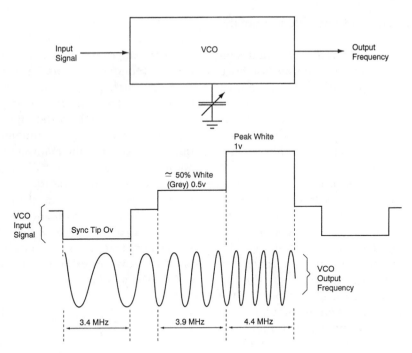

Fig. 5.21. *A frequency modulator is a special purpose VCO.*

sponds to the peak white video levels in the picture. Thus the total FM deviation range extends from the sync tip to the peak white levels. In a conventional VHS machine the sync tip frequency is 3.4 MHz and the peak white frequency is 4.4 MHz. Every video level occurring in the picture will fall between the sync tip and peak white levels. This constrains the FM deviation frequencies to a range between 3.4 MHz and 4.4 MHz.

For example, assume that the video signal voltage input to the frequency modulator ranges between zero volts for sync tip and one volt for peak white. Furthermore, assume that the VCO sync tip frequency is 3.4 MHz and the total deviation output for peak white (one volt) is 4.4 MHz. When the VCO in Fig. 5.21 senses the sync tip voltage at its input the sync tip (carrier) output frequency is 3.4 MHz. When a voltage level of 0.5 volts is sensed by the VCO input (half its input voltage range), the device produces a frequency of 3.9 MHz. Since the input voltage and output frequency

are directly proportional, the output frequency increases 500 kHz—one-half of the 1 MHz total deviation—changing the 3.4 MHz to 3.9 MHz. Finally, when the VCO senses 1 volt on its input, the maximum output frequency is produced (4.4 MHz). The total output frequency varies between 3.4 MHz and 4.4 MHz (1 MHz).

Video information broadcast from a TV station or coming from a video camera is measured in units of amplitude and consists of many complex signals. These signals include color or chroma information, brightness or luminance (Y) information, horizontal and vertical synchronization pulses (sync), and a short, high-frequency oscillation (burst) signal that is used to synchronize the color circuits of the receiving device with those at the video source.

For the moment, consider the luminance, or Y signal, as containing the brightness information, picture detail, and the horizontal and vertical synchronization pulses. Horizontal sync pulses phase- and frequency-lock a receiver's horizontal circuits with those from the broadcast station.

In a TV receiver, these horizontal synchronization pulses are responsible for horizontally locking the picture to the broadcast signal. Vertical sync pulses perform a similar role in the vertical circuitry of the TV and prevent the TV picture from rolling in a vertical direction. Vertical sync is also used in the servo circuitry of the VCR.

Picture detail (resolution) is a result of rapid video signal changes from one brightness level to another. For example, consider a video scene containing a dark telephone pole standing in front of a brightly lit building. The changes in video level relate to a change in voltage level within the video signal. The frequency modulator must make a rapid change in its output frequency when the different voltages corresponding to the white building and dark telephone pole are detected at the VCO input.

In a frequency modulation system, detail is dependent upon the modulator's ability to rapidly adjust the output frequency according to the rapidly changing input video signal voltage levels. If the VCO in the FM circuit changes frequencies slowly as the transitions between the dark telephone pole and the white building are reached, the video reproduced would be blurred and not produce sharp edges at the luminance transition points. The video would appear to transcend through various shades of gray until the new video level is finally achieved. Our eyes would observe these gradual transitions as smearing or blurring and we'd notice soft edges on the object. The detail (resolution) would be lost.

In the conventional 8mm system, the Y portion of the TV signal is frequency modulated with a signal between 4.2 MHz and 5.4 MHz. The horizontal and vertical sync tip pulses correspond to the 4.2 MHz portion of the frequency spectrum. Maximum brightness in the TV image corresponds to 5.4 MHz. The FM sync tip and deviation frequencies used vary according to the VHS, SVHS, 8mm, or Hi 8mm format being used.

As the picture brightness varies from one part of the TV display to another, the signal frequency changes within the specified deviation range. Frequency modulation produces many complex sidebands, which extend above and below the deviation frequency. These sidebands are recorded on the tape. Most of the sideband signal energy is contained in the frequencies closest to the main signal in the deviation passband. Fine detail (high-frequency information) in the picture is contained primarily in the sidebands. During playback, enough sideband information is recovered to produce a close replica of the original video signal. For example, in VHS machines the sideband information extends from about 1 MHz to 5.1 MHz as shown in Fig. 5.22.

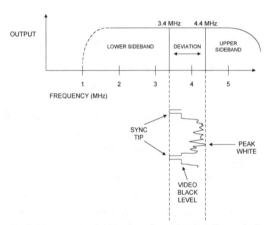

Fig. 5.22. *Frequency spectrum for FM VHS video includes a narrow 3.4–4.4 MHz deviation band and upper and lower sidebands.*

The upper sideband extends higher than 5.1 MHz, but the video tape cannot retain the higher frequencies, and the high-frequency losses created by the head cause the upper sideband information to be lost during the playback process. In consumer VCRs, only the lower sidebands are recovered.

The amount of lower sideband signal that is recovered is important in reproducing the fine detail in the original signal. The higher resolution capability of SVHS and Hi 8mm formats is a direct result of extending the lower sideband passbands.

Table 5.1 lists the upper and lower deviation frequencies for several formats. Figure 5.23 compares the VHS and SVHS sidebands. Notice that some of the higher resolution formats also have extended deviation ranges as compared to standard conventional models. When the lower sideband is extended, the deviation range is typically increased to maintain good video signal-to-noise characteristics. All of the higher resolution formats have extended lower sideband bandwidths when compared to their predecessor conventional format.

Table 5.1
Frequency and Resolution Specifications
for Various Consumer Formats.

	Sync Tip Frequency	Peak White Frequency	Total FM Deviation	Luminance Resolution Format (MHz)
VHS	3.4	4.4	1.0	240
SVHS	5.4	7.0	1.6	400
8mm	4.2	5.4	1.2	270
Hi 8mm	5.7	7.7	2.0	430

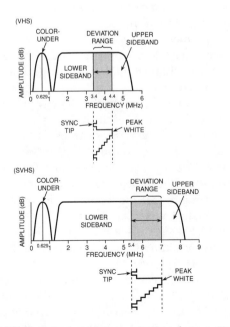

Fig. 5.23. *Frequency spectrum comparison between VHS and SVHS formats.*

Frequency modulating the video signal allows the entire video bandwidth of more than 16 octaves to be recorded within less than four octaves. The modulated lower sideband containing the fine detail is also part of that modulated four octave bandwidth. Thus, modulating a wideband signal and using sidebands overcomes the difficult task of equalizing over 16 octaves of the original signal.

The use of FM overcomes another obstacle to video recording. A video signal can contain DC information. For example, if a blank picture (having no detail and covering the entire screen, such as a blank white piece of paper) were transmitted from the TV station, the video signal would contain no alternating current information. This represents a constant DC signal. If this DC signal were recorded on tape using conventional recording methods, no playback information could be retrieved from the tape. For video heads to extract tape information, the magnetic lines of flux must be changing polarity. However, if this DC signal is converted to a frequency that represents that DC signal, it can be recorded on tape, played back, and converted once again to the original DC signal. This is accomplished using FM recording.

FM recording is ideal because it solves tape dynamic range limitations. Also, the system can be designed to ignore minor playback amplitude variations of the FM signal caused by imperfections in tape magnetic coatings and mechanical errors in the machine. During playback, the VCR monitors variations (deviation) from a set modulation frequency. Upon detection, the machine converts these signals into the original received signal. Because the system detects variations in frequency and not amplitude, it isn't sensitive to minor amplitude variations in playback signal strength.

The FM process also permits the recorded signal to saturate the video tape without causing adverse playback video signal distortion. This permits the FM signal to be directly recorded onto the tape without using bias. During playback, the FM signal coming off the tape is distorted, but, because the FM demodulator is only looking for variations in frequency, the original signal can be adequately restored. The ability to saturate the tape during recording produces stronger playback signals and decreases the chance of tape or VCR mechanical imperfections affecting the playback signal.

All consumer formats use two-head helical-scan recording. Two heads are used to record one full frame of video. These heads are mounted on a rotating head drum assembly 180 degrees apart as shown in Fig. 5.24. Each head records one field or one-half of the TV frame.

Fig. 5.24. Two heads are mounted on a rotating head drum assembly.

One head records the first field and the other head records the second field. With the video tape wrapped slightly more than 180 degrees around the head drum, an extra three horizontal lines of video are recorded from the alternate field at the beginning and end of each sweep of the head. This produces six lines of identical video information during playback. The two video heads have the record signal simultaneously applied at all times.

If you compare a video frame to a single frame of movie film, you will notice that if the movie film is stopped at a particular frame, an image is projected on the screen that resembles a 35mm slide picture. Movie film is com-

prised of a series of still pictures which are moved past a projection lens at a rate of 24 frames per second. The velocity of this motion is just fast enough to cause our brain to sense continual motion in the projected image without flickering or discontinuity. This is analogous to the image from a video recorder system.

The video picture on a TV screen is divided into 30 frames. Each of these frames is further divided into 525 horizontal lines. Line 1 is at the very top of the TV screen and line 525 is at the very bottom of the screen. The 525 lines are categorized into two fields. Even-numbered lines make up one field; the odd-numbered lines make up the other field. To reduce flicker on the TV screen, these fields are projected onto the face of the TV cathode ray tube (CRT) screen at alternating intervals. The field containing the odd-numbered lines begins by scanning the odd lines on the CRT from left to right and top to bottom. As soon as this is completed, the beam is quickly returned to the top of the screen and the even-numbered lines are scanned onto the screen. This procedure is called 2-to-1 interlace. The horizontal lines of video are interlaced on your screen using a procedure that is repeated 59.94 times per second (vertical field rate). Since each field contains 262.5 horizontal lines, one frame contains two vertical fields or 525 horizontal lines.

As mentioned, VHS units have the video tape wrapped around the head drum assembly slightly more than 180 degrees, although the video heads are mounted exactly 180 degrees apart. The tape is wrapped just far enough around the head drum assembly that one head is starting to exit the tape after the other head has made contact as shown in Fig. 5.25. The record FM signal is constantly applied to both heads simultaneously so there is a duplication of recorded material on the tape for slightly more than three horizontal lines. As described in Fig. 5.26, when video head B is starting to exit the tape, it is recording the same information as video head A, which is just starting to touch the tape on the other side of the drum.

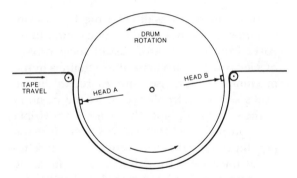

Fig. 5.25. *As one VHS video head is lifting away from the tape, the 180 degree opposite head has already made contact.*

This duplication of recorded material exists for three to four horizontal lines, yet the recording process causes the video head passing over the tape first to make contact approximately 10 horizontal lines before the vertical sync period occurs.

During playback, the video heads are electronically switched on and off to produce a normal picture and are timed to avoid playing back the duplicated three horizontal lines as shown in Fig. 5.27. The video heads must be switched on and off during playback to prevent video noise generated by the head, which

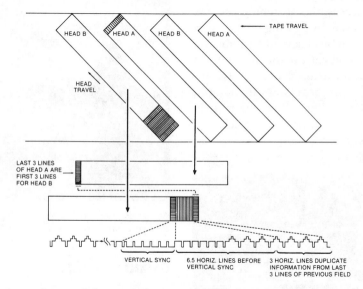

Fig. 5.26. *Video heads A and B record the same information for three horizontal lines.*

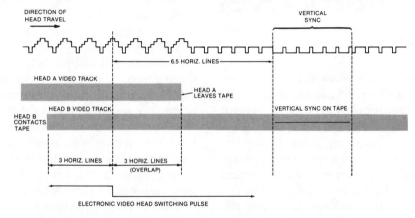

Fig. 5.27. *The video head switches 6.5 horizontal lines before vertical sync (playback mode).*

is not in contact with the tape. Each video head is connected to its own high-gain amplifying circuit called a preamplifier. If the preamplifier isn't gated off when the video head is not in contact with the tape, it would produce snow on the screen. Only the amplifier for the head in contact with the tape should be passing a signal. Video head switching occurs 6.5 horizontal lines prior to the vertical sync pulse at the very bottom of the video screen.

Figure 5.28 is an oscilloscope display showing the relationship between a playback video waveform and the head switching pulse. On a TV monitor with properly adjusted vertical height and linearity, this switching point is below the normal screen viewing area so it is not visible. But it can be observed by adjusting the vertical hold control on the monitor to bring the vertical synchronization pulse up into the display viewing area. You will then notice a very fine line 6.5 lines above the vertical sync pulse. This is the time location where the video heads are switched.

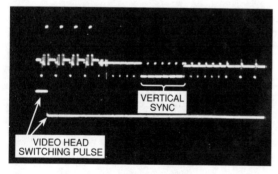

Fig. 5.28. *A playback waveform with vertical sync and the video head switching pulse.*

A special monitor called a pulse cross monitor can display the head switching point located 6.5 horizontal lines above vertical sync.

Video head switching occurs before the vertical sync pulse to prevent noise interference that could desynchronize the TV vertical circuits causing the picture to roll, become unstable, or jump up and down. Designing the switching point so it occurs 6.5 lines before

vertical sync provides time for the TV horizontal automatic frequency control circuitry to recover from timing error disturbances caused by electronic or mechanical instabilities in the machine. Very precise timing circuitry in the VCR's servo electronics controls the position of the switching point. The point is not visible on a properly adjusted video screen. It is important to note that the vertical sync pulse interval must be recorded at exactly the same place on the video head record track with each sweep of the head. See Fig. 5.29.

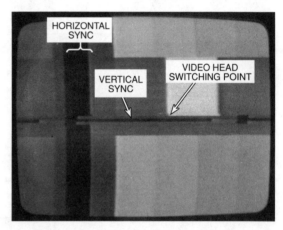

Fig. 5.29. *The video head switching point and vertical and horizontal sync are shown on a pulse cross monitor.*

GUARD BANDS

Video tape is passed across the video drum assembly so the video information is recorded at an approximate five degree angle referenced from the edge of the video tape. Some of the very first helical-scan recorders had open reel-to-reel tapes with a space between each of the video record tracks known as guard bands. Video guard bands are thin areas on the tape where no signal is recorded. Guard bands were used to prevent interference on the TV screen caused by information from adjacent recorded tracks being detected by the head during playback. This occurs if head A detects information from video track B. This crosstalk produces very fine, closely spaced lines of interference on the screen. Video guard bands are spaced to prevent a video head from picking up adjacent track information. Crosstalk is also called cross modulation.

The use of guard bands represents wasted tape so engineers developed a technique for preventing Y signal crosstalk and eliminating the need for a guard band. This technique has been termed zero guard band.

Azimuth Recording

The principle of azimuth recording is important in eliminating crosstalk and the need for guard bands. From magnetic tape recording theory, a designer knows that maximum head output signal occurs when exactly one-half the recorded wavelength is within the boundaries of the video head gap such that the gap is exactly 90 degrees perpendicular to the recorded signal.

An example will illustrate this point. Audio technicians know that when an audio tape recorder has high-frequency losses from tapes recorded on another machine, yet correctly plays back its own recordings with good high-frequency response, the probable cause is the angle or azimuth of the audio head itself. If the audio recorder head gap is not mounted exactly 90 degrees perpendicular to the recorded tape signal as shown in Fig. 5.30, a reduction in high-frequency response occurs. It is now no longer possible for exactly one-half the recorded wavelength to fit in the head

gap. What occurs instead is a deterioration of high-frequency playback signal when magnetic domains of opposite polarity are sensed by the head. This process cancels some of the correct signal field intensity.

The video head in the VCR is mounted so its head gap is perpendicular to the tape information plus or minus the azimuth angle offset. This allows elimination of the guard band while still providing adequate isolation for the luminance portion of the recorded signal. The video head is constructed with the gap at a slight angle. In VHS machines, one head has a gap placed at +6 degrees while the angle of gap for the other head is at –6 degrees as shown in Fig. 5.31. The effective azimuth difference between the two heads is 12 degrees. Heads in 8mm machines have gap angles of plus and minus 10 degrees for a total of 20 degrees azimuth difference.

VHS video head A records a track of information at a 6 degree angle to the perpendicular. Then when head A reads that same track during playback, it picks up its own signal and not that of the other track. However, as shown in Fig. 5.32, video head B, with an az-

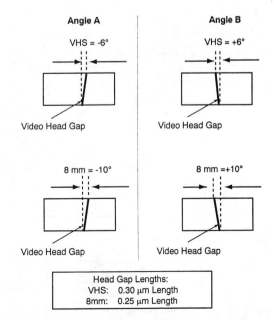

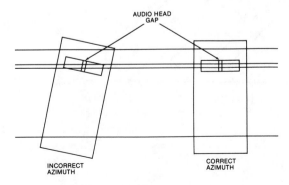

Fig. 5.30. *Correct and incorrect audio head azimuth.*

Fig. 5.31. *Video head gaps have different azimuths.*

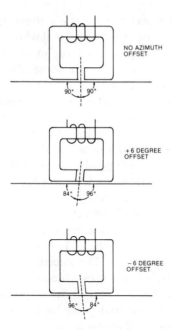

Fig. 5.32. Azimuth offset enables a video head to detect only the signal it recorded.

band and provides sufficient video track isolation to prevent high-frequency crosstalk. No guard bands are used in 8mm SP, 8mm LP, and VHS LP and SLP/EP modes, but some VHS models do develop guard bands on the tape when recording in the SP mode. This occurs because of the head core width used in the design.

You can analyze the presence of guard bands in some VHS models by looking at some early production consumer models. Figure 5.33 describes the use of guard bands in the Panasonic PV1100 operating in the SP mode. Each of the two video head cores are 38 microns wide with a video head gap of 3 microns. The two video head gaps are designed at ±6-degree azimuth offsets. The video record tracks on the tape are equal to the 38 micron width of the video head core material and each video head track has a 20-micron-wide guard band in the standard play mode.

Chroma or color information is recorded at a lower frequency making the azimuth angles insufficient to reduce the chroma crosstalk, so another means must be used for isolating the chroma signals on two recorded tracks. This will be discussed later.

imuth offset of 12 degrees from head A, detects virtually none of the recorded luminance information from track A. Therefore, azimuth recording eliminates the need for a guard

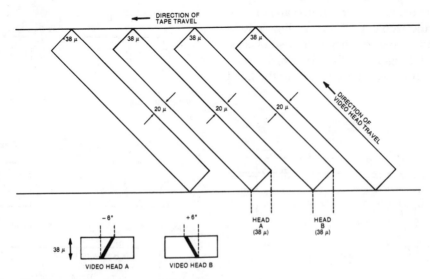

Fig. 5.33. Guard band spacing in SP (Panasonic PV1100).

High-Density Recording and Special Effects Playback

To maximize the use of video tape recording surface while eliminating the need for guard bands, the video record tracks are partially overlapped for slower tape speeds. Positioning the tracks on the tape in this manner is a form of high-density recording. This should not be confused with the increased resolution produced from recording shorter wavelengths on the tape, which is a separate form of high-density recording.

Figure 5.34 shows what happens when the PV1100 begins recording in LP mode. The video head tracks (commonly called RF tracks for radio frequency tracks) are caused to overlap by 9 microns. During the record process, video head A writes information on the tape as the capstan motor advances the tape, and video head B begins to write its information partially recording over information previously written by video head A. This procedure alternates with head A overwriting a nine-micron-wide strip of track B, and vice versa. The information first recorded by video head A loses some of the effective width when video head B rerecords over track A because video head B reorients the magnetic domains in the 9 micron strip. During playback, video head A reads its previously recorded track and recognizes enough of its recorded information to

faithfully reproduce the original signal, although the width of the video head playback track has been reduced from 38 microns to 29 microns.

Engineers discovered that it was possible to further slow tape speed and use a smaller video head core to pack the video RF tracks even closer. This resulted in the introduction of SLP or EP (super long play or extended play) speeds. Early versions of machines capable of these speeds have video heads which are only 30 microns wide using video record overlap technology.

However, a trade-off developed between track width and overlap. Reduction in video head size caused a slight reduction in video picture quality, which prompted the introduction of video tape recorders using four heads. Two heads are required for record and playback, so one pair of heads was designed to operate only during the SP mode, thus offering optimum picture quality. The second pair was dedicated to LP and EP modes and was designed for maximum tape economy. The different linear tape speeds meant the use of unequal size head pairs between SP and LP/EP modes. These four-head VCRs provide a higher quality video image in SP because wider heads are used during this speed. Using narrower heads in LP and EP was economical because it provided longer record/playback time on the same length of tape.

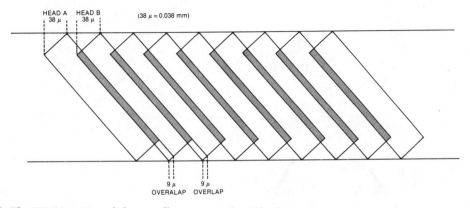

Fig. 5.34. The PV1100 in LP mode has an effective RF track width of 29 microns (0.029 mm).

Video tape recorders with four heads can produce improved special effects in playback because designers could select the video heads which give the best playback picture for slow motion and noise-free still.

Engineers also found that by selecting video heads with unequal widths, they could achieve better special effects. Figure 5.35 shows the record track for an early-model two-head Panasonic PV5000. The video head sizes in this model differ by 5 microns. Video head A is 26 microns wide and video head B is 31 microns wide. In SP speed, the video heads have two different guard band widths—32 microns and 27 microns. A similar technology exists in many current model video head designs.

As shown in Fig. 5.36, operation in LP produces a 3 micron guard band and a 2 micron head overlap of the adjacent track.

In SLP mode, as described in Fig. 5.37, no guard band exists, and the overlaps alternate between 6.7 and 11.7 microns. However, the effective playback video head track width is 19.3 microns wide for each of the video playback tracks. This is significant because the relationship between the width of the video head

core and the effective video playback track directly affects the quality of playback during special effects such as search forward and reverse, still playback, and slow motion playback.

Figure 5.38 describes the relationship between video head A and B and their respective recording tracks. The RF envelope produced by the video heads generates a noise-free TV image.

Figure 5.38 also describes the relative angle changes when the video head touches the tape and when the tape is stopped. During still frame playback, the video tape stops and the moving drum assembly causes video head A to begin its excursion up the video head track. It first encounters track B information, which it doesn't recognize due to the azimuth difference. The small amount of track A information present at the same point is recognized and becomes the radio frequency (RF) output waveform envelope. As head A continues up the tape, it detects more of the recorded A track and the RF amplitude increases. At all times during still frame playback, some track A information is detected so the coil in head A always produces some RF output.

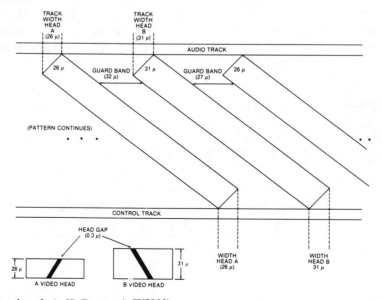

Fig 5.35. *Video head tracks in SP (Panasonic PV5000).*

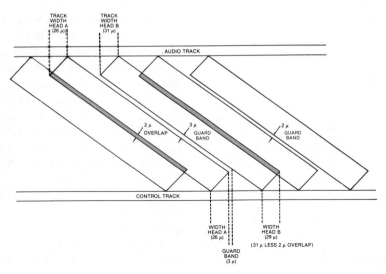

Fig. 5.36. *Video head tracks in LP (Panasonic PV5000).*

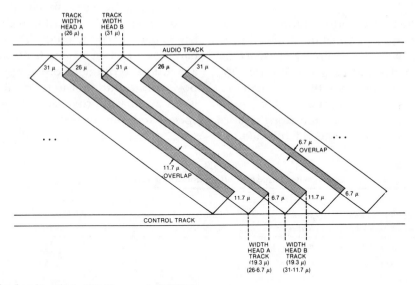

Fig. 5.37. *Video head tracks in SLP (Panasonic PV5000).*

When head B begins to move up the tape, it covers and detects a large amount of track B record information, and its output signal is high. But the RF information begins to deteriorate as the head moves up the track because it detects less track B information and more track A recorded information.

Head B is 5 microns wider than head A, so it detects slightly more RF information as it passes over the tape. If the B head wasn't 5 microns wider, it would run out of recorded track B information during a pass and produce no RF output near the last part of its excursion over the tape. No output occurs when a head

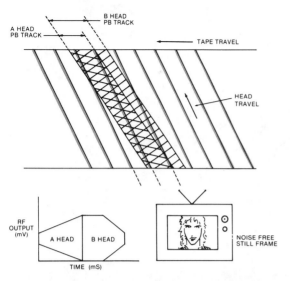

Fig. 5.38. *Video output during SLP still frame playback (Panasonic PV5000).*

detects no recorded video information or when the recorded information is traced by a head with the wrong azimuth. The absence of an output signal causes snow on the TV screen. The snow appears at the portion of the scene where no RF is present.

The place where the tape stops is also important in positioning the video heads for proper scanning of the recorded track information and for producing a noise-free still picture. Stopping the tape at the correct position is controlled by the capstan servo and will be discussed in the next chapter.

Table 5.2 compares the video head sizes for a number of Panasonic VCRs.

Early production VCRs with only one video head width couldn't produce noiseless still playback because the video head size was too large to properly scan the recorded RF tracks. The TV screen displays were so poor that designers decided to mute, or blank out, the video during playback pause. Figure 5.39 describes the playback RF envelope and the noisy TV picture during still frame. It was not until head drums with varying head widths were introduced that noise bands could be reduced in search modes and eliminated in still/pause modes.

During the search functions (also known as cue-and-review), all machines generate noise bands that are visible on the TV screen. These bands occur when a video head crosses either a guard band or the recorded track of the opposite head. During search functions, the tape is advanced at approximately 10 times normal speed and the respective read angle with which the video head crosses the tape changes

Table 5.2.
Early Version Panasonic VCRs and Their Head Sizes

Record Playback Speeds Available	Model Number	Number of Video Heads	Video Head Core Width (in microns)	
			A Head	B Head
SP/LP/SLP	PV-1200 series	2	30	30
SP/LP/SLP	PV-1300 Series	2	30	30
SP/LP/SLP	PV-1400 Series	2	30	30
SP/LP/SLP	PV-1600	2	30	30
SP/LP/SLP	PV-1650	2	70	90
SP/LP/SLP	PV-1750	2	26	30
SP/LP/SLP	PV-1770	2	26	30
SP/LP/SLP	PV-1780	4	30 & 32	45 & 32
SP/LP/SLP	PV-3000	2	26	31
SP/LP/SLP	PV-4000	2	26	31
SP/LP/SLP	PV-5000	2	26	31
SP/LP	PV-1000A	2	38	38
SP/LP	PV-1100	2	38	38
SP/LP	PV-1500	2	38	38

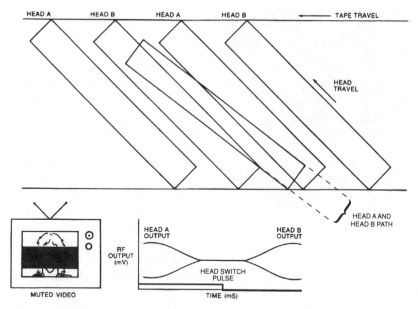

Fig. 5.39. *Playback video muted on early VCRs during pause because of poor picture quality. During playback still, the A head and B head travel up the same path because tape movement has stopped.*

dramatically. In these modes the heads cross several recorded tracks during each scan cycle. As the head crosses a guard band or a recorded track with opposite azimuth angle, the output from the read head is zero resulting in bands of noise or snow across the TV screen. Figure 5.40 is a picture of snow bands occur-

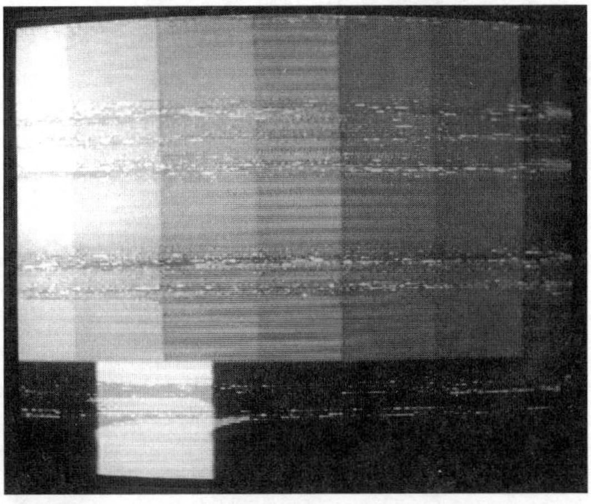

Fig. 5.40. *If, during search, the video head crosses an RF track of opposite azimuth or a guard band, snow bands like these occur.*

ring on a current model VCR in search mode. The number of noise bands on the TV screen closely approximates the speed with which the tape is being advanced.

For example, if nine bands of snow are counted on the screen, the VCR is advancing at approximately 10 times the normal speed because the head is passing over 10 individual RF tracks for every rotation of the head drum assembly. These bands vary in width according to the width of the head core. The original version VHS units with only one head width generated wide noise bands in search. Recent designs display much narrower bands due to the variance in head widths. Figure 5.41 is an oscilloscope display of the measured RF envelope and the head switching pulse during search.

The output waveform and TV screen image for forward and reverse search modes are shown in Figs. 5.42 and 5.43. These figures also show how the noise bands are produced on the screen. In the PV5000 model, different size video heads act to reduce visible snow lines on the screen.

The 31 micron head produces an effective track width of 19.3 microns in SLP record

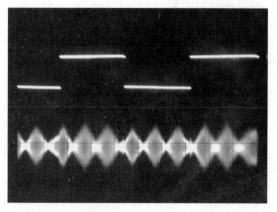

Fig. 5.41. *Video head switching pulse and the RF envelope during search.*

mode causing head B output at all times in the search mode. This substantially reduces the size of the visible noise bands. When searching on an SP-recorded tape, the guard band becomes quite visible and the noise bands on the screen are much wider.

The two-dimensional representations of video head and recorded tape track angle rela-

tionships shown in Figs. 5.42 and 5.43 may mislead you without more explanation. In the illustrations, it appears that the relative azimuth angles between the head and the tape track are too far off to reproduce a usable RF output signal because of the azimuth angle losses in the head gap. Actually, the azimuth angle at which the head approaches the tracks changes very little. The drawings distort this fact because they are only intended to show how the head scans different recorded tracks in high-speed search modes.

The head does scan several tracks in search modes so in the figures, its angle was tilted (as referenced to the tape edge) to show the multiple track scanning from a two-dimensional perspective. In search mode, the head scans several recorded tracks during one revolution because the tape is moving around the drum at a faster longitudinal speed. The video heads rotate at the same speed and approach the tape at the same helix angle (as referenced to the tape edge), but since the tape is moving faster than normal, more tracks are pulled be-

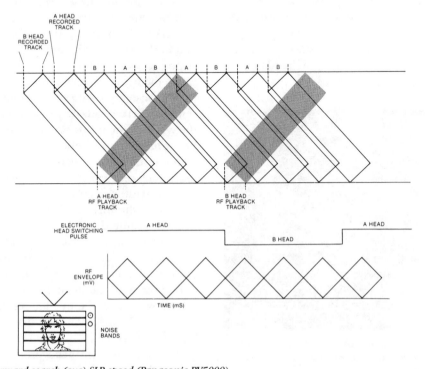

Fig. 5.42. *Forward search (cue) SLP speed (Panasonic PV5000).*

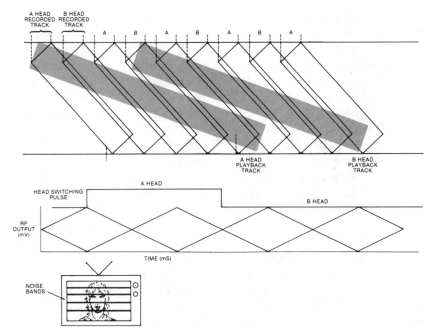

Fig. 5.43. Reverse search (review) SLP speed (Panasonic PV5000).

neath the head and, hence, get scanned. Just as the head starts to read up one track, the capstan motor pulls another track beneath the rotating head. The faster the tape moves, the more tracks are pulled past the rotating head during each scan. This multiple track scan causes noise bands to appear in the picture during search. The relative azimuth angle between the recorded tracks and the head gap is close enough that most of the RF signal recorded on the tracks is recovered.

Special Purpose Playback Heads

Some VCRs include a designated playback head to minimize guard band noise in slow motion and still/pause playback modes. Figure 5.44 shows a typical three head configuration with one head used for guard band noise minimization. This special playback head (designated CH1′ for channel 1 prime in Fig. 5.44) may also be identified as an "SS" head for still/slow head or "trick play head" all of which indicate its special function.

The trick playback head is strategically mounted on the drum so that it can trace a specific RF track as the tape speed is slowed down for slow motion playback or stopped for a still display. In Fig. 5.44, the CH1′ head traces up the channel 1 track. To read channel 1 information, the azimuth of the CH1′ head gap is the same as the normal channel 1 record/play head. Special switching circuits introduce the RF signal detected by the trick play head into the playback processing circuits when slow

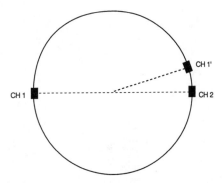

Fig. 5.44. Trick play head drums contain more than two heads.

motion or still/pause modes are engaged. VCR servo circuits play an important role in enabling the trick play head to produce a noise free still/pause and slow motion video. Servo operation is described in Chapter 8.

In the still/pause mode, the tape is carefully parked in such as way that the CH1′ head can trace over one of the CH1 RF tracks. The CH1′ head scans the same track over and over again until the still/pause mode is released. In this mode, the video out of the machine is a single field displayed continually on the television screen.

Previously we learned that the relative angle between the track recorded on the tape and the angle of the video head scan changes when the tape speed is something other than normal play (i.e., still and slow motion forward and slow motion reverse). The CH1′ head has a wider head core than the normal record/playback head so it can detect some RF under its gap when playing back under these abnormal conditions. The relationship of the wider CH1′ head core to the recorded track in the still/pause mode is shown in Fig. 5.45.

During slow motion operation, the tape does not travel smoothly through the tape path. Instead, it moves in a stepped fashion. The tape is pulled then parked, pulled then parked, over and over again to give the effect of slow motion. A careful examination of the display screen in this mode reveals that the slow motion video is not a continuous and smooth slow motion movement, but a stepping action from one field of video to another. This stepping movement is a result of the pull and park action of the tape during this mode.

In the slow motion mode, the CH1′ head operates similar to the still mode. The trick head scans up the CH1 RF track each time the tape is parked by the mechanism. The CH1′ head continues to trace over the same track until the tape is jogged and positions another CH1 RF track under the head's scanning path. The number of scans that the head makes over the same track is dependent on the slow motion speed selected by the VCR operator. A reduced slow motion speed will cause the CH1′ head to trace over the same track more times than during a faster slow motion speed before a new track is pulled into position.

Not all VCRs use a trick play head to achieve special effects playback. Some of the four-head designs use the wider SP head cores

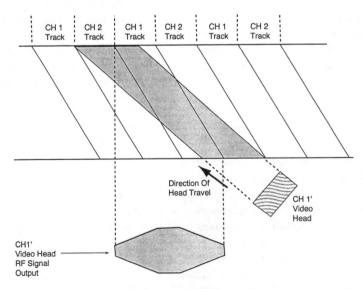

Fig. 5.45. *The CH1′ head has a wide enough head core to maintain an RF output as it scans the CH1 RF track in the still/pause mode.*

to provide guard-band-free slow motion and still pictures when playing back recordings made in the SLP/EP speed. Noise-free playback is possible because the wider core of the SP heads can trace and detect information up the narrower record track written by the SLP/EP recording heads. The operation is similar to the third head (trick head) function shown in Fig 5.45. The difference is that the four-head units use the SP heads rather than a CH1′ head as the trick playback head to perform the special effect function.

Some RF signal is always present in the SP head gap in the still/pause and slow motion modes. Some models with this four-head configuration may only offer guard-band-free special effects when playing back SLP/EP recordings. For these units, entering slow motion or still/pause while playing back SP and LP recordings will still produce guard band interference in the picture. Other models use a combination of varying head core widths in the four-head design to provide noise-free slow motion and still/pause in both the SP and SLP/EP modes. Some manufacturers offer a CH1′ head in addition to the four heads. These five head machines offer good special effects for all tape speeds. Two-head and four-head VHS models with LP recording modes typically don't offer good special effects at the LP tape speed.

WRITING AND READING THE 8MM SIGNAL

The 8mm format specifies that a recorded track width must not exceed 20.5 microns. Like the VHS units, the track width on the 8mm tape is determined by the width of the head core writing the signal and not the head gap length. Some 8mm models may use head core widths measuring less than 20.5 microns to provide a slower recording speed (LP) and to reduce the guard band noise in still/pause, slow motion, and search modes. In fact, some 8mm models use heads with a core width of just 15 micrometers.

Just like select VHS models, some 8mm units use a trick play head to obtain optimal

slow motion and still/pause playback displays. With the exception of the drum size, 8mm head drums with a third (trick) play head look like the one shown in Fig. 5.44.

There is a noticeable difference between the 8mm and VHS formats in how much tape wraps around the head drum. An 8mm recorder/player wraps the tape at least 221 degrees around the drum. VHS units wrap the tape around the drum about 186 degrees. The VHS tape wrap consists of 180 degrees for the video track and an additional 3 degrees on both the entrance and exit sides of the drum to allow for some mechanical tolerance and some video information redundancy at both ends of the recorded tracks as previously explained.

Extending the 8mm wrap to a minimum of 221 degrees provides space on the recorded RF track for additional information to be written along with the FM video signal. This information includes PCM audio, ATF, RC time code, and DATA code information. PCM audio is a form of digital sound. It's an optional means of recording a stereo audio program. All 8mm units record monaural FM audio along with the video signal as their primary sound track. ATF (automatic track find) is a signal used by the drum and capstan servo circuits to perform tracking automatically. The 8mm units don't provide a manual tracking control like VHS machines do. RC time code and DATA code information are optional signals that provide frame number identification (time code) and recorded time and date (DATA code) displays, respectively. The time code information is useful in editing a tape. The date and time display are useful for archival information. With the exception of the ATF signal that is required on all 8mm equipment, each of the additional signals is optional. To provide interchangeability between models, all 8mm machines must make space on the recorded RF track for each of these signals whether they are recorded or not. The theory supporting these new signals is described in Chapter 6.

The "footprint" produced with the extended tape wrap is shown in Fig. 5.46. Notice that the format also specifies that sections of

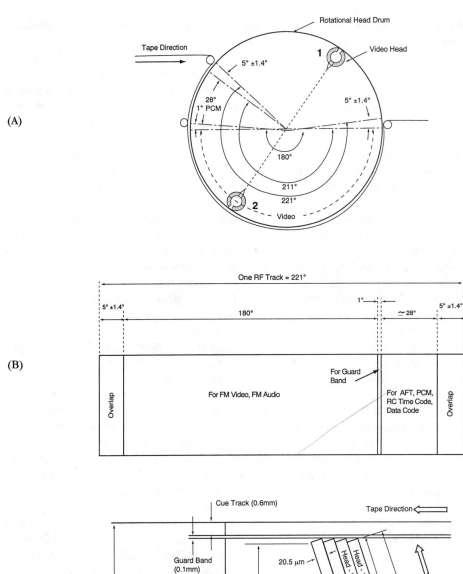

(A)

(B)

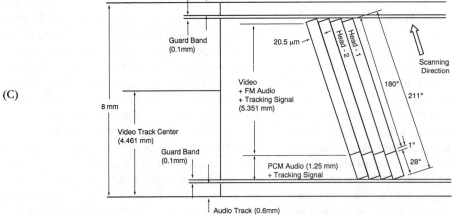

(C)

Fig. 5.46. *The 8mm format specifies a tape wrap of 221 degrees. (A) The wrap around the head. (B) A longitudinal view of the tape wrap. (C) The recording pattern (footprint) of the tracks on the tape.*

the tape be preserved for a cue track and a longitudinal audio track. Although all machines must preserve these linear track regions in the recording process, they are not presently being used in any consumer model.

Figure 5.46 shows that the FM video signal is recorded for 180 degrees of wrap just as with the VHS video signal. The extra 5 degrees of wrap shown at either end of the drum provides many of the same features that the extra 3 degrees of wrap offers in the VHS format. Among these benefits is the assurance of interchangeability of tapes even when slight mechanical tolerance differences might exist between various models and equipment manufacturers.

The 8mm format specifications also require a minimum of 1 degree of space between the section occupied by the FM audio and video signals and the 28 degree space reserved for the PCM and ATF information. This 1 degree separation is another form of guard band. It's placed on the tape to provide spatial isolation between the separate packets of information. Models that record and play back the RC time code and DATA code signals write and read the information in this 1 degree guard band zone. Figure 5.46C shows that the additional wrap of the tape around the head (approximately 360 degrees) containing the RC time code, data code, PCM, and ATF signals occupies 1.25mm of the 8mm-wide tape.

Not all models reproduce PCM audio. However, all models are required to reserve this 1.25 mm space for the PCM signal. Every 8mm unit records audio using a monaural audio FM technique. Therefore, even units offering PCM audio will simultaneously record sound using the FM audio method to assure compatibility with non-PCM models. The FM audio shares some of the frequency spectrum used by the FM video signal and is recorded along with the FM video signal on the same (180 degree) portion of the track.

Figure 5.46C shows a longitudinal audio track at the bottom of the 8mm tape and a longitudinal cue track at the top edge. Both of these longitudinal tracks are separated from the RF tracks by a 0.1 mm guard band. These guard bands provide isolation from the RF signals and help prevent signal crosstalk. Neither of these longitudinal tracks is currently used, but both are preserved for use in future models.

VHS TAPE PATH CONFIGURATIONS

Figure 5.47 describes the tape path for a typical VHS machine. Figure 5.48 is a close-up of the VHS tape path as it travels from left to right past the full track erase head, around the drum and past the audio erase and record heads. The full track erase head removes all previously recorded audio, video, and control track information. It uses the bias frequency generated for recording the normal audio signal as a high amplitude erase signal and demagnetizes the tape before the tape passes around the video head drum. This head is deenergized during audio or video insert edit modes.

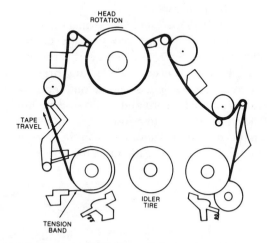

Fig. 5.47. *Typical VHS tape path layout (top view).*

As the tape moves past the full track erase head, it passes through a series of guides that position it over the rotating head drum assembly. The tape is drawn across the head drum assembly at a slight angle to enable helical

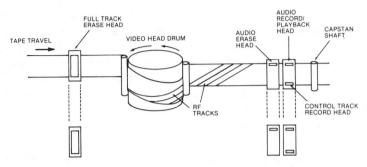

Fig. 5.48. *VHS tape path (front view).*

scan recording. To achieve accuracy, the tape must be carefully guided along the head drum assembly by the entry and exit guides adjacent to the head drum. Proper tape path movement around the scanner (head drum assembly) is controlled by a lip called a rabbet located around a small ledge on the lower part of the drum assembly. The rabbet keeps the tape on a precise path around the drum. VHS and 8mm machines keep tape travel in the rabbet by using an upper rotating drum that is slightly larger than the lower stationary portion. The difference in size causes the tape to press lightly into the rabbet.

In a VHS machine, the normal track audio head assemblies have two separate heads as shown in Fig. 5.49. The first head is the audio erase head. It is used to delete any previously recorded audio signals and is only energized during normal record and audio insert editing. The audio erase head removes the audio signal on the longitudinal audio track in the same way the full track erase head erases the entire tape.

The second head in the stack, the audio/control head, is divided into two head sections—the audio head and the control track head. The audio head (on top) is for recording the normal track (also known as the linear track) audio signal. It has two record channels if the unit has the stereo linear audio option. Channel 2 (the right channel) is at the very top edge of the tape; channel 1 (the left channel) is at the lower part of the audio section of the tape. In the case where the normal audio track is not stereo, the head uses the full longitudinal audio track width for the monaural audio information.

The lower portion of the audio/control track head contains the control track head. This head is used to record the control track pulse on the tape. Control track pulses let the machine identify where to place the video head during playback. The procedure for this operation will be described later.

Figure 5.50A describes the video tape format for VHS machines. Figure 5.50B shows

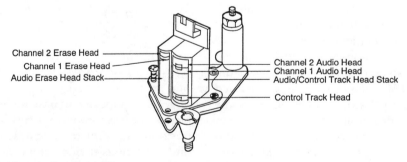

Fig. 5.49. *The VHS audio head assembly.*

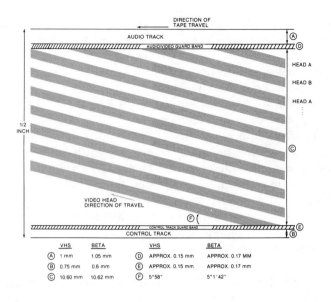

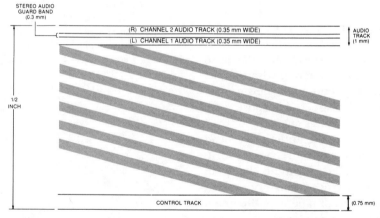

Fig. 5.50. *(A) VHS tape format for the SP speed and a monaural normal track audio head. (B) In a VHS recorder with stereo linear audio, the audio/video guard band is eliminated and a 0.3 mm space is used as a stereo audio guard band to separate right and left audio channels.*

the audio track tape sections produced by a linear track stereo VCR. There are slight differences between the monaural and linear stereo video tape formats.

The effect of the full track erase head used in a VHS machine can be seen on a TV monitor when a previously recorded tape is rerecorded. For the first few seconds of tape travel, a rainbow pattern is often visible on the TV screen. This rainbow pattern zigzags back and forth across the entire screen. The de-gaussing effect of the full track erase head removes this rainbow pattern at the beginning of each new recording. We see the effect of the erasure process as a wiping away of the rainbow pattern starting at the top of the screen and progressing to the bottom until the rainbow eventually disappears. What you are seeing is the remnants of the previous video program. The video head is able to record over and erase most of the previous video but some of the color information remains on the tape because the video record signal is not strong enough to completely realign the polarities of all the previous magnetic domains.

Figure 5.48 shows that on a VHS unit, a few inches of tape exist between the full track erase head and the video head drum assembly. These few inches of tape cause the rainbow pattern that eventually disappears as the completely erased tape passes the full track erase head and moves across the video head drum assembly. The duration of this rainbow pattern is determined by the speed with which the new recording is made. It takes longer for the completely erased tape to get from the full track erase head to the video heads on the drum in SLP than it does in SP. SLP is one-third the speed of SP so the rainbow pattern will appear on the screen three times as long.

Some of the more feature-rich VHS models mount a video track erase head on the head disk to stop the previously recorded video from bleeding through and interfering with the picture for that first few seconds of the new recording. The operation of this special purpose erase head is explained later in this chapter.

In a VHS machine, the audio erase head is close in front of the audio record head to prevent the same effect from occurring in audio. If an audio erase head were not present in the machine, a strip of tape would contain both old and new audio programs.

When recording on a previously used tape, the mixture of previous and newly recorded sound would be heard when playing back the start of the new recording. This "sound-on-sound" effect would be present until the portion of tape erased by the full track erase head reached the audio/control head assembly. From that point forward, only the new audio would be played back. This sound-on-sound phenomenon would occur for several seconds each time a new recording was initiated on top of a previously recorded tape.

So, to quickly eliminate previous audio programming, an audio erase head was placed in the tape path just prior to the audio

record/playback head. The normal audio signal is then recorded using conventional audio record methods.

8MM TAPE PATH CONFIGURATIONS

The tape path configurations for 8mm units are shown in Fig. 5.51a and b. Although the tape paths for 8mm units appear quite different, tapes produced on both units are interchangeable.

Figure 5.52 shows a front view of the 8mm tape path as the tape is pulled from left

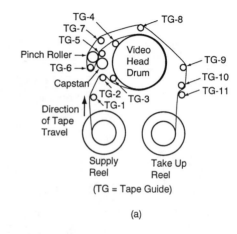

(a)

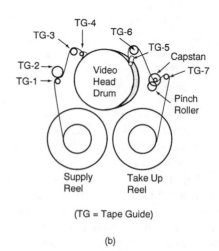

(b)

Fig. 5.51 a,b. Two common 8mm tape path layout configurations (top view)

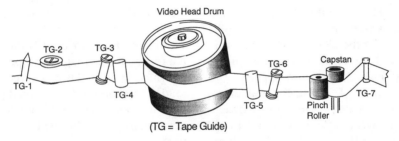

Fig. 5.52. *8mm tape path (front view).*

to right across the head drum. Notice that the 8mm machines do not have a full track erase head, a normal track audio/control track head, or an audio erase head in the tape path. This is because all audio, video, and servo control signals are recorded on the RF track by the video heads. A study of Fig. 5.46A, B, and C shows that all audio, video, and servo information is recorded by the video heads as part of the 221 degree wrap around the video head drum. The FM audio is recorded in the same portion of the RF track as the FM video signal. The ATF signal, used for servo tracking control, is recorded as part of the portion of FM video and in the additional 30 degree segment of the tape wrap. The PCM audio (if used) is also recorded on the extra 30 degree wrap portion of the RF track. The extra amount of tape wrap required in 8mm units provides the space on the RF track for this nonvideo information and eliminates the requirement of the normal track audio and control track heads necessary in the VHS format.

Figure 5.52 also reveals that no full track erase head is used in the 8mm tape path. Erasure of previously recorded information is performed in 8mm models by a flying erase head or rotary erase head mounted on the head drum. This erase head is positioned on the rotating head drum so that it will contact the tape just prior to the recording video head and erase the previously recorded tracks on the tape before the recording head starts to write the new material. This provides an accurate and instantaneous erasure of any previously recorded material existing on the tape when a

new recording is initiated. Flying erase heads eliminate the previous program bleed through that is common in VHS models for the first few seconds when a new recording is written on top of an existing recording. Some of the more expensive VHS models include a flying erase head to eliminate this phenomenon.

Figure 5.53A shows an 8mm head disk configuration with two video heads and a flying erase head attached. Figure 5.53B displays a four video head disk with a flying erase head as

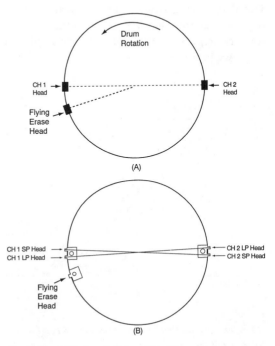

Fig. 5.53. *8mm head disks with a flying erase head. (A) A two video head disk with a flying erase head. (B) A four video head disk with a flying erase head.*

used in 8mm machines. In this design, one pair of video heads is used for the SP mode while the second set is for the LP speed. On this multihead disk, the flying erase head erases previous recordings for both recording speeds.

Only one pass of a single flying erase head is required to erase the two recorded tracks previously written by the channel 1 and channel 2 video heads. Figure 5.54 shows how one head can accomplish this in a single rotation of the head. The flying erase head core is 41 microns wide. This is wide enough to allow one trace of the flying erase head to remove the two 20.5 micron-wide RF tracks recorded by the channel 1 and channel 2 video heads. The flying erase head is mounted at the correct height (referenced to the lower edge of the head disk) and rotational position (referenced to the location of the channel 1 recording head) to precisely travel over the previously recorded tracks and erase them when a new recording is initiated. The flying erase action continues throughout the recording process. The flying erase head operates during every record session regardless of whether a new (blank) or previously recorded tape is placed in the VCR.

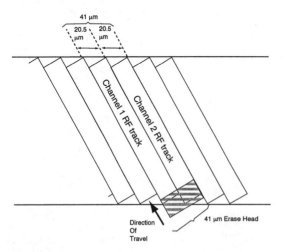

Fig. 5.54. *One swipe of a single flying erase head will erase two RF tracks.*

SUMMARY

This chapter has covered audio and video signal recording on magnetic tape and the means by which the signal is recovered. You have also read about the recorded signal pattern on the tape for both VHS and 8mm formats. In the next chapter you'll explore VCR operating theory.

CHAPTER REVIEW QUESTIONS

1. What is the function of the audio bias oscillation used in writing a normal track audio signal in a VHS recorder?

2. What is the typical operating frequency range of a bias oscillator in a VHS VCR?

3. Other than mixing with the record audio signal, bias oscillation is frequently used for one other purpose. What is it?

4. What is FM deviation?

5. (a) What is the head gap azimuth angle for an 8mm format VCR?

   (b) What is the head gap azimuth angle for a VHS unit?

6. Why should a video tape be stored vertically?

7. What is dropout?

8. What is brown stain?

9. Why must a metal head be used to write a signal onto a metal tape?

10. Video tape is designed to create some wear on the video head. Why?

11. List three symptoms of a defective video tape.

12. Why doesn't the 8mm format require a control track head?

Chapter 6

VCR Color Processing Theory

In the last chapter, you discovered how magnetic principles are used to record and recover information stored on a special oxide-coated plastic length of tape. Chapter 6 describes the complex circuitry that surrounds the magnetics and enables audio and video to be sensed and converted into analog and modulated signals to produce a beautiful color program on a TV or monitor screen. This portion of the book begins with a look at color signal processing and time-base error.

TIME-BASE ERROR

Mechanical VCR instabilities during record and playback cause timing differences or time-base errors, which occur in the horizontal, vertical, and color synchronization pulses. Synchronization pulses must occur at precise intervals and the sync pulses must remain for very specific periods of time. Any slight variation in synchronization pulse intervals can cause vertical

jumping, horizontal back-and-forth wavering, or color hue and saturation differences.

One time-base error is caused by expansion and contraction of the video tape itself. If the tape expands or contracts after recording, the physical distance between the recorded information on the tape changes, causing video synchronization pulses to occur at different times during playback. The result is an unstable TV picture.

Another form of time-base error is caused by the tension on the tape as it wraps around the video head drum assembly. VCRs require that the tape be held under a specified tension as it passes by the video head drum assembly. This tension is mechanically controlled by break bands, or, in the case of direct drive reel motors, by the supply reel motor itself. If the video tape is recorded under one holdback tension and then played back on a machine whose holdback tension is different, time-base errors result.

Figure 4.11 described severe time-base error. Figure 6.1 shows a slight shifting (horizontal time-base error) of the horizontal sync pulse at the video head switching point as seen on the monitor. This is caused by a small (few grams) difference in forward holdback tension between record and playback. The tape tension servo maintains the tape at a specified amount of forward holdback tension at the point where the video tape enters the head drum assembly (25 grams/cm for VHS). Meanwhile, the capstan and head drum servo circuits ensure that the video head contacts the recorded tape at a prescribed location.

As the tape is drawn around the drum, however, the tape tension gradually escalates due to the increased friction coefficient between the moving tape and the stationary portion of the lower drum assembly. This increase in tape tension causes the tape to stretch slightly as it travels around the drum. Since tape tension rises as it winds around the drum, the tape stretching also elevates proportionally. During record, the horizontal sync pulses are written onto the tape as it moves around the video head drum. The synchronization pulses are recorded on the tape as it is being stretched. Since the tape is stretched less at the initial head-to-tape contact point than it is

at the drum exit point, the distance between the various horizontal sync pulses on the tape can be altered if the tape is not played back under the same conditions as it was recorded. This is because the elastic nature of the tape causes it to return to its normal (unstretched) condition after it has passed through the machine. In order for the recorded horizontal sync pulses to be played back at the same interval as recorded, the tape must be returned to the same length that it was during record. Tape tension, and the condition of upper and lower video head drum assembly (which also affects the tension coefficient) must be consistent between record and playback modes to prevent serious horizontal time-base errors.

Assuming that the upper and lower sections of the video head drum assembly are working correctly, and the tape tensions between the record and playback modes are the same, the distance between the horizontal sync pulses on the tape will appear consistent to the monitor as they are read by the playback head. If not, the timing between horizontal sync pulses will vary as the video head scans up the RF track.

To illustrate, let's assume that a tape is recorded with lower than specified forward holdback tension and then played back on a

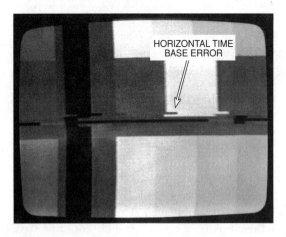

Fig. 6.1. Horizontal time-base error caused by minor forward holdback tension differences between record and playback.

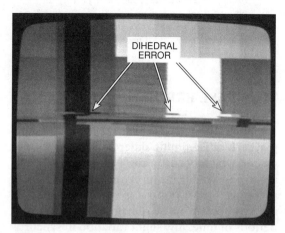

Fig. 6.2. Dihedral error causes horizontal time-base error. (Compare with Fig. 6.1 which has no dihedral error.)

second machine with correct tension settings. (For this example, assume that while the tape tension is lower than it is supposed to be, it is high enough to hold the tape against the spinning head cylinder for recording.) During playback, the detected horizontal synchronization pulses appear to stretch farther apart as the head scans the RF track. This occurs because the tape is stretched more as it passes around the drum on the second VCR than it was during the recording process. The monitor attempts to compensate for the playback horizontal timing error using its horizontal automatic frequency control (HAFC) circuit. As the video head exits the tape, the horizontal synchronization pulses reach a point where they are most separated. At this time, however, the next head begins to contact the video tape at the video tape entrance point and detects horizontal synchronization pulses that are closer together. It is at the point where the video heads switch to the next field (the next RF track) where the HAFC circuit in the TV monitor receiver must rapidly compensate for the new horizontal pulse read off the tape. The HAFC circuit makes a dramatic jump between horizontal time-base errors occurring at the end of one RF track (the drum exit point) and the horizontal time-base errors occurring at the beginning of the next RF track (the tape entrance point). The result is a bending or tearing of vertical objects in the upper portion of the TV picture (see Figures 4.11 and 6.1).

If the tape recorded with the incorrect forward holdback tension is played back on the same VCR with the same incorrect tension, the differences between the record and playback horizontal sync pulse timing will be minimal and the HAFC circuit will not have to make the dramatic jumps between fields previously described. The bending at the top of the picture will thus be gone. As long as the tape recorded with the incorrect tension is played back in the same machine, no tearing will be seen in the displayed picture.

Yet another form of time-base error can occur as a result of misalignment of the video

heads. Video heads must be positioned exactly 180 degrees apart. If one head is slightly advanced relative to the other, the advanced head will start playing back video signals prematurely. This produces synchronization signals that occur early causing the synchronization and picture information to take a slight jump forward in time at the video head switching point. As soon as the second head begins to contact the tape, the video signal and horizontal synchronization pulses jump backward in time. This form of time-base error is called *dihedral error*. It is corrected by realigning the position of the video heads on the disk. This adjustment is not considered a field adjustment, however, and should only be done at the factory. Dihedral error produces a horizontal jumping back and forth at the video head switching point between the two vertical fields, as shown in Fig. 6.2.

Time-base errors can also be caused by friction between the tape and the guides in the video tape path, and by mechanical vibrations of the video tape itself. High humidity or a failure of the tape binder's lubricant can cause the tape to adhere to the metal guides and drum assembly. Other time-base errors can be caused by instabilities in the video tape speed as the tape is drawn through the machine.

These errors are most noticeable in the higher frequency synchronization pulses, including the horizontal and chroma sync pulses. The low-frequency sync pulse (vertical rate) is least affected. Horizontal time-base errors generally appear as bending or flagging of vertical objects on the upper portion of the video screen.

The TV horizontal automatic frequency control circuit plays an important part in reducing the effects of horizontal time-base errors. Until the early 1980s, American-made TV sets were designed to operate only with extremely stable TV broadcast signals. The horizontal automatic frequency control (HAFC) circuits were deliberately designed to slowly compensate for timing errors and avoid the time-base errors caused by external random

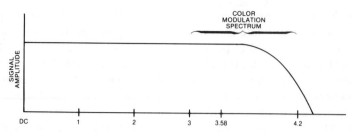

Fig. 6.3. *The video frequency bandwidth.*

impulse noise. Such noise could be caused by a passing vehicle or a nearby electric motor. With the introduction of VCRs, American televisions needed HAFC circuits that provided faster compensation time. Because older TVs are not able to respond quickly enough to the horizontal time-base errors introduced from the VCR, a VCR connected to an older TV may cause severe bending or flagging at the top of the picture yet appear to operate normally on a newer TV set. Figure 4.11 in Chapter 4 shows flagging.

Consumer video tape recorders don't record color subcarrier frequencies directly on the tape. Only the expensive broadcast-quality one inch "C" format helical-scan VTRs can directly record color frequencies. Color recording in consumer machines is done by a process called *color-under*, an economical way to record chroma signals and still eliminate the visual effects from time-base errors.

Color or chroma information is broadcast by TV stations using a subcarrier frequency of 3.579545 MHz (3.58 MHz). The higher the sync frequency the more sensitive the synchronization circuits become to minor time-base errors. Consumer VCRs don't record chroma information onto the tape at 3.58 MHz. Instead, the chroma subcarrier is converted to a frequency below that to make room for the luminance FM information (thus the term color-under). The color-under process was developed to reproduce color TV programs while minimizing time-base errors produced by the VCR during record and playback. A common form of chroma time-base error is seen on the video screen as hue (tint) changes, and in severe instances, as loss of chroma

synchronization or loss of all color. Loss of chroma synchronization can cause either a rainbow effect or a total loss of color.

THE BROADCAST TELEVISION SIGNAL

Knowing how TV signals are broadcast helps you understand the color-under process. Video information on the TV screen has a bandwidth of approximately 4.2 MHz, as shown in Fig. 6.3.

Picture detail is contained in the black and white portion of the video signal. This is called *luminance*, and is designated Y. The Y signal varies in intensity from black through shades of gray to 100 percent white and contains all of the fine detail in the picture scene, as well as the vertical and horizontal sync information. Figure 6.4 is an oscilloscope display of the Y signal without the presence of color information.

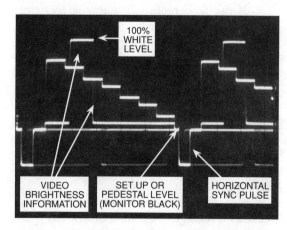

Fig. 6.4. *Video (Y) signal only (no color information).*

Color information is modulated using a 3.58 MHz subcarrier and "painted" on the black and white detail in the video picture. Figure 6.5 shows how chroma is "painted" over the black and white detail so the resultant broadcast video contains both Y and chroma information. Figure 6.6 is a broadcast video signal that contains brightness detail, burst, chroma, and horizontal sync information.

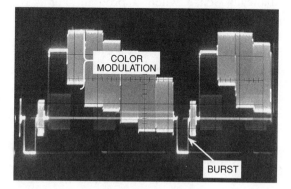

Fig. 6.5. *Video (Y) and color information*

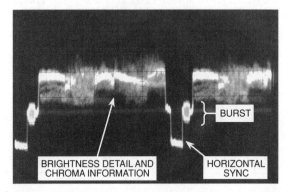

Fig. 6.6. *A broadcast video signal containing brightness detail, burst, chroma, and horizontal sync information.*

Colors are represented by different phases and amplitudes of the 3.58 MHz subcarrier as referenced by a color sync signal called *burst*. The phase of the subcarrier, referenced to burst, produces the color hue, while the amplitude of the subcarrier represents the color intensity (saturation). The color subcarrier signal consists of sidebands of approximately 600 kHz that extend above and below the main 3.58 MHz subcarrier frequency. A frequency of 3.58 MHz is selected because it is low enough to allow the 600 kHz sideband to fall within the 4.2 MHz bandwidth upper limitation and yet retain the color subcarrier frequency in the upper portion of the video frequencies. On the other hand, the 3.58 MHz chroma subcarrier frequency is high enough that it reduces the size of the dot pattern produced on the TV screen generated by the subcarrier frequency modulation, thus minimizing its visual effects on the display.

The audio in the TV broadcast signal is modulated around a 4.5 MHz carrier frequency. The proximity of the color and audio subcarriers required that the color subcarrier frequency be selected to minimize interference with the audio carrier. The effects of the chroma subcarrier frequency appear as small dots and are less noticeable on the TV screen because the 3.58 MHz color information has a fixed relationship with the horizontal picture line rate (15,734 Hz). The subcarrier is not an even multiple of the horizontal frequency and appears opposite in phase on alternating lines in the same field on the TV screen. The screen dot pattern from the chroma subcarrier is less noticeable because it occurs at a 45-degree diagonal on alternating lines. Figure 6.7 shows

Fig. 6.7. *Color modulation. Adjacent even or odd lines have opposite subcarrier phase relationship.*

the relationship between the chroma subcarrier frequency and the dot pattern produced on the TV screen.

Notice that the phase of the subcarrier frequency is not flipped on adjacent lines but instead is a continuous oscillation. Each line of color information contains a series of full cycles minus a half cycle so alternating lines appear to be 180 degrees out of phase. Figure 6.8 is the dot pattern display on a high-resolution monochrome monitor screen.

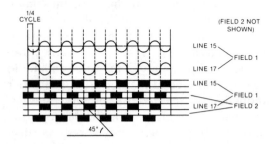

Fig. 6.8. *Color subcarrier dot pattern on a TV screen.*

COLOR-UNDER PLAYBACK

Color-under processing is designed to fulfill two tasks. First, it shifts the recorded chroma information to a lower frequency. This is necessary because the original broadcast chroma signal is 3.58 MHz, which falls within the luminance FM deviation passband of conventional consumer formats. The chroma signal must be shifted to a new frequency or beat interference will result from the mixing of the luminance FM and 3.58 MHz chroma signals. If a higher luminance FM frequency was used and the chroma frequency was left at its original spectrum point, higher writing speeds would be needed to handle the higher frequencies. This would increase tape speed requirements, equipment complexity, and tape costs. Secondly, color-under is designed to reduce chroma phase errors resulting from chroma time-base error without expensive broadcast-grade time-base stabilizing equipment.

In VHS and 8mm machines, the color subcarrier and sidebands are converted to

743 kHz for 8mm, and to 629 kHz for VHS. To record color information at the lower end of the frequency spectrum, some of the lower FM sideband is given up, as shown in Fig. 6.9. This produces some loss in black and white detail because the fine detail is contained in the lower sideband. The extreme lower portion of the lower FM sideband is electronically chopped via filter networks.

Some VCR models are designed to automatically sense when a program being recorded is not in color. When black and white (monochrome) signals are detected by these units, the filters that cut off the lower part of the lower FM sideband are bypassed to allow full lower sideband operation. This provides greater picture detail when recording black and white TV programs. The extension of the lower sideband for monochrome broadcasts was prevalent in early models, but recently, this has been reserved for more expensive models. This feature is sacrificed to reduce cost in many less expensive models.

The color-under frequencies of 629 and 743 kHz were selected to allow enough room for the upper and lower sidebands of approximately 500 kHz for the color-under frequencies. Color-under frequency conversion is accomplished by heterodyning two signals of differ-

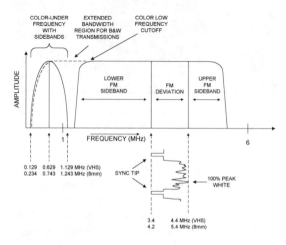

Fig. 6.9. *Frequency allocations for VHS and 8mm video cassette recorders.*

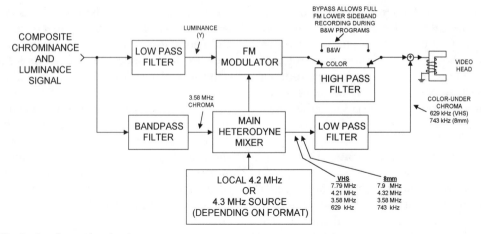

Fig. 6.10. Basic color-under circuitry.

ent frequencies to form a resultant frequency that is the sum of the two base frequencies— the difference of the two base frequencies and the original base frequencies. The color-under frequencies in VHS and 8mm are derived by adding the standard 3.58 MHz color subcarrier frequency to a stable frequency generated inside the VCR, as shown in Fig. 6.10. For 8mm units this frequency is approximately 4.32 MHz. VHS units use a frequency of approximately 4.21 MHz.

Color information is separated from the video luminance information, as shown by the oscilloscope photograph in Fig. 6.11. Next it passes through band-pass filters and combines with locally generated frequencies. In 8mm, the 4.32 MHz local frequency is combined with the 3.58 MHz chroma frequency to produce 7.9 MHz, 4.32 MHz, 3.58 MHz, and 743 kHz signals. The VHS system produces 7.79 MHz, 4.21 MHz, 3.58 MHz, and 629 kHz signals.

Figure 6.12 is an oscilloscope display of the color-under and 3.58 MHz separated chroma signal before and after frequency conversion. A low-pass filter passes only the color-under signal to a mixing circuit where it is combined with the frequency modulated Y signal. The combined color-under and FM Y signals

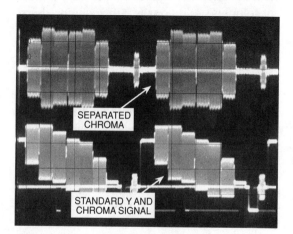

Fig. 6.11. Chroma is separated from luminance for color processing.

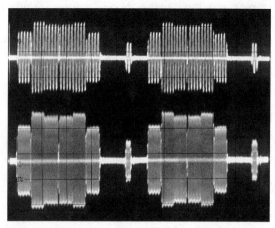

Fig. 6.12. The color-under chroma (top) compared with the original 3.58 MHz signal (bottom).

are then recorded on the tape by the video heads. The color-under chroma signal is recorded on the tape using the same principles as that for audio recording. Both audio and color signals require AC bias during recording to avoid distortion. The FM Y signal becomes the bias during recording of the color-under signal.

Color-Under and Time-Base Error

During playback, the color and luminance FM signals are read off the tape, separated by frequency, and processed individually. The color-under and luminance signals contain time-base instabilities created as a result of the record and playback process.

Luminance time-base errors are not corrected in consumer formats. Under normal circumstances, luminance time-base errors are within stability correction tolerances of the television receiver's deflection circuitry. The VCR luminance (including the horizontal and vertical sync pulses) time-base errors appear reasonably stable on the screen because the TV makes the necessary corrections. However, it's important to note that the time-base errors are not corrected. They are merely disguised by the television circuitry.

Chroma time-base errors are more severe than errors observed in the luminance because the color signal occurs at a substantially higher frequency than the luminance scan rate. The television cannot disguise the larger time-base related chromaticity errors so circuitry must be designed into the VCR to improve the playback chroma time-base stability.

A major function of the color-under circuit is to reduce the chroma time-base error to a level comparable to the luminance time-base instability. It causes the chroma errors to track the luminance time-base jitter. This places the chroma instability within the correction window of the television's chroma correction circuitry.

Cotracking the luminance and chroma time-base errors minimizes the visibility of the remaining jitter effects that would otherwise be very obvious on the screen. When luminance and chroma time-base errors occur in unison, they are less noticeable. For example, a very slight shift of a vertical object to the right or left on the screen is less noticeable if the color saturation and hue also shift so the color information is "painted" directly on top of the object. If the chroma time-base error weren't reduced and caused to track the luminance jitter, the luminance could shift slightly right while the chroma hue changes could shift left producing a distracting luminance/chrominance time-base error display.

8mm color-under playback and chroma time-base error correction

The 8mm color-under system will be examined first. Many of the principles learned in this section will help you comprehend the more complex VHS color-under circuitry.

Color-under circuits are designed to reduce unwanted chroma phase (color hue) changes that occur in the playback picture due to time-base errors. In our TV system, color hue is determined by the phase relationship existing between the reference burst signal and the phase of the 3.58 MHz subcarrier signal contained in the reproduced video scene. Each change in the subcarrier phase, relative to a stable burst reference signal, will produce a different color hue. Additionally, if the reference burst signal is not stable in phase and frequency, the hue in the reproduced picture will not be accurate. Since time-base errors affect the stability of both the burst and chroma subcarrier signals, they also disrupt the delicate frequency and phase relationship between burst and the subcarrier information that is critical for accurate hue reproduction. In a VCR we see these errors as incorrect colors. Typically the hue errors are very short in duration and only exist for a few microseconds. These errors are most easily seen on the screen when viewing solid, heavily saturated colors such as red or magenta. The errors show up as fine incorrectly colored streaks that extend horizontally across the screen.

Chroma saturation (color intensity) is determined by the amplitude of the subcarrier relative to the reference burst level. Time-base errors do not typically effect chroma amplitude (color saturation). Chroma saturation errors do exist in a VCR, but are handled by another part of the playback process circuitry and will be described later.

Figure 6.13 describes how a color-under circuit is capable of reducing phase-related time-base errors during the playback process. During the playback recovery process, the 743 kHz 8mm playback color-under signal contains time-base errors as it enters the frequency conversion block for up-conversion back to the original 3.58 MHz frequency. The time-base errors manifest themselves as phase and frequency instabilities in the 743 kHz oscillation recovered off the tape. If we were to view the color-under chroma arriving at the input to the up-converter, we would see that the 743 kHz signal would have an unstable jittering appearance. If the 4.32 MHz oscillation entering the frequency conversion box was stable and did not track the time-base errors off the tape, the frequency output from the up-converter would be 3.58 MHz ± the jitter component (labeled as delta f or Df). For example, if the 743 kHz jitter caused the frequency to shift up by 15 Hz, the Df value would be +15 Hz and the 3.58 MHz output from the up-converter would be off frequency by 15 Hz as well. If, however, the generated 4.32 MHz oscillation frequency is also caused to shift up by 15 Hz at the same time, the up-converted frequency output will still be 3.58 MHz because

4.32 MHz (+15 Hz Df) − 743 kHz (+15 Hz Df) = 3.58 MHz (without Df)

As long as the 4.32 MHz generator is able to track the phase and frequency chroma errors present in the 743 kHz input signal, the frequency and phase of the signal output from the up-converter will always equal the 3.58 MHz difference frequency and will be relatively stable. Some instabilities will still be present after the up-conversion is completed due to the system's lag in responding to the incoming errors off the tape and the circuit's limited ability to rapidly track and correct for large time-base errors that exist in any format. The residual time-base errors are normally small enough that the circuits in the TV can still reproduce an acceptable color picture even with the small phase errors that remain. As long as the color-under circuit has reduced the chroma time-base errors to the point that the television can reproduce good color with only minor phase errors, it has performed its task.

The signal from the 4.32 MHz oscillator is constantly altered by a controlling error voltage coming from the error detection block in Fig. 6.13. The error detector operates by monitoring the frequency and phase of the 3.58 MHz up-converted signal coming from the heterodyne block and by comparing that signal to a highly stable 3.58 MHz crystal oscillator as shown in Fig. 6.13. Changes in the heterodyned 3.58 MHz output produce an error voltage at the error detector that is proportional to the difference in frequency between the crystal reference oscillation and the 3.58 MHz heterodyned frequency. The error voltage out of the error detector acts to correct the phase and frequency of the 4.32 MHz voltage controlled oscillator (VCO). The end result is a stabilized 3.58 MHz heterodyned frequency with good chroma phase and frequency characteristics.

8mm chroma playback automatic phase control (APC)

Figure 6.14 shows a complete operational block of an 8mm heterodyne circuit and the control-

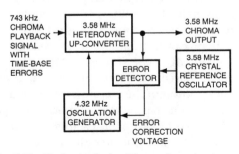

Fig. 6.13. The basic playback heterodyne circuit.

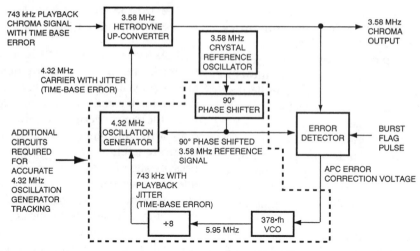

Fig. 6.14. *The 8mm APC circuit.*

ling VCO. The individual sections in this figure combine to form a processing section in the VCR called the *APC* (automatic phase control) circuit.

Figure 6.14 contains the up-converter and error detector previously shown in Fig. 6.13. The 4.32 MHz oscillation generator section in Fig. 6.13 has been expanded within the dashed lines of Fig. 6.14 to show additional circuits required for accurate 4.32 MHz tracking. The functions of the new sections in Fig. 6.14 will be explained next.

The enhanced view of the 4.32 MHz oscillation generator shows that the generator is not directly controlled by the error detector block. Instead it is driven by a 378 fh VCO and a divide-by-eight circuit. The 378 fh VCO generates an oscillation 378 times the frequency of the horizontal line rate (fh) or 5.95 MHz. Instead of a crystal controlled oscillator, the error correction voltage directs a VCO to provide a wide time-base error correction range. The phase and frequency of a crystal controlled oscillator can also be corrected to some extent by an error control voltage to form a *VCXO* (voltage controlled crystal oscillator). A VCXO has a much narrower correction range than a VCO and as such is not suitable for the wider correction window required to correct chroma time-base errors. The VCO operates at

a high frequency (378 fh or 5.95 MHz) to yield a wide time-base error correction window that is accurate and can rapidly respond to phase and frequency errors. The 378 fh multiple of the horizontal line rate is selected because it can be easily divided down to the 743 kHz frequency by a common divide-by-eight circuit while still providing the accurate tracking benefits of a high-frequency VCO.

The frequency of the VCO is controlled by the error correction circuit, which alters its correction voltage according to the up-converted 3.58 MHz frequency and phase errors coming out of the up-converter. Frequency and phase monitoring in the error detector is performed during the brief chroma burst interval present on each horizontal line. The burst period is identified by a burst flag pulse coming from the demodulated playback luminance signal. The color burst oscillation is used for error detection because it does not vary in phase and amplitude as changes in scene content occur. The 3.58 MHz chroma subcarrier oscillation present in the active viewing portion of the picture cannot be used to monitor time-base errors because it changes amplitude and phase with each new color in the scene. It does not provide the necessary stable reference that burst does. On the other hand, changes in the burst reference frequency and phase result only from playback time-base errors and not the normal changes in the color content.

The 378 fh VCO produces a 5.95 MHz oscillation which is phase- and frequency-altered by the correction voltage coming out of the error detector. The VCO frequency is divided by eight to produce a 743 kHz signal which contains a quantity of jitter proportionally equal to the time-base error present in the off-tape playback chroma signal. As the frequency and phase of 3.58 MHz oscillation from the up-converter shifts away from the standard frequency, the error detector circuit senses the change and produces a correction voltage which in turn shifts the phase and frequency of the 378 fh VCO.

The internally generated 743 kHz oscillation from the VCO is fed to the 4.32 MHz frequency generator. Notice that the second input to the 4.32 MHz oscillation generator is a stable 3.58 MHz reference signal. When the manufactured 743 kHz from the VCO, with its representative time-base instabilities, is mixed with the stable 3.58 MHz reference, an output frequency that also contains time-base errors is produced by the 4.32 MHz oscillator. The 4.32 MHz oscillation contains jitter because the frequency output from the 4.32 oscillation generator is the summation of the stable 3.58 MHz reference, the 743 kHz oscillation from the VCO and divide-by-eight circuits, and the time-base error (or jitter content) present in the 743 kHz signal. Recall that the jitter content is the *delta frequency* (Df). The 4.32 MHz time-base errors are thus representative of the jitter content present in the output of the heterodyne up-converter. When the 4.32 MHz generator tracks the 743 kHz frequency time-base instabilities coming off the tape and adjusts accordingly, the difference frequency out of the heterodyne converter will always be corrected to a relatively stable 3.58 MHz.

Notice in Fig. 6.14 that the reference 3.58 MHz frequency is shifted by 90 degrees. Shifting the phase of the standard oscillator aids the error detector and 4.32 MHz oscillation generator in their respective functions. For example, it is easier for the error detector to monitor and produce an error correction voltage when comparing two signals of the same frequency that are slightly out of phase with each other. Since the heterodyned frequency and the standard reference frequency are the same, the phase of one frequency is shifted to assist in accurate monitoring.

8mm automatic phase control identification (APC ID)

If the chroma playback system contained only APC correction for time-base errors, some shortcomings would still exist. These are:

1. Full time-base compensation will not always be possible during variable speed playback modes because the time-base errors are too large for the error correction and VCO circuits to handle. This would result in a complete loss of color during these modes.

2. When a high-frequency, wide tracking range VCO is used, frequency mis-locking may occur during normal playback. Frequency mis-locking is a situation where the VCO and error detector circuits lock onto a frequency other than the correct one. The mis-locking typically occurs at some multiple of the correct frequency. In a VCR mis-locking may occur because:

   a. The VCO may erroneously lock onto a harmonic frequency produced by a mixture of the burst and horizontal line frequencies. This harmonic frequency is expressed as

   $$fsc \pm n \times fh$$

   where fsc = the subcarrier frequency, n = the horizontal line number, and fh = the frequency of the horizontal line rate (15,734 Hz).

   b. The VCO may erroneously lock onto a frequency of fsc $\pm \frac{1}{2}$ fh. This frequency produces a quasi-stable VCO frequency, which will mislead the VCO error detector to believe it has correctly locked onto the correct frequency. The VCO output frequency will be stable but will cause the up-converter to operate at a frequency of $\pm\frac{1}{2}$ fh from the required 3.58 MHz frequency.

c. If adjacent track crosstalk is detected by the playback head, the VCO may incorrectly lock onto the adjacent track's chroma signal. This will cause the VCO to produce an incorrectly phased signal. This frequency is fsc $\pm\frac{1}{2}(2n + 1)$fh.

When the chroma frequency from the up-converter deviates more than a few hertz from the standard, a circuit known as the *automatic color killer* (ACK) prevents the subcarrier and burst signals from being passed to the VCR output jacks. This kills all chroma in the picture and is why a VCR with a mis-locked VCO may produce no color. If the mis-locking VCO produces a color frequency close to the 3.58 MHz standard, the ACK may not detect the error, thus allowing the incorrect up-converted frequency to be passed from the machine to the TV or display monitor. In this case, incorrect color hues, or in some cases, a rainbow (*barber pole*) effect will be present in the reproduced picture.

The automatic phase control identification (APC ID) circuit is part of the 378 fh VCO and exists to prevent mis-locking. Figure 6.15 explains how one APC ID circuit operates. The output of the 378 fh VCO is input to the APC ID block where the frequency of the VCO is monitored and compared to a reference frequency. If the VCO frequency wanders off frequency by more than ± 7.5 kHz ($\pm\frac{1}{2}$ fh) the APC ID interrupts the error detector voltage input and brings the VCO back into the correct frequency range for the error detector to function. The APC ID only functions when the VCO frequency moves beyond the ± 7.5 kHz

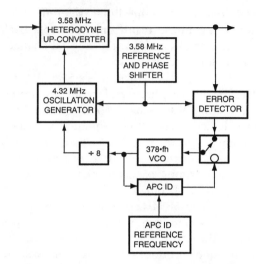

Fig. 6.15. *APC ID circuit operation.*

correction range of the error detector. APC ID functions as an *AFC* (automatic frequency control) to correct gross frequency errors while the APC works to refine the less drastic phase and frequency errors. Phase correction is the more precise adjustment made by the circuit. Frequency control is a coarse adjustment. Phase correction cannot occur unless the frequency is very close to the correct value.

Thus, the APC ID circuit functions to keep the APC circuit operating at the correct frequency. When time-base errors create a frequency error beyond the APC's ability to track the signal, the APC ID engages to make the required corrections. Once the APC ID has restored the normal VCO frequency, VCO control is returned back to the error detector. The APC ID does not function to correct phase errors. It only operates to keep the playback frequency errors within a ± 7.5 kHz window so that phase correction can take place.

The frequency correction range of the 8mm APC circuit is ± 7.5 kHz. The phase correction window of the APC is expressed in degrees and is limited to ± 90 degrees out of a possible 360 degrees. It is possible for time-base errors to possess the correct frequency but have the wrong phase. When phase errors coming out of the up-converter extend beyond ±90 degrees from the correct phase, the APC circuit does not perform phase correction efficiently. The response time for correct VCO phase lock-up is increased as the VCO makes a second attempt to lock-up and track the large phase errors. In this case, the increased lock-up time causes incorrect hue changes to become much more apparent on the TV screen than they would otherwise be and minor chroma time-base errors that used to be acceptable to the viewer will now be much more apparent and distracting.

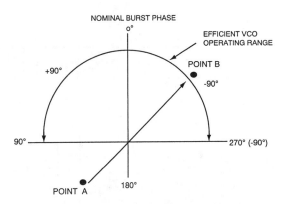

Fig. 6.16. *The Burst ID circuit keeps the playback subcarrier phase within ±90 degrees of the correct phase.*

8mm burst identification (Burst ID)

The burst identification circuit accelerates the response time of APC phase-locking by inverting the chroma phase when the playback color signal phase is outside the ±90 degree range of the APC. A slow phase-locking response time will produce color hue errors that streak horizontally across the playback picture. Figure 6.16 shows how this works.

The correct chroma phase is located at the top of the drawing (zero degrees). The APC functions correctly as long as the time-base error does not cause the playback chroma to extend beyond ±90 degrees (shown as the 90 and 270 degree points in the figure). If the phase errors increase beyond this window, the Burst ID circuit inverts the phase and brings it back within the 90 degree range. Point A in Fig. 6.16 represents a time-base-generated phase error of 135 degrees (90 degrees plus 45 degrees). Since this is outside the APC's efficient operating range, the Burst ID circuit inverts the phase of the carrier and places it at point B in the drawing. This 315 degree point (–45 degrees) places the subcarrier well within the efficient tracking window of the APC circuit. The Burst ID circuit sits idle until the phase of the playback signal extends beyond the ±90 window.

Figure 6.17 shows a composite APC playback circuit with the Burst ID blocks and some

of the components that we have described thus far. The other blocks shown in this figure will be covered shortly.

Overall APC function review

Before addressing the other circuits in Fig. 6.17, let's review the functional responsibilities of the APC circuits previously discussed.

The APC ID circuit performs the most coarse correction of the playback chroma frequency. It only activates when the up-converted frequency of the playback chroma signal exceeds ±7.5 kHz. The APC ID is responsible for bringing the frequency into the range of the APC error correction circuit.

The Burst ID circuit is the next stage in coarse time-base error correction. It maintains the phase of the up-converted signal within ±90 degrees. It prepares the chroma signal for the APC's phase correction stage, which functions most accurately when the phase correction requirements are within this window.

APC is the final adjustment in the time-base error correction process. APC can only operate if the APC ID and Burst ID sections have properly minimized coarse time-base errors. This fine adjustment is responsible for the correct phase (hue) of the playback color picture. Phase corrections can only be made if the playback chroma frequency is held at

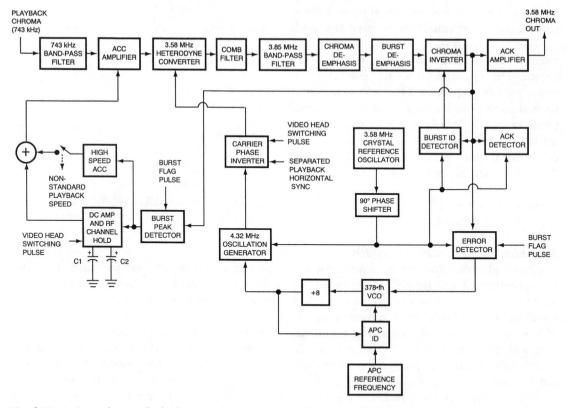

Fig. 6.17. *An 8mm chroma playback circuit showing many of the major components used for playback APC.*

3.58 MHz and the coarse phase maintained within a ±90 degree window.

Chroma Playback Automatic Color Control (ACC)

During playback, the level of the color-under signal detected by the playback heads changes amplitude. Some causes of tape playback level fluctuation are poor head-to-tape contact, differences in playback head gains and tape surface defects. These unwanted changes in the color-under amplitude will affect the playback chroma saturation, if not corrected. The symptom appears on the TV screen as a fluctuating chroma level. The duration of this chroma flicker depends on the duration of the incor-

rect chroma level. If an all red screen was being reproduced and the chroma saturation error exists for the duration of only a few horizontal lines, the saturation errors are noticed as horizontal streaks flashing across the screen in various intensities of red. If the chroma intensity only changes from one head to another, the full field of red will flicker at a 30 Hz rate because this error only manifests itself between the different heads as each one scans its respective field.

In Fig. 6.17 there is an automatic color control (ACC) block on the input line of the 743 kHz playback signal. This circuit stabilizes chroma amplitude fluctuations that occur during playback and eliminates the color saturation errors that would otherwise be present on the TV screen. The ACC circuit's first function is to eliminate chroma level differences that naturally occur as part of the playback process. Its other function is to equalize chroma level differences between the video heads resulting from variations in head sensitivity.

The ACC circuit operates to keep the playback burst signal coming off the tape at a constant level. Since chroma subcarrier levels are supposed to change in the active portion of the video according to color saturation, the subcarrier in the video cannot be used as a level reference. Thus, chroma burst is the only stable level reference in the video signal. Unwanted level changes occurring during the playback process impact both the burst level and the chroma saturation in the picture proportionally. Since burst is present on every line of video displayed on the screen, it can be used to provide an accurate representation of chroma level errors occurring on each horizontal line that we view. When the ACC acts to adjust incorrect burst levels, the incorrect chroma levels in the picture are also restored.

The ACC in Fig. 6.17 keeps the burst levels consistent on a line-by-line basis. On the other hand, level differences existing between video heads are corrected by switching a smoothing capacitor (C1 or C2) into the channel hold circuit by means of a video head switching pulse. This head switching pulse turns on the correct video head as the signal is read off the tape. This pulse is also used to switch in the correct smoothing capacitor. One smoothing capacitor is used for each head. The capacitors receive and hold a detected voltage representing the peak burst level coming from the processed chroma playback signal. Separate capacitors are used for each video head to provide independent level compensation for each video field. The time constant for each video field is dependent on the channel hold capacitor switched into the circuit. These time constants must be long enough to eliminate chroma flicker between heads yet short enough to provide a quick re-sponse time to rapid chroma amplitude changes without introducing chroma noise.

High-speed ACC

The demands placed on the ACC are not the same for all modes of operation. During search mode for example, there are many short duration chroma level changes that would not otherwise be present. The chroma amplitude can change three or four times in the search mode. Since the time constant for each field in the normal video playback mode is dependent on the channel hold capacitors, and because the special playback modes place unusual demands on the ACC time constant, the channel hold capacitors C1 and C2 are switched out of the ACC circuit during the special playback modes. When a nonstandard playback speed is selected, the burst peak hold output is amplified and added directly to the ACC amplifier. The channel hold capacitors are not in the circuit at this time. This shortens the ACC time constant substantially, although chroma noise is increased to some extent during this time.

Automatic Color Killer (ACK)

The automatic color killer (ACK) circuit prevents chroma noise from leaving the color processing circuit when chroma up-conversion phase lock is not obtained in the playback mode. The absence of chroma phase lock produces incorrect hues when a low level mis-lock occurs. Large mis-lock errors may cause various symptoms including a complete loss of color, color banding, color streaking, color rainbowing (barber poling), color flashes, and random color speckles throughout the playback picture. The ACK prevents erroneous and noisy color signals from appearing on the screen when up-conversion phase mis-lock occurs.

The ACK detection circuit monitors the phase of the color up-conversion output signal and compares that with an internal 3.58 MHz crystal reference oscillator. If the phase of the up-converted subcarrier signal exceeds permissible bounds, the ACK detector initiates a voltage that activates the ACK block. The ACK block acts to prevent any color signal from passing to the output line. When frequency errors occur in the APC section, proper phase lock cannot be achieved in the APC and the ACK is also activated. Thus, by monitoring the phase of the reproduced chroma, the ACK detector is able to sense both frequency and APC troubles.

Miscellaneous 8mm Color Processing Circuits

Other circuits within the playback color processing section shown in Fig. 6.17 include band-pass filters, chroma, and burst de-emphasis circuitry, and a comb filter. These will be described next.

Band-pass filters
The 743 kHz band-pass filter (BPF), shown in the upper left corner of Fig. 6.17, is responsible for filtering out all unwanted frequencies coming from the playback heads. These unwanted signals include the playback luminance FM, FM audio, PCM audio (if present) ATF servo signals, and noise. Only the 743 kHz chroma signal should be input to the color processing circuit.

The 3.58 MHz band-pass filter on the output of the up-converter eliminates all heterodyned frequencies other than the desired 3.58 MHz signal. Whenever two frequencies are beat together (heterodyned), sum and difference frequencies and related harmonic frequencies are produced. If these additional heterodyned frequencies are output to the television set, fine diagonal beat patterns will be present on the screen. The 3.58 MHz BPF blocks out (filters out) these extra frequencies eliminating this interference from affecting the reproduced picture.

Chroma and burst de-emphasis
The chroma de-emphasis block, near the upper right hand corner of Fig. 6.17, returns the recovered chroma playback levels back to their original prerecord processing values. During record, the chroma levels input to the VCR are amplified for recording. This color boosting action is called *chroma emphasis*. The playback circuits must de-emphasize the boosted levels to restore the correct amount of chroma saturation. The emphasis and de-emphasis actions reduce the appearance of chroma noise created in the chroma color-under processing path. We'll revisit this subject again in the record chroma section later in this chapter.

Also in the upper right hand corner of the drawing is a burst de-emphasis circuit. During record, the burst signal is identified and separately amplified. This process is called *burst emphasis*. Later, the playback circuits must de-emphasize or return the increased burst level back to its standard value so that it can provide a correct level reference for the television set. Burst is boosted (amplified) in the record circuitry to reduce the effects of noise on the burst signal during the frequency conversion and record/playback write and read processes. Noise in the burst signal tends to interfere with the playback APC circuit's ability to accurately reproduce color hues. Burst emphasis helps immunize the playback APC electronics from the effects of this noise.

The comb filter
The comb filter following the up-converter in Fig. 6.17 and the phase inverter located between the 4.32 MHz oscillation generator and the 3.58 MHz heterodyne up-converter work together to eliminate chroma crosstalk. The chroma crosstalk, detected by the playback head, is 743 kHz color information coming from the adjacent RF track on the tape. This process will be described later in the chapter.

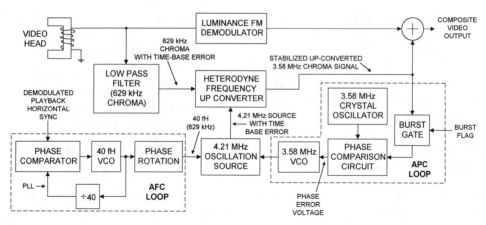

Fig. 6.18. VHS playback chroma frequency and phase control circuits reduce time-base error.

VHS Color-Under Playback and Chroma Time-Base Error Correction

Playback automatic frequency control (AFC)

As shown in Fig. 6.18, many of the same basic principles for minimizing color time-base errors are applied in the VHS chroma processing system. There are some differences in how color time-base errors are minimized between the VHS and 8 mm formats, however. These differences include separate color-under frequencies, the use of AFC (automatic frequency control) rather than the APC ID used by 8mm, and a phase rotation scheme to reduce chroma crosstalk rather than the phase inversion method used in 8mm. The VHS system uses a form of chroma phase rotation to modify the recorded color phase. This phase rotation is applied to the signal in the AFC circuitry during record. The 8mm system inverts the chroma phase of only specific lines in one of the two fields. It accomplishes phase inversion in the 4.32 MHz generator block rather than in the AFC section where VHS systems modify the signal phase. An operational description of the VHS chroma phase rotation circuitry shown in Fig. 6.18 will be covered later during the detailed description of the chroma crosstalk reduction methods used in both formats.

Notice that the color-under frequencies in the VHS block are different than those in 8mm units. In this block diagram, the VHS AFC loop operates at 40 fh (629 kHz). The particular AFC circuit in Fig. 6.18 is from one of the original VHS designs popular several years ago. The color-under frequency impressed on VHS tape is 629 kHz (precisely 629.371 kHz) instead of the 743 kHz frequency used in the 8mm format. In this figure the APC section controls a VCO that operates at 3.58 MHz. Recall that 8mm units use a higher frequency VCO (378 fh or 5.95 MHz) for APC control. Recent VHS designs have an APC section that closely resembles the higher frequency VCOs found in the 8mm APC circuits.

The complete playback signal containing the VHS color-under subcarrier and the modulated luminance information enters the low-pass filter where the color-under signal is separated from the higher frequency luminance FM signal. The 629 kHz chroma signal contains jitter from the tape and is mixed with the output of 4.21 MHz oscillator. The 4.21 MHz oscillator's output also contains jitter. The source of the 4.21 MHz signal jitter is found at the input of the AFC circuit in the lower left hand corner of Fig. 6.18. This jitter is the separated horizontal synchronization pulses (H sync) coming from the demodulated playback luminance signal. This H sync input represents the time-base error of the playback luminance signal.

In the 8mm units, there is no playback chroma AFC functioning. These units apply an APC ID correction to keep the coarse chroma playback frequency under control. In VHS units, there is a playback AFC circuit that is constantly active. The color playback AFC circuit has two fundamental responsibilities.

1. To correct coarse color-under frequency errors resulting from time-base error. The AFC section must stabilize the playback color frequency to a point where the APC can perform the phase adjustments.

2. To reverse the chroma phase rotation imposed upon the color-under chroma record signal for chroma crosstalk elimination.

The playback AFC in this VHS circuit contains a VCO that runs at 40 times the frequency of H sync (40 fh) or 629 kHz. The VCO output is sampled, divided by 40, and passed to a phase comparator that locks the VCO's frequency and phase to the incoming H sync. The output of the AFC phase-locked-loop (PLL) is phase-rotated to undo the phase rotation modifications made during the record process. This AFC output is then passed to the 4.21 MHz oscillation source where it is mixed with a 3.58 MHz signal from the APC to produce the 4.21 MHz source. The 4.21 MHz signal contains the same amount of jitter as the off-tape playback luminance signal because the jitter was present in the separated H sync at the input to the AFC. When the 629 kHz color-under signal off the tape (which contains time-base error Df) and the 4.21 MHz oscillation signal containing the same amount of error (Df) are mixed, the sum and difference frequencies output from the main heterodyne mixer are always the same. They will be the same as long as the two frequencies input to the heterodyne mixer track one another. This is because:

[4.21 MHz + the jitter component (Df)] − [629 kHz plus + Df]
= 3.58 MHz without Df

The output of the 4.21 MHz source drives a heterodyne frequency converter that up-converts the 629 kHz playback chroma signal back to its original 3.58 MHz frequency.

VHS playback automatic phase correction (APC)

Although the output of the AFC tracks the time-base frequency changes in playback color fairly well and produces a stabilized 3.58 MHz playback chroma oscillation, the AFC circuit supplying the same generator corrects for frequency errors only. Phase correction must still be applied to reproduce correct color hues. VHS playback APC operation corrects for phase discrepancies in the playback color, which disrupts correct hue reproduction. Once the off-tape color signal has been up-converted, the phase must be monitored and corrected to remove the remaining time-base error not eliminated by the AFC playback circuit. The APC circuitry makes fine adjustments to the 4.21 MHz oscillation generator to correct the remaining chroma playback time-base errors.

Figure 6.18 shows that APC is accomplished by sampling the phase of the up-converted 3.58 MHz burst signal from the main heterodyne converter. A burst gate, controlled by a burst flag pulse, samples the 8 to 11 cycles of up-converted 3.58 MHz oscillation which make up the burst signal. The sampled playback burst signal is passed to a burst phase comparator where it's mapped against a stable 3.58 MHz crystal oscillator reference. Notice that the reference 3.58 MHz frequency in Fig. 6.19 is shifted 90 degrees. Shifting the phase of the standard oscillator helps the error detector recognize unwanted changes in phase. It's easier for the error detector to monitor and produce an error correction voltage when comparing two signals of the same frequency that are slightly out of phase relative to each other. Since the heterodyned frequency and the standard reference frequency are the same, the phase of one frequency is shifted to assist in accurate monitoring.

Phase changes in the burst signal off the tape are sensed by the phase comparator, which generates a correction voltage at its output. The level of this correction voltage shifts as it tracks phase errors in the playback chroma. These correction voltage changes cause the 3.58 MHz VCO to change phase. The related phase shift changes in the VCO circuitry fine tune the 4.21 MHz oscillation generator, whose frequency has been coarsely adjusted by the AFC correction voltage input. The resulting output of the 4.21 MHz oscillation source is a signal containing phase and frequency shifts (jitter) which, when heterodyned with the 629 kHz off-tape color-under signal, will produce an up-converted 3.58 MHz subcarrier oscillation with reduced time-base error.

After studying the APC block diagram in Fig. 6.18, one might wonder if it wouldn't be simpler to have the phase correction occur at 629 kHz, since it would eliminate one of the mixers in the circuit. This isn't done because the color-under burst contains only two cycles, and this is not enough to perform phase comparison. Normal burst at 3.58 MHz contains between 8 and 11 cycles of oscillation.

VHS burst identification (Burst ID)

The APC section in VHS equipment monitors and corrects playback chroma phase errors that exist after the AFC circuit has done its job. The VHS APC can efficiently track and correct phase errors of up to ±90 degrees. Phase errors beyond these limits require the additional assistance of Burst ID correction. Figure 6.19 shows the location of the Burst ID detection circuit in the block diagram.

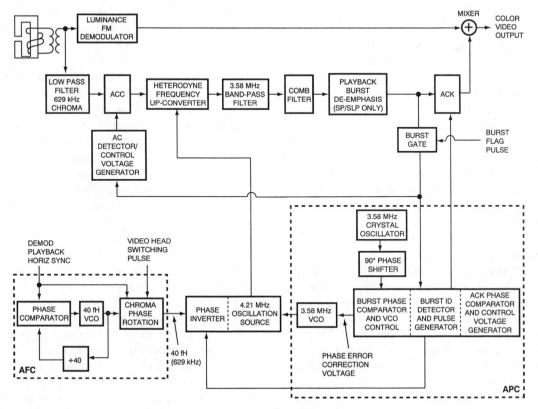

Fig. 6.19. *Additional circuits used in the VHS chroma playback section.*

Just as in 8mm machines, VHS Burst ID only corrects phase errors greater than ±90 degrees. The Burst ID is necessary to accelerate the response of the APC phase locking time by inverting the chroma phase when the playback color signal phase is outside the efficient ±90 degree operating window of the APC. A slow phase-locking response time will produce color hue errors that streak horizontally across the playback picture.

A look back at Fig. 6.16 shows how this works. The correct chroma phase is at zero degrees and is located at the top of the drawing. The APC functions correctly as long as the time-base error does not cause the playback chroma to press beyond the ±90 degree limits (shown as the 90 degree and 270 degree points in the figure). If the phase errors extend beyond this window, the Burst ID circuit acts to invert the phase of the up-converted 3.58 MHz signal by altering the 4.21 MHz phase. This action brings the up-converted signal's phase back within the ±90 degree range. Point A in Fig. 6.16 represents a time-base generated phase error of 135 degrees (90 degrees plus 45 degrees). Since this is outside of the APC's efficient operating range, the Burst ID circuit inverts the phase of the subcarrier and places it at point B. This is the 315 degree point (minus 45 degrees), which places the subcarrier well within the efficient tracking window of the APC circuit. The Burst ID circuit then sits idle until the phase of the up-converted playback signal again extends beyond the ±90 window.

A look at Fig. 6.19 shows that the VHS Burst ID inverts the phase of the AFC 629 kHz oscillation. This in turn causes the 4.21 MHz to be inverted, which shifts the phase of the up-converted 3.58 MHz signal by 180 degrees. The effect of the Burst ID in the VHS system is the same as that of the Burst ID in 8mm systems where the phase of the 3.58 MHz signal is inverted after it has been up-converted (see Fig. 6.17).

The VHS Burst ID functions as a part of the APC section. Just because Burst ID shifts the phase of the 629 kHz oscillation coming from the AFC section, it should not be confused with the phase rotation action in the AFC circuitry. Remember that the playback AFC does not correct for phase errors remaining in the playback color oscillation resulting from time-base instabilities. That is the responsibility of the APC and Burst ID blocks. Thus, the Burst ID section is sensitive to large time-base related phase errors and is ready to initiate a correction command to the 4.21 MHz oscillator if required.

VHS burst ID and chroma phase rotation

There is an advantage in using Burst ID to invert the 629 kHz signal coming out of the AFC section. It relates to the masking of occasional phase rotation errors that occur in the playback AFC phase rotation section.

During record, the VHS color-under system modifies the phase of the 629 kHz chroma subcarrier on a selected line-by-line basis. This makes it possible for circuits in the playback section to eliminate chroma crosstalk picked up by the heads from the adjacent RF tracks. The VHS phase modification process is called *chroma phase rotation.* During record, the chroma rotation circuit shifts the original phase of the color-under oscillation in predetermined quantities. The original phase is rotated either clockwise or counterclockwise with respect to its original value depending on which head is in contact with the tape. During playback, the rotated chroma phase must be returned to its original phase for correct color hue reproduction. The phase rotation process and its benefits will be explained in detail later. For now, it's sufficient to know that the playback chroma AFC section is only responsible for correcting frequency errors and for undoing the phase rotation imposed upon the color-under signal during record.

To restore the correct color hue for the television display, the playback rotation circuit must shift the phase back to its original position. It's critical that the playback chroma phase rotation circuit shift the phase of each modified line by the correct amount and in the correct direction so that the original playback phase can be reestablished. Obviously the line that was rotated with a minus 90 degree phase shift during record must be unrotated by shifting it plus 90 degrees during playback to return the chroma signal back to its proper phase. If the playback rotation circuit gets out of sequence, it could rotate the phase in the wrong direction—rotating another minus 90 degrees instead of shifting it back by applying the required plus 90 degrees. Assuming that the ACK is not preventing the color signal from exiting the unit (thus allowing only a black and white picture to be produced), the screen would show fine horizontal streaks of the wrong color. Each horizontal line would display incorrect, different-colored hues from the previous line. If the AFC rotation circuit produces errors like these, the Burst ID will sense these errors at the 3.58 MHz up-converter output, recognize them as time-base related phase errors, and act to correct them in the APC circuit on a line-by-line basis.

The chroma phase rotation circuit in the playback AFC block is responsible for rotating the playback chroma phase back to its original prerecorded value. This is part of the coarse AFC signal correction applied to the playback color-under signal done in preparation for action by the APC circuitry.

During playback, pulses supplied to the playback AFC circuit identify each horizontal line and vertical field so that the phase rotation electronics recognizes just how much to rotate the playback signal and which way to rotate it to reestablish the original prerecorded chroma phase. The signals used to identify horizontal lines with modified phase are the playback H sync and video head switching pulses. Playback rotation circuits can lose track and get out of sequence. When this happens, the horizontal line that is supposed to have the rotated phase returned to normal will still have a large phase discrepancy as compared to its original value. For example, when dropout occurs in a tape

being played back, the playback AFC circuit may lose sequence because a long time period passes with missing H sync pulses. If the input H sync pulse is not detected, it's possible for the phase rotation circuit to lose track of what line it's on during playback. The result would be a series of fine, colored horizontal streaks flashing across the screen. We would see these streaks until the ACK circuit recognized the gross phase error and blocked the chroma from leaving the circuit.

Playback AFC sections do contain a VCO circuit that will continue to oscillate close to the desired frequency when the H sync pulses are missed due to dropout (see Fig. 6.19). This design helps keep the rotation circuit in step when the H sync signal is missing for a brief period. However, chroma phase rotation inaccuracies are still possible when dropout occurs. The Burst ID section is ready to help correct the gross phase errors resulting from dropout if the VCO wanders too far from the correct point.

Another potential for incorrect playback color phase rotation occurs when the playback mode is initiated. When playback is first started, it's possible that the VCR's rotational phase circuit, used to undo the rotated signal, may be wrong as referenced to the signal on the tape. This mismatching of phase rotation between the VCR's circuits and the recorded chroma signal is a random possibility every time the playback mode is first engaged. When this happens, it takes a brief period of time for the rotation circuit to match up to the signal on the tape. During this time, the Burst ID operates to bring the phase of the playback color into the operational window of the APC until the next video head switching pulse comes along and resets the chroma rotation circuit to correct operation.

Whatever the cause, phase errors greater than ±90 degrees will result in a pulse from the Burst ID generator in the APC. This pulse causes the 4.21 MHz generator source block to invert the phase of the signal output by 180 degrees. This returns the up-converted 3.58 MHz chroma to the proper ±90 degree range for APC lock-on and control allowing good color reproduction. Whenever the Burst ID is working to correct chroma phase errors resulting from an incorrect chroma phase rotational sequence in the AFC section, the Burst ID correction pulse is only required until the next head switching pulse arrives. At that time, the rotation circuit is reset by the head switching pulse and normal operation resumes.

The Burst ID circuit is internal to most VHS APC ICs and is contained in the color killer and phase comparator section as shown in Fig. 6.20. Many VHS designs use a color processing IC similar to the one shown.

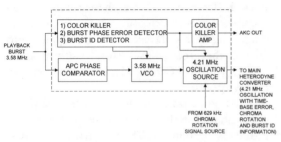

Fig. 6.20. VHS chroma Burst ID block showing color killer and burst phase comparator circuitry.

VHS Burst Boost

One difference between VHS and 8mm designs is how burst is handled. For 8mm machines, burst is boosted in both SP and LP recording speeds. VHS machines amplify the burst signal 6 dB in only the SP and SLP (EP) record modes. Burst is not amplified in LP.

The color signal-to-noise ratio decreases when burst is amplified (boosted) during the VHS LP record speed thus defeating the signal-to-noise increases caused by boosting the recorded burst level. In LP mode, the color signal-to-noise ratio deteriorates because of frequency beating between some of the synchronization pulses in the adjacent RF tracks on the tape. The FM frequencies representing the horizontal sync and the color-under frequency of the burst signals present on adjacent tracks beat together to produce a form of crosstalk interference. This beat interference produces finely-pitched line patterns in the playback video. Some of the line patterns can be colored. These beat patterns are present in all VHS machines, but in the SP and SLP (EP) modes they occur at locations off the visible portion of the screen. Thus, during SP and SLP recordings, VHS designs take advantage of the increased signal-to-noise provided by the boosted burst without concern for the appearance of the beat pattern, which is simultaneously increased by the boost circuit.

Regardless of the VHS recording speed, during playback, these particular beat patterns only occur where horizontal sync and burst appear in the video signal. Thus this interference only manifests itself within the horizontal blanking interval of the video signal. If we could see magnified views of the signals on recorded SP and SLP (EP) tapes, we would notice that the horizontal and burst synchronization pulses are perfectly aligned with their respective counterparts on the adjacent RF tracks. Thus the beat patterns and the associated crosstalk will always be located in the same position for every line of every field when SP and SLP (EP) recordings are made. Since the horizontal blanking interval occurs off the screen and the synchronization pulses are aligned with respect to the adjacent tracks, any beating or crosstalk created by these pulses will never be seen in the picture regardless of the line being displayed in any field.

However, if we could see a magnified view of signals on the LP recorded RF tracks, we would observe no sync alignment between tracks. Thus, the adjacent channel horizontal sync and burst beat and the associated crosstalk patterns no longer occur in an area of the screen that is out of view. Instead, the beat and related crosstalk can appear within the picture information itself. To keep the presence of the LP beat pattern as low as possible in the playback picture, the burst level is not boosted during LP record. This reduces the intensity of the colored beat pattern in the LP playback picture.

Fluctuations in the playback chroma level, caused by playback head-to-tape contact flaws, typically don't create noticeable beat pattern level fluctuations. The intensity of the LP beat pattern is relatively consistent during playback because the ACC circuits work to stabilize unwanted variances in playback chroma levels. Since the LP burst signal is not boosted, the persistent and unwanted chroma level fluctuations not eliminated by the ACC and that cause the LP beat pattern to fluctuate in intensity are minimized when the LP burst is not boosted. If the LP mode burst signal was boosted during record, minor variances in the chroma playback level, not caught and corrected by the ACC, would cause the beat pattern to fluctuate in amplified amounts drawing even more attention to this screen interference. Therefore, the chroma signal-to-noise improvement produced in the color processing circuitry by boosting the burst is insignificant in LP when the sync beat patterns are also increased.

Not all VHS manufacturers have elected to provide an LP record speed. Models without this record speed will play back LP tapes recorded on other units, but will record only in the SP and SLP (EP) speeds. These manufacturers have elected not to record in the middle (LP) speed because of the inferior picture quality produced from this mode.

VHS Automatic Color Control (ACC)

The block diagram in Fig. 6.19 shows that VHS ACC signal processing is similar to that described for 8mm. However, the method used to achieve ACC in VHS machines is different for the LP speed than it is for the SP and SLP (EP) speeds. In LP, the burst level is sampled and then averaged as it is processed by the ACC circuitry. Peak detection is used in the SP and SLP (EP) speeds. LP ACC requires burst

level averaging as part of the detection to assure that the burst level is held constant to reduce the beat pattern from sync pulse misalignment described earlier. Averaging the burst signal level changes generates a slower and smoother chroma level error correction action and minimizes the distracting beat level fluctuations that occur in the LP speed.

In the SP and SLP modes, ACC detects peak chroma levels and rapidly adjusts the signal for unwanted chroma level fluctuations. The purpose in using burst peak detection or burst signal averaging to accomplish ACC in the various speeds is to provide maximum signal-to-noise ratio in color processing. Peak level detection provides the best signal-to-noise ratio for chroma processing and offers the most rapid correction to the unwanted level changes so it is used whenever possible.

Burst is the part of the chroma signal that is peak detected during SP and SLP and averaged in the LP mode because it often represents the highest level of incoming chroma during all playback modes. Chroma levels are rarely saturated in the broadcast signal and burst is typically the highest chroma level found in the color picture.

There is another reason that burst is used as the level monitoring reference. Burst levels do not change with scene content. Unwanted chroma level changes will simultaneously affect both the chroma in the picture and the burst amplitude. Since burst is present on each line in the screen picture, it can be used to provide an accurate representation of undesirable color level changes occurring on each line. When the ACC acts to correct faulty burst levels, incorrect chroma levels are also returned to their correct levels.

The available VHS design documents do not describe the use of a separate high-speed ACC operation. If VHS units require separate ACC processing for the nonstandard playback modes (such as search and slow motion) this takes place inside the playback color processing IC and is not discussed in manufacturer's literature. It is also possible that the VHS ACC circuit was designed to compromise between ACC response times for normal and nonstandard playback speeds, instead of including separate circuits like the 8mm units have to optimize the ACC response for the various modes.

VHS Automatic Color Killer (ACK)

The automatic color kill (ACK) function is performed in the VHS APC circuitry. ACK prevents erroneous color information from exiting the machine when the AFC and APC circuits are not phase locked. Without ACK interruption, wrong colors and chroma noise will show up on the screen when the color up-conversion is not correct. The lack of phase lock in the APC produces incorrect hues. When APC errors cause the ACK to activate, a black and white picture results. Chroma frequency errors caused by incorrect AFC action produce many of the symptoms associated with the loss of APC phase lock. AFC errors may also produce other symptoms including a complete loss of color, color banding, color streaking, color rainbowing (barber poling), color flashes, and random color speckles throughout the playback picture. Incorrectly reproduced color hues and excessive chroma noise are frequently the result of a malfunctioning AFC or APC section.

The ACK detection circuit monitors the phase of the color up-conversion output signal and compares that with the internal 3.58 MHz crystal reference oscillation. If the phase of the up-converted subcarrier signal exceeds permissible bounds, the ACK detector initiates a voltage that activates the ACK circuitry. The ACK electronics then prevents any color signal from continuing down the output line. When frequency errors occur in the AFC section, proper phase lock cannot be achieved in the APC and the ACK is also activated. Thus, by monitoring the phase of the reproduced chroma, the ACK detector is able to sense both AFC and APC troubles.

Miscellaneous VHS Color Processing Circuits

The other circuits within the playback color processing section shown in Fig 6.19 will be described next.

Low-pass and band-pass filters

The 629 kHz low-pass filter, located at the upper left corner of Fig. 6.19 is responsible for stripping out all unwanted frequencies coming from the playback heads. These unwanted signals include the playback luminance FM, FM audio, and noise. Only the 629 kHz chroma should be input to the color processing circuit.

The 3.58 MHz band-pass filter (BPF) following the up-converter is responsible for eliminating all heterodyned frequencies other than the desired 3.58 MHz difference signal. Whenever two frequencies are beat together, sum and difference frequencies and related harmonic frequencies are produced. If these additional heterodyned frequencies were allowed to enter the television set, fine diagonal beat patterns would be present on the screen. The 3.58 MHz BPF filters out these extra frequencies eliminating this interference from the reproduced picture.

Burst de-emphasis

The burst de-emphasis circuit block, near the upper-right hand corner of Fig. 6.19, returns the recovered colored burst playback levels back to their prerecord processing values. During record, the burst signal is identified and separately amplified by 6dB for recording. This boosting action is called *burst boost* or *burst emphasis*. Burst is amplified in the record circuits to reduce the effects of noise placed on the burst signal during the frequency conversion and record/playback write and read processes. Noise in the burst signal tends to interfere with the playback APC circuit's ability to accurately reproduce color hues. Burst emphasis helps immunize the playback APC electronics from the effects of this noise. The playback circuits must de-emphasize or return the increased burst level back to its standard value so that it can provide a correct level reference for the television set. Note that the de-emphasis circuit only operates in the SP and SLP (EP) modes. More on this in the record chroma discussion later in this chapter.

It's worth noting here that the VHS format does not use a separate chroma emphasis in conjunction with the burst boost like the 8mm format does. VHS units only provide burst emphasis and de-emphasis.

The comb filter

The comb filter following the 3.58 MHz up-converter in Fig. 6.19 is used to eliminate chroma crosstalk. Chroma crosstalk is 629 kHz color information detected by the playback head coming from the adjacent RF track on the tape. The theory of this process will be described later in this chapter.

VHS Chroma Playback Circuit Summary

Before we address the color playback circuitry of current VHS models, let's review some of the principles covered so far. The VHS AFC circuit stabilizes large time-base errors in the playback color-under signal. It also performs the chroma phase rotation required to undo the phase rotation imposed on the signal during record. The AFC however, does not correct time-base related phase errors. These are handled by the APC circuit. The AFC's chroma phase rotation restoration is performed in fixed amounts and directions based upon the predetermined modifications placed on the signal during record. During record, the record color-under signal is phase-rotated to assist the playback electronics in eliminating the color-under crosstalk signal coming from adjacent RF tracks. Since the AFC's chroma

phase rotation is accomplished in fixed amounts, it is not able to correct for phase instabilities resulting from time-base error. Once these processes have been completed, the signal is ready for the APC correction.

APC corrects the remaining time-base related phase errors. Phase errors are detected by comparing the up-converted 3.58 MHz chroma burst signal with the signal from a fixed crystal reference oscillator. Differences between the reference oscillation and the up-converted signal produce a voltage that changes with the amount of detected difference. This voltage is used as a fine correction voltage in the 4.21 MHz oscillator for precise control of its output.

Burst ID assists the phase error correction by issuing a control signal to the 4.21 MHz oscillator when phase errors exceed ±90 degrees. Burst ID also helps correct hue reproduction when the chroma phase rotation circuit in the playback AFC section gets out of sequence.

Current VHS Chroma Playback Circuit Designs

Over the years, VHS chroma circuits have remained relatively unchanged although higher frequency oscillators in the AFC section and digital technology have replaced the previous circuits. Even with the introduction of SVHS, the fundamental principles used in the chroma circuits have remained the same. Figure 6.21 shows one circuit where the AFC VCO frequency has been increased to 160 fh rather than the 40 fh frequency previously used. Notice that the final AFC frequency arriving at the 4.21 MHz oscillator source is still 629 kHz. This 629 kHz AFC frequency is still required to heterodyne with the 3.58 MHz frequency from the APC section to produce the 4.21 MHz signal with an amount of jitter proportional to the off-tape time-base error. The use of a higher frequency VCO in the AFC section pro-

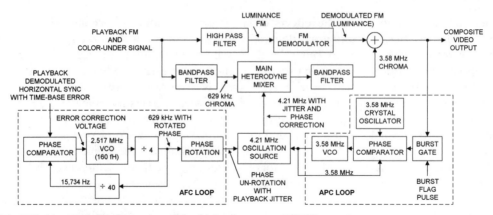

Fig 6.21. *VHS chroma playback circuitry with a higher frequency AFC VCO.*

vides improved tracking and a faster response time to correct existing errors.

In the most recent designs, digital circuits like the one outlined in Fig. 6.22 have been used in the chroma processing section. This enables lower production costs while improving the accuracy of the digital circuits, making these state of the art designs very popular. The block diagram shows a typical current playback digital chroma processing method.

By applying what you learned from the earlier circuit designs to this current model, you can see that the AFC and APC sections still exist. Although they do appear in a little different form, it is still possible to distinguish the AFC and APC sections.

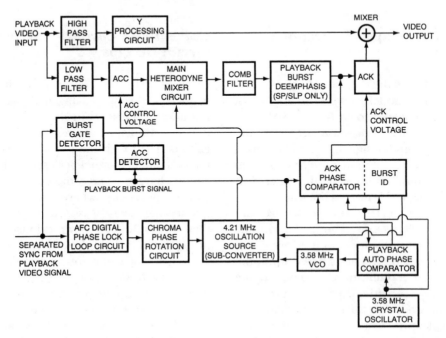

Fig. 6.22. *Current VHS chroma playback is similar to earlier methods except digital processing has replaced analog processing.*

Let's begin analyzing the operation of the digital chroma playback by looking at the AFC section. The digital chroma playback AFC circuit operates as a digital phase-locked-loop (D-PLL) as shown in the lower left-hand corner of the figure. Playback horizontal sync containing the playback jitter (Df) enters the D-PLL where the AFC circuitry acts to track the time-base error.

The D-PLL acts as a buffer oscillator in the event playback horizontal sync is missing just like the AFC VCO did in a previously described model. Here the D-PLL can be considered an artificial sync pulse generator that functions when incoming sync is missing or incorrect.

The D-PLL frequency is phase rotated before being applied to the 4.21 MHz oscillator generator. The AFC D-PLL operates at 5.03 MHz or 320 fh. It is important to realize that all AFC circuits operate at some multiple of the horizontal line frequency (fh). In the first design that we presented, the AFC VCO operates at 40 fh. In a later model, you saw that the AFC VCO operates at 160 fh (or four times higher in frequency than the 40 fh model). The current AFC designs operate at an even higher multiple of 320 fh.

Regardless of the oscillating frequency of the AFC design, the signal must eventually be divided down to 629 kHz to mix with the 3.58 MHz VCO in the APC section so that the required 4.21 MHz frequency can be produced.

Also notice that the digital circuit in Fig. 6.22 still requires ACC, ACK, and Burst ID. These sections are used in all VHS VCRs regardless of the vintage. Many models use large scale integrated circuits that house the entire record and playback chroma circuitry. Often, the repair schematics and block diagrams supplied by VCR manufacturers reveal little detail concerning the internal workings of these complex ICs. Items typically excluded from the drawings include the ACC, ACK, and Burst ID sections. However, these sections do exist in all models.

In the APC section described in Fig 6.22 the APC circuit seems similar to the block diagrams previously described. Current models still use a 3.58 MHz VCO that is controlled by a correction input coming from a phase comparator.

Chroma Crosstalk Elimination

Now that we've explained the playback jitter correction techniques used in both formats, we'll describe how the system eliminates the color-under signal crosstalk that is picked up by the heads during playback.

In previous sections you read how the playback luminance FM signal crosstalk can be eliminated by using azimuth recording techniques. Figure 6.23 shows that the luminance FM crosstalk signal can be eliminated by recording the signal using high-frequency FM for the Y portion with offset azimuths in the gap.

Azimuth recording techniques work effectively only for high-frequency signals but have little effect on crosstalk elimination for frequencies below 1.5 MHz because some adjacent track information is detected by the op-

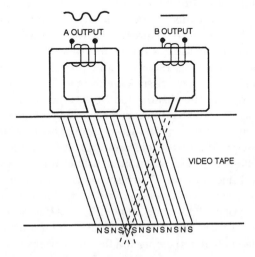

Fig. 6.23. *Azimuth recording techniques prevent head B from detecting the head A Y FM signal.*

posite head as shown in Fig. 6.24. Therefore, when color-under frequencies well under 1 MHz are used, azimuth recording techniques provide little isolation for the color portion in adjacent channels.

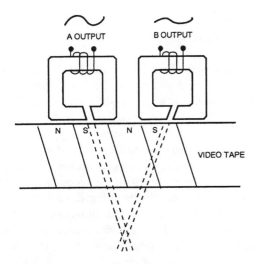

A OUTPUT   B OUTPUT

VIDEO TAPE

Fig. 6.24. *Head B detects more N than S, but since N isn't canceled by an equal amount of S, some A signal is detected by head B.*

Complex electronic circuitry is used to eliminate chroma crosstalk on adjacent RF tracks, and this is one area where 8mm and VHS formats differ greatly. However, the end result is the same. Chroma information on adjacent tracks is recorded and played back in such a way that crosstalk is eliminated. Chroma crosstalk cancellation for both tape formats uses the fact that the chroma subcarrier signals at any given point on two sequential horizontal lines in the same field are opposite in phase. Referring to Fig. 6.8, note that line number 15 and line number 17 are sequential horizontal lines in field number one, and in the lower portion of the figure line number 15 and line number 17 are the resultant screen dot pattern. Also note that the dots and the subcarrier waveform are exactly 180 degrees out of phase.

Chroma crosstalk cancellation circuitry for both tape formats electronically modifies

this phase relationship during record. The modification is performed in record mode so the crosstalk component present during playback can be eliminated by a comb filter in the playback circuitry. The comb filter is located in the playback circuits following the 3.58 MHz up-converter block (Figs. 6.17 and 6.19).

During playback, the original chroma phase relationship is restored. As part of this process, the crosstalk component, which is moving through the same circuitry as the desired signal, is also reconfigured by the same circuit. The result is a crosstalk component that has been electronically manipulated so it can be canceled in the comb filter, letting the original signal pass through unaffected. Although the two formats differ in the method used to modify the subcarrier phase relationship, a special circuit called a *comb filter* is the common link between the two processes.

Record chroma phase inversion and color-under crosstalk elimination for the 8mm format

Comb filters are important for eliminating chroma crosstalk caused by the use of zero guard bands. Before the comb filter circuitry can function, the color-under chroma signal phase relationship must be modified during the record process. During record, this chroma subcarrier modification positions the crosstalk component so that during playback it will be detected by each head in a specified phase relationship on a sequential line-by-line basis. The phase of the chroma crosstalk component must have a relationship unique and different from that of the desired signal so it can be selected and eliminated. Without this modification, crosstalk couldn't be detected and canceled. Since the 8mm system uses a simpler approach to modify the subcarrier phase relationship, it will be described first.

During record, the 8mm system alters the subcarrier phase relationship of selected horizontal lines occurring within a given field. This phase modification only happens on every other horizontal line within one of the two vertical fields. This phase modification occurs only when the channel 2 video head is touching the tape. The phase of the 4.32 MHz oscillation used to create the 743 kHz color-under record signal is modified. The phase of the 4.32 MHz oscillation source is passed unmodified when the channel 1 head is recording the signal. Phase modification is accomplished by the circuitry between the 4.32 MHz oscillation source and the main heterodyne mixer.

Figure 6.25 is a simplified block diagram showing how the 4.32 MHz oscillation source is derived. Horizontal sync that has been separated from the incoming record video source enters the AFC section where it is used to produce a 743 kHz oscillation that follows any changes that may occur in the record input video color signal. In the 8mm format, the AFC section is unique to the record mode. Recall that no AFC circuit is in the 8mm chroma playback path. Instead, 8mm units use APC ID to correct for large frequency errors in playback. In record, the 8mm AFC operates like a VCO. The output of the 8mm record AFC enters a 4.32 MHz heterodyne submixer circuit where it combines with a very stable 3.58 MHz signal from the crystal controlled oscillator. The 3.58 MHz and 743 kHz signals are mixed to-

gether to produce 4.32 MHz, the sum of the two frequencies.

The 4.32 MHz output becomes one input to the phase inverter selection circuit. The same signal is inverted and becomes the second input to the circuit. During the time that the channel 2 video head is recording, the phase inversion selection circuit is receiving horizontal synchronization pulses. These pulses toggle a switch between the inverted and noninverted 4.32 MHz oscillator inputs. The output is a 4.32 MHz oscillation signal that is phase inverted every other horizontal line while the channel 2 video head is recording.

During the time that the channel 1 head is recording, the phase inverter selection circuit toggles to the noninverted input causing the 3.58 MHz chroma subcarrier signal to be converted to 743 kHz in the main heterodyne mixer circuit with an unmodified subcarrier phase relationship. The horizontal lines recorded with the channel 2 head have modified a chroma subcarrier phase relationship that is exactly in phase with each sequential line. The subcarrier phase relationship recorded with the channel 1 head maintains the original phase alternating relationship of the original color video signal. The only difference between the

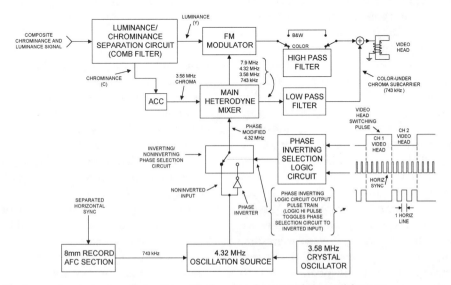

Fig. 6.25. The 8mm color-under record circuitry with chroma subcarrier phase modification.

channel 1 head recorded chroma signal and the original chroma subcarrier is that the frequency has changed.

The 4.32 MHz oscillation signal is phase-locked to the incoming video's color horizontal sync pulses. This means that the color-under chroma record signal is always phase locked to the incoming horizontal sync multiplied by 47.25 (743 kHz), and that the 743 kHz color-under signal maintains the exact phase relationship with the input video's original 3.58 MHz subcarrier frequency.

The phase inverter selection circuit is driven by a logic circuit that has two inputs. One input is a 30 Hz (actually 29.97 Hz) video head switching pulse derived from the video head itself. This signal will be described in the chapter on servos. The other logic circuit input is a sequence of horizontal synchronization pulses that cause the phase inverter selection circuitry to alternate between inverting and noninverting inputs. Thus, in the field that the channel 2 head has recorded, the phase of every other line is inverted. The resultant output is such that each sequential horizontal line of the field recorded with the channel 2 head has an in-phase subcarrier. During the time that the channel 1 head is recording, the 30 Hz head-switching pulse forces the phase inverter selection circuit to remain in the noninverting input position, which allows the subcarrier phase to alternate during each sequential line. The phase of each line alternates at this time because color video signal convention normally produces an out of phase relationship between adjacent lines (see Fig. 6.8). Figure 6.26 shows the subcarrier phase relationship for each horizontal line of the RF track for the channel 1 and channel 2 heads.

During playback, the channel 2 head subcarrier phase must be restored to its original alternating phase for display on the TV screen. In most consumer models, playback phase restoration is done by the same circuitry that performed the modification during record. During playback the phase inversion circuitry performs in a manner similar to that

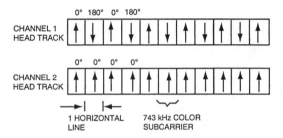

Fig. 6.26. Phase relationship of adjacent RF tracks when the 743 kHz color subcarrier is modified.

shown in Fig. 6.25. The differences in playback center around the signals supplied to this circuitry. The playback horizontal sync pulses supplied to the phase inverter are now coming from the playback demodulation section rather than from the input video sync separator. The 4.32 MHz oscillation coming from the oscillation generator contains time-base related jitter in playback. The 3.58 MHz chroma signal into the heterodyne converter during record came from the input video source. But, once the unit is put into playback, the input signal is 743 kHz and contains time-base error and a crosstalk component picked up by the playback head from the adjacent tracks. Figure 6.27 shows the subcarrier phase relationship of the recorded signal and the restored playback subcarrier during playback. Section A of the figure compares the channel 1 head and the channel 2 head recorded tracks. The color-under subcarrier phase is modified for the channel 2 head so each sequential line for the channel 2 head field is in phase. The channel 1 head color-under subcarrier phase isn't modified so each horizontal line alternates in phase.

Figure 6.27B describes the playback phase relationship of the color-under signal from the video heads. Here the channel 1 head chroma information contains the phase-alternating 743 kHz color-under subcarrier and the chroma crosstalk from the channel 2 head track whose phase was modified and does not alternate. Crosstalk information is indicated by the dotted arrows. The channel 2 head playback energy contains a 743 kHz signal whose phase is exactly as recorded on the tape (modified), plus some signal detected from the adjacent recorded track laid down by the channel 1 head. The chroma energy from the channel 1 track and detected by the channel 2 head is the crosstalk component, whose phase is alternating. Crosstalk information is again indicated by the dotted arrows in the figure.

Figure 6.27C shows what happens to the subcarrier phase relationship when the color-under signal is restored to its original alternating phase condition. Only the channel 2 head information was modified during record so in playback only the channel 2 head information must be restored.

Let's analyze what happens to the channel 2 head chroma information in playback. The desired chroma information and the crosstalk component are picked up by the channel 2 head and are conducted down the same processing path. Since the channel 2 head signal was modified in the record process, it must be undone in playback. Restoring the channel 2 chroma signal to its original phase by inverting every other line also inverts the crosstalk component at the same time. The channel 1 track information was not phase

modified during record, so its phase was altered 180 degrees according to NTSC (National Television Systems Committee) color video signal design specifications. During playback, restoring the channel 2 track chroma phase also alters the phase of the channel 1 track chroma crosstalk component. This is why Fig. 6.27C shows the channel 1 crosstalk signal for the channel 2 head in the same phase for all adjacent lines. The crosstalk has actually been modified by the playback circuits while the original signal phase was restored. It's important to remember that the only reason the chroma is phase-modified during record is to establish a specific phase relationship for the crosstalk component so the comb filter can cancel it out during playback.

Figure 6.27C also shows the chroma phase relationship for the channel 1 head track, and the channel 2 track crosstalk that the channel 1 head detects. During record, the channel 1 head chroma signal wasn't modified so it doesn't require restructuring during playback. The channel 2 track crosstalk component picked up with the desired channel 1 track information also is not changed during playback. The channel 2 chroma signal was phase modified during record. Since it was

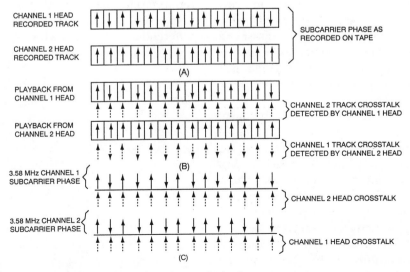

Fig. 6.27. Chroma subcarrier phase relationships during playback.

recorded at the same phase on each line, the channel 2 crosstalk detected by the channel 1 head will also be at the same phase for each line. The channel 1 chroma signal phase needs no restoration so the channel 2 chroma crosstalk passes through as recorded in its modified fashion.

In playback, the channel 1 and channel 2 head color-under signals and their respective crosstalk components are converted back to the 3.58 MHz color subcarrier frequency. The crosstalk component from each of the heads is also converted to 3.58 MHz, but each adjacent horizontal line is of the same phase. The phase of the desired subcarrier signal is returned to NTSC specifications so that it alternates 180 degrees out of phase on adjacent lines in each field. Once it has been up-converted to 3.58 MHz, the crosstalk component and the desired signal are passed to the comb filter circuitry for crosstalk elimination.

The comb filter

The comb filter received its name from the combing action that it performs on an electronic signal. It acts like an ordinary hair comb in that unwanted portions of a signal are combed out of the information passing through the circuit. Some recent designs use digital comb filter circuits, but earlier units and many less expensive models use conventional analog circuitry. These analog comb filters consist of a glass delay line and a summation circuit, as shown in Fig. 6.28A and B. The glass delay line retards the signal one horizontal line (63.5 microseconds).

When a glass delay line is used, the signal delay is caused by the time it takes for a signal to pass through a measured length of glass. Two transducers are attached at each end of the glass delay line. A transducer acts like a speaker or a microphone depending on whether it is sending or receiving a signal. The delay line's input transducer receives the electronic signal and converts it to acoustical waves that travel through the glass reflecting off the glass edges. When the acoustical waves

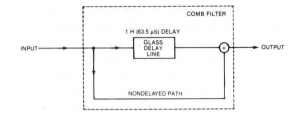

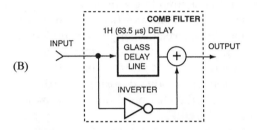

Fig. 6.28. *(A) The comb filter uses a glass delay line and a summation circuit to remove (cancel out) crosstalk. (B) The same comb filter circuit with an inverter in the nondelayed path.*

reach the output transducer, they are converted back into an electronic signal by the output transducer as shown in Fig. 6.29.

In a VCR, the comb filter is the circuit that cancels out the unwanted chroma crosstalk component of the playback signal. It does this by taking advantage of the frequency interlaced chroma oscillation present in the composite video signal. The comb filter can differentiate the desired chroma subcarrier os-

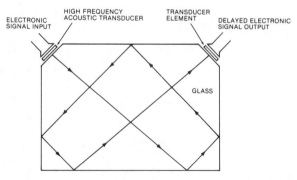

Fig. 6.29. *A 63.5 microsecond glass delay line in the comb filter.*

cillation from the crosstalk signal because of the phase differences between these two signals. In a color video program, the 3.58 MHz subcarrier oscillation present on each sequential horizontal line in the same vertical field has a chroma subcarrier phase exactly 180 degrees out of phase from the same points on the previous line. This relationship was designed to reduce the visual effects of the chroma subcarrier on the TV screen.

The delay line that is part of the comb filter allows the device to compare the color signal that is in one line of the video with the chroma content on the previous line. Since the subcarrier oscillation from the previous line was 180 degrees out of phase with the present line, the comb filter is able to recognize the desired chroma information. Since the comb filter in Fig. 6.28A is comparing two lines that contain subcarrier signals of the opposite phase, it will cancel out the oscillation in its summation circuit. If an inverter is placed in the nondelayed signal path as shown in Fig. 6.28B, the comb filter summation circuit will double the subcarrier output level because the phase of the subcarrier on the previous and current lines is the same phase as they arrive at the summation circuit, and when these two values are added together, the output is doubled. Thus, the comb filter is capable of either adding to, or subtracting from, the signal on one line by comparing it to the information on a previous line depending on how the circuit is designed. In this VCR circuit, it's desirable to keep the chroma information in the signal, so the comb filter is configured so that the chroma signal is doubled in amplitude and allowed to pass.

The comb filter and playback color crosstalk elimination for the 8mm format
Now let's see what effect the comb filter has on the unwanted crosstalk component that is part of the subcarrier oscillation. Recall that during record, the subcarrier phase of every other line in the field recorded by the channel 2 head was modified by inverting it. This puts

the subcarrier oscillation in phase with the adjacent horizontal lines for all lines of that field (see Fig. 6.26).

Now look back at Fig. 6.27C. Notice the phase relationship of the crosstalk and the desired subcarrier when the color-under signal is restored to its original alternating phase condition. The comb filter circuitry differentiates between the desired alternating subcarrier phase and the nonalternating crosstalk subcarrier phase. It identifies and combs the crosstalk component out of the chroma information.

Figure 6.30 describes how the comb filter circuitry actually cancels out crosstalk information. The 3.58 MHz up-converted chroma information enters the comb filter circuitry from the left side of the figure. The chroma information takes two separate paths. The first path is through the nondelayed line and the second path is through the delay line. In the delay line path, the incoming 3.58 MHz up-converted chroma signal is converted into acoustical vibrations that travel through the glass. This signal stream is delayed by 63.5 microseconds (the time period of one horizontal line) before entering the summation circuit.

Chroma crosstalk cancellation occurs inside a summation circuit—typically a resistor network. Cancellation occurs when two signals of equal amplitude and opposite phase meet and are summed together. The desired signal is of the same phase and is summed with the delayed signal. This is how the level of the wanted subcarrier is doubled.

In Fig. 6.30 the wanted and unwanted signal characteristics are portrayed at significant locations of the circuit. The phase of the desired signal is shown in the solid arrows and the crosstalk component is depicted with dashed arrows. The double lined arrows in sections D and E represent the doubled chroma subcarrier amplitudes. The X's indicate the crosstalk component that has been canceled out of the signal.

The inverter closest to point C in the figure is used to set up the wanted chroma signal phase for doubling in the summation circuit. It

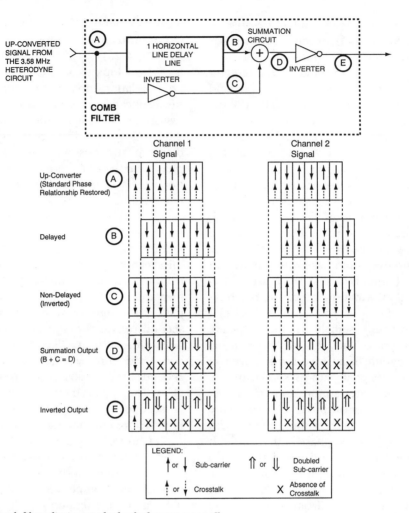

Fig. 6.30. The comb filter eliminates playback chroma crosstalk.

also configures the crosstalk phase for cancellation as part of the summation process.

The signal output from the summation circuit is the product of adding the delayed signal and the nondelayed (inverted) signal paths together (B + C = D). There is still some crosstalk component in the first line of the summation output. Also, the desired chroma signal was not doubled for that same line. This is because there was no signal coming from the delay line (point B) at that moment. In this instant of time D = C + 0 because the delay line retards the arrival of the first line by 63.5

microseconds (one horizontal line) just as it does any other line. Therefore, the first line does not arrive in time for the correct product to be produced. The desired result does not occur until the second line appears.

The inverter at the output of the summation circuit between points D and E is used to return the phase of the inverted subcarrier back to the value originally input to the comb filter.

For complete crosstalk cancellation to occur, the comb filter must have a color crosstalk signal that is precisely the same amplitude on both of the two sequential horizontal lines being compared. Also, a precise phase relationship of the crosstalk component between the two lines is required. If the crosstalk signal phase between adjacent lines is not equal, complete cancellation will not occur in the summation circuit. Even though proper operation of the comb filter requires that color crosstalk information be equal in amplitude with a precise phase relationship between any two lines of a given field, this is not always the case. When these relationships are not identical, some crosstalk interference is present on the screen. Because color information isn't always identical between adjacent lines, crosstalk is substantially reduced, but not completely canceled. Yet, even under these conditions, comb filters reduce the crosstalk component well below the typical viewer's perception.

Another important point is that comb filter operation requires two lines of information for cancellation. This means that chroma crosstalk is not canceled on the first and last lines of a field on the TV screen. For example, notice that the leftmost arrows representing the summation output signal at point D in Fig. 6.30 indicate that crosstalk is still present on the first line of video output from the summation circuit. This occurs because signal path C has no comparison signal to enable cancellation during the time of the first horizontal line. The signal from the first horizontal line following the path at point B has been delayed 63.5 microseconds and will not be arriving until one horizontal line time later. There are two reasons that the residual crosstalk on the first and last lines of a field has little effect on what the viewer sees. First, the horizontal lines are extremely fine on a TV screen—525 lines make up the entire picture. Secondly, the first and last lines of video information are respectively above and below the normal viewing area of the screen.

Comb filters have another flaw. They typically do not completely double the wanted chroma amplitude. In order for doubling to take effect, the amplitude of the chroma oscillation on adjacent lines must be identical in amplitude and exactly 180 degrees out of phase. This rarely occurs in a normal color program. Also, some minor distortion occurs as the chroma signal passes through the delay line. Both the lack of perfect doubling and the absence of complete crosstalk cancellation in the summation circuit allows some of the adjacent channel crosstalk to pass through the comb filter. However, the residual crosstalk and the minor chroma distortions imposed upon the wanted chroma signal represent a very small percentage of the color quantity in the picture. Typically, the distortions in the chroma signal and the remaining crosstalk information itself are not detectable on the TV. On the other hand, if a comb filter were not used in the playback processing, the crosstalk component would be very visible on the screen and the distracting, colored beat pattern overlaid on top of the video scene would be quite distracting.

Some VCR models have digital circuitry that improves comb filter operation by using charge coupled devices (CCDs) as delay lines. CCDs are more efficient and occupy less space than traditional glass delay lines. A CCD functions by sampling and holding electrical charges for a period of time until commanded to release (dump) the charges. In a comb filter, the CCD samples the color signal, stores the signal as electrical charges for 63.5 microseconds, and then releases the information. The CCD is the major component in a digital comb filter.

CCDs have a major advantage over glass delay lines, which attenuate the input signal severely. This attenuation reduces the strength of the signal output from the glass device. CCDs are not perfect, however; they produce some chroma signal distortions of their own. These distortions relate to noise produced in the CCD by quantization errors that occur in the digital sampling process and by linearity errors. The linear response of the device can be maintained by proper CCD biasing. The sampling noise and quantization errors can be minimized by using faster sampling rates. Using good circuit design, VCR manufacturers are able to keep CCD noise and signal distortions well below our ability to perceive these unwanted defects.

VHS color processing and crosstalk elimination

Phase rotation in a VHS machine is much more complex than the phase-inverting method used in the 8mm format, but an understanding of 8mm operation helps when analyzing the VHS technique for color processing and crosstalk elimination. Recalling 8mm theory will help because the VHS system also uses phase cancellation in a comb filter to eliminate adjacent track chroma crosstalk during playback. Therefore, both the VHS and 8mm formats must process the color signal so a common comb filter can remove the undesired signals. However, there are major differences in the way the signal applied to the comb filter is prepared in each format. First, the conversion frequencies are different. Second, there is a major difference in the way that the subcarrier phase is modified in the color-under conversion process. In the 8mm format, the phase of specific lines is inverted only for the field recorded by the channel 2 video head. In the VHS system, the record chroma subcarrier is rotated in 90 degree increments for each line of every field. This is why this method is called *chroma phase rotation*. When analyzing VHS chroma crosstalk, remember that the rotation process is required to establish a specific phase relationship between the chroma crosstalk components during playback so the comb filter can cancel the undesired signals.

VHS circuitry modifies the subcarrier phase on each line of both the A head field and the B head field (known as the channel 1 and channel 2 video heads respectively in the 8mm format). In VHS, the phase of the recorded color-under signal is advanced 90 degrees from each previous horizontal line for every line in the A head field, and is retarded 90 degrees with reference to each previous horizontal line for every line in the B head field. This is achieved by rotating the phase of the 629 kHz signal out of

the record AFC circuit before it heterodynes with the 3.58 MHz signal from the APC. This produces a 4.21 MHz oscillation signal with a rotating phase that in turn phase-rotates the 629 kHz chroma signal recorded on the tape.

VHS phase rotation

The phase rotation process is performed during record and occurs inside the record AFC loop. Figure 6.31 shows a VHS rotation design using an AFC that operates with a 160 fh VCO. Incoming separated horizontal sync pulses enter the AFC phase comparator and are matched with a 15.734 kHz input to generate a DC control voltage for the 2.517 MHz (160 fH) VCO. The 160 fH VCO output is passed to a ring counter where it is divided by four (40 fh) and each output shifted 90 degrees (0, 90, 180, 270 degrees). The outputs, described as signals phase 1, phase 2, phase 3, and phase 4, are phase-shifted 90 degrees from each other. The phase 2 output is divided by 40 to produce a 15.734 kHz (the horizontal sync rate) input to the AFC phase comparator. This provides a feedback sample for locking the 2.517 MHz VCO on frequency and phase.

The outputs from the divide-by-four ring counter are fed into individual two-input AND gates that produce an output only when each input is a logic level 1 (high). Thus the 629 kHz phase-shifted signal does not pass through one of the AND gates until its other output is also high. The second input to the AND gates comes from a pulse generator enabling circuit. The AND gate pulse generator selects which of the four output lines to enable according to the phase-locked horizontal sync and the 30 Hz head switching pulse applied. Since each of the AND gates at the ring counter output has a phase-shifted 629 kHz signal on one input at all times, all that is needed to allow that signal to pass through is a logic high from the appropriate gate in the AND gate pulse generator.

The enabling or firing order from the AND gate pulse generator is shown in the lower right of Fig. 6.31. The 30 Hz head switching

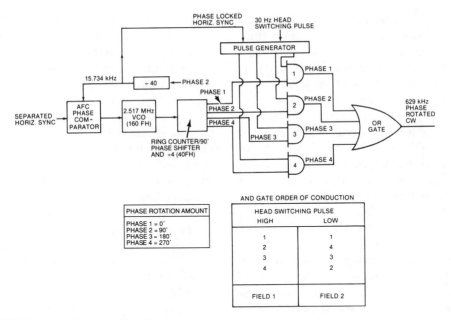

Fig. 6.31. VHS 90 degree phase rotation circuitry.

pulse is used to alter the order of firing. The horizontal sync input causes the AND gate pulse generator to produce an output at the horizontal scan rate.

The outputs from the four AND gates are fed into a four-input OR gate that passes any signal present at any input. It doesn't care if it has a signal present at one input or all of them. It will pass one or a composite through to its output. In this case, however, the AND gates preceding the OR gate will only provide one pulse at a time to the OR gate. While the A head is recording the first horizontal line, AND gate 1 is turned on by signals at both of its inputs, and it passes a zero degree phase-shifted signal to the input of the OR gate. All of the other AND gates are disabled. The OR gate passes the signal from AND gate 1 to the 4.21 MHz subconverter and the 3.58 MHz chroma signal is phase-altered accordingly as it is down-converted. In this case, the phase of the original subcarrier signal is not phase-shifted because the output of the OR gate is set to zero degrees.

The next horizontal line comes along and enables AND gate 2. At this firing of the pulse generator, AND gate 2 outputs a +90 degree phase-shifted signal through the four-input OR gate to the 4.21 MHz oscillation source causing the 3.58 MHz subcarrier phase of the input signal to alter +90 as it is converted to the color-under frequency. This causes the phase of the second line to be set to 270 degrees. It is placed at 270 degrees because the original input subcarrier was already 180 degrees out of phase from the previous line before the signal was shifted an additional 90 degrees. For the next line, AND gate 3 is enabled and the phase of the color-under conversion signal is modified to 180 degrees from its original value. This modification places the converted subcarrier phase at zero degrees. The phase of the third line is set at zero degrees because a 180 degree phase shift was imposed on the original subcarrier signal, which was already 180 degrees out of phase with the previous line. (Recall that the NTSC television standard requires that the subcarrier phase of every other line be 180 degrees out of phase with the previous line.) The phase of the third line in the sequence is changed by 180 degrees from its original value because the rotation sequence advances the phase by 90 degrees from the previous line (90 degrees plus 90 degrees equals 180 degrees). Therefore, it is the combination of the rotation signal and the phase of the original NTSC signal phase that produces the final converted color-under subcarrier.

This process may appear confusing at first, but it will be easier to understand the rotational phase sequencing if you recall that for the input signal, the chroma phase of each adjacent line transmitted from the TV station is always 180 degrees out of phase from the previous line. To calculate the recorded phase of the modified signal for a given line, you must first know the original phase of the subcarrier (zero degrees or 180 degrees) and the modified phase applied to the OR gate in Fig. 6.31.

Examine the truth tables in Figs. 6.31 and 6.32B. Two things must be determined in order to define the modified phase of the down-converted color signal. First, the direction of the phase rotation (clockwise or counterclockwise) must be determined. Second, the phase of the original unmodified signal (zero or 180 degrees) must also be known. In Fig. 6.32B the original phase of the input recorded chroma signal is indicated by the open arrows; the solid arrows indicate the phase of the modified signal to be recorded. Assume that the circuit modifying the down-converted signal phase is rotating clockwise. On the first line of field one, the input subcarrier signal is zero degrees. During the time that line one is passing through the chroma down-converter, the 629 kHz oscillation from the AFC ring counter also remains at zero degrees phase reference. This in turn causes the 4.21 MHz oscillation generator to output a 4.21 MHz signal at zero degrees phase reference. At this time the phase of the down-converter's output will remain at zero degrees phase reference, so the first line is considered unaltered.

In the second line, the phase is rotated +90 degrees because the phase 2 AND gate signal is applied to the OR gate. At first glance, the solid phase arrow of the second line may appear rotated too much (270 degrees). The arrow for line 2 is pointed in this direction for two reasons: (1) the phase of the original chroma signal in line 2 is 180 degrees out of phase with the previous line (NTSC convention), and (2) the phase is rotated +90 degrees from its original position by the VHS rotation circuitry. Since the original input chroma signal was already 180 degrees out of phase with the previous signal, the rotation circuit rotates it another 90 degrees to 270 degrees.

Now look at line 3 in Fig. 6.32B. The phase of the original input signal, which would have normally been at zero phase, is now rotated +180 degrees (90 degrees for line 2 and an additional 90 degrees for line 3). This places the modified signal 180 degrees out of phase with the original signal. (This may be considered 270 degrees from the previously modified signal because the modified phase of line 2 was 270 degrees, and the phase of line 3 is an additional 90 degrees from the rotation circuit, plus the original input signal shift of 180 degrees.)

In line 4 the input chroma signal is 180 degrees out of phase with the first line. The phase 4 signal from the rotation generator imposes a 270 degree shift on the signal to produce a modified phase as shown in Fig. 6.32B. The result is that the modified line 4 signal is +90 degrees out of phase with the original chroma signal (180 degrees + 270 degrees = 90 degrees).

When the next line of information comes into the main converter, its normal phase is at zero degrees and the rotation circuit signal outputs a signal shifted zero degrees. This matches the phase of the original signal, and it is recorded unaltered. At this point, the cycle starts over again. This process continues until the B head comes in contact with the tape.

During the head B record time, the same process takes place, but now the AND gates

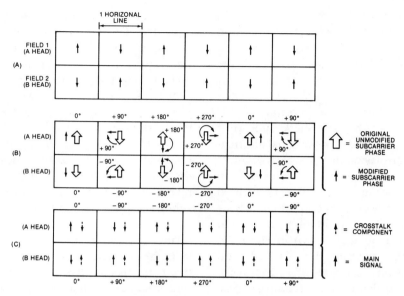

Fig. 6.32. *VHS chroma phase relationship before and after record phase rotation.*

are enabled in a different order because the 30 Hz head switching pulse has changed polarity. This causes a different sequence of commands from the AND gate pulse generator, and the firing order is such that the phase is retarded 90 degrees rather than advanced.

Once again, to calculate the recorded phase of the modified signal for a given line, you must first know the phase of the original recorded subcarrier signal (zero or 180 degrees) and the modified phase of the previous line.

As shown in Fig. 6.32B, the phase rotation circuit retards the color-under phase by 90 degrees during the time that the B head is recording. This rotation occurs in a direction opposite to that for the A head chroma information. Once again, the phase of line 1 is established at zero degrees (unmodified). Since a zero degree 629 kHz phase oscillation is passed to the OR gate from the ring counter, the OR gate's output will result in recording a signal with zero degrees (unmodified) phase difference. However, the unmodified phase for line 1 of the B head field is starting out at 180 degrees rather than zero degrees as the first

line for the A head did. This follows NTSC specifications for the color system. The first line in the sequence is considered unaltered because the rotation circuit does not change the input signal.

Now consider the second line to be recorded. Its phase is rotated –90 degrees. The solid phasor arrow for head B shown in Fig. 6.32B indicates 270 degrees because the phase angle of the original signal was zero (180 degrees from the previous line) and 90 degrees counterclockwise from this is 270 degrees. Note that this phase corresponds to that of the corresponding line of the A head.

In line 3, the phase of the original input signal has been rotated –90 more degrees to –180 degrees. This places the modified signal 180 degrees out of phase with the original input. (This may also be considered –270 degrees phase difference from the previously modified signal.)

On line 4, the input chroma signal phase is 180 degrees out of phase from that in the first line of the B head. The rotation generator shifts the original signal a total of –270 degrees to +90 degrees.

When the next line comes into the main converter, its normal phase is 180 degrees and the rotation circuit outputs a signal with zero degrees phase difference. At this point the cycle starts over.

VHS playback chroma phase rotation

During playback, the modified subcarrier phase must be restored to its original alternating phase for display on the TV screen. In most consumer models, playback phase restoration is done by the same circuitry that performed the modification during record. In playback mode, the phase rotation circuitry performs in a manner similar to that shown in Fig. 6.21. The differences between the playback and record processing techniques relate to the signals supplied to this circuitry (compare Figs. 6.21 and 6.31). During playback, the horizontal sync pulses supplied to the phase rotation section in Fig. 6.31 are now coming from the playback demodulation section rather than the input video sync separator. The 4.21 MHz signal coming from the oscillation generator contains time-base related jitter as in playback. The chroma signal fed into the main heterodyne converter during record was from the

input video source and was at a frequency of 3.58 MHz. Once the unit is put into playback, the input signal will be 629 kHz and contain time-base error and a crosstalk component picked up by the playback head from the adjacent tracks. Figure 6.33 shows the subcarrier phase relationship of the recorded signal and the restored playback subcarrier during playback. The figure compares the subcarrier phase of selected horizontal lines recorded by the A head and B head. Unlike the 8mm format, the VHS system modifies the color-under subcarrier phase for both heads.

Figure 6.33B describes the playback phase relationship of the color-under signal from the video heads. Here the A head chroma information contains the phase-rotated 692 kHz color-under subcarrier that was modified in +90 degree increments in the clockwise direction and the chroma crosstalk from the A head. It is important to recall that the B track crosstalk component is the 629 kHz color-under signal modified in the record processing by shifting it in –90 degree increments (counterclockwise). Crosstalk information detected by both heads is indicated by the dotted arrows. The A head playback energy contains the 629 kHz signal

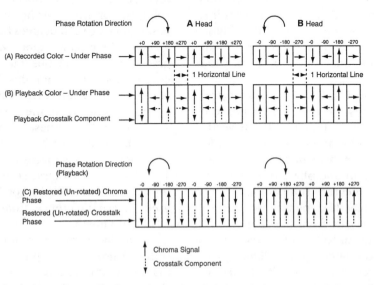

Fig. 6.33. VHS playback chroma crosstalk phase relationships before and after the record chroma phase rotation has been removed.

whose phase is exactly as recorded on the tape (rotated clockwise), plus some signal detected from the adjacent recorded track laid down by the B head (rotated counter clockwise). The chroma energy from the B track and detected by the A head is the crosstalk component. Crosstalk information is again indicated by the dotted arrows in the figure.

Part (C) of Fig. 6.33 shows what happens to the subcarrier phase relationship when the color-under signal is restored to its original NTSC phase alternating condition. Since the subcarrier phase recorded by both heads was rotated, the playback chroma received from both heads must be restored.

During playback, the desired chroma information and the crosstalk component are picked up by the A head and are conducted down the same processing path. Since the A head signal was rotated in 90 degree increments in the clockwise direction during the record process, it must be undone in playback by rotating it in the counterclockwise direction by the same amount. Restoring the A head's detected chroma signal to its original phase also rotates the crosstalk component that was detected by that head in the counterclockwise direction.

The B track chrominance was rotated in the counterclockwise direction during record, so its phase must be shifted back in playback by moving it in the clockwise direction. During playback, restoring the B track chroma phase also alters the phase of the A track chroma crosstalk component picked up by the B head. This is why Fig. 6.33C displays the crosstalk signal for a given head at the same phase for all adjacent horizontal lines. Notice also that the crosstalk component is 180 degrees out of phase with the adjacent track. The end result of the complete record and playback phase rotation process is that the crosstalk component has been modified for cancellation in the comb filter by the playback circuits while the original signal phase was restored. It's important to remember that the only reason the chroma is phase-modified during record is to establish a specific phase relationship for the crosstalk component so the comb filter can cancel it out during playback.

In playback, the A and B head color-under signals and their respective crosstalk components are converted back to the 3.58 MHz color subcarrier frequency. The crosstalk component from each of the heads is also converted to 3.58 MHz, but each adjacent horizontal line is of the same phase. The phase of the desired subcarrier signal is returned to NTSC specifications so that it alternates 180 degrees out of phase on adjacent lines in each field. Once it has been up-converted to 3.58 MHz, the crosstalk component and the desired signal are passed to the comb filter circuitry for crosstalk elimination.

The comb filter and playback color crosstalk elimination for the VHS format
The combination of the record and playback phase rotation processing has created a crosstalk phase relationship that the comb filter can identify and eliminate. Figure 6.34 shows how this is accomplished. The 3.58 MHz up-converted chroma information enters the comb filter circuitry from the left side of the drawing. The chroma information takes two separate paths. The first path is through the nondelayed line and the second is through the delay line. Here the incoming 3.58 MHz up-converted chroma is delayed by 63.5 microseconds (the time period of one horizontal line) before entering the summation circuit.

Chroma crosstalk cancellation occurs inside a summation circuit that is typically a resistor network. Cancellation occurs when two currents of the opposite phase and amplitude meet and are summed together. The desired signal is of the same phase and is also summed together in the same circuit. This is how the level of the wanted subcarrier is doubled.

In Fig. 6.34 the wanted and unwanted signal characteristics are shown at significant locations of the circuit. The phase of the desired signal is shown in the solid arrows. The crosstalk component is depicted with the

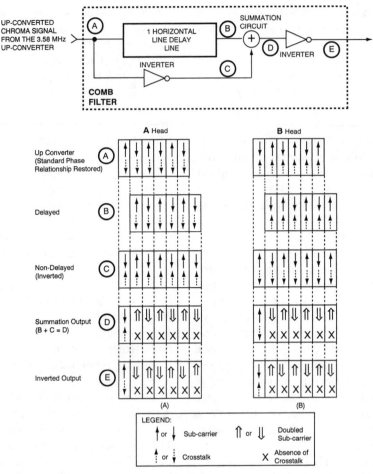

Fig. 6.34. *VHS comb filter operation: (A) when the A head signal is applied to the comb filter, and (B) when the B head signal is applied to the comb filter.*

dashed arrows. The double lined arrows in sections D and E represent the doubled chroma subcarrier amplitudes. The X's indicate the crosstalk component that has been canceled out of the signal.

The inverter closest to point C is used to establish the wanted chroma signal phase for doubling in the summation circuit. It also configures the crosstalk phase for cancellation as part of the summation process.

The signal output from the summation circuit is the product of adding the delayed signal and the nondelayed (and inverted) signal

paths together (B + C = D). Notice that there is still some crosstalk component in the first line of the summation output and that the desired chroma signal was not doubled for that same line. This is because there was no signal coming from the delay line (point B) at that moment. In this instant of time D = C + 0 because the delay line retards the arrival of the first line by 63.5 microseconds (one horizontal line) just as it does any other line. Therefore, the first line does not arrive in time for the correct product to be produced. The desired result does not occur until the second line appears.

The inverter at the output of the summation circuit between points D and E is used to return the phase of the inverted subcarrier back to the value originally input to the comb filter.

For complete crosstalk cancellation, the comb filter must have a color crosstalk signal that is exactly the same amplitude on both of the two sequential horizontal lines being compared. Also, a precise phase relationship of the crosstalk component between the two lines is required. If the crosstalk signal phase between adjacent lines is not precisely the same, complete cancellation will not occur in the summation circuit. While proper operation of the comb filter requires that color crosstalk information be equal in amplitude and maintain a precise relationship between any two lines of a given field, this is not always the case. When these relationships are not perfect, some crosstalk interference will still be present on the screen. Therefore complete crosstalk cancellation is not normally achieved because color information isn't always identical between adjacent lines. In this case, crosstalk is substantially reduced, but not completely canceled. Even under these conditions, comb filters are capable of reducing the crosstalk component well below the typical viewer's perception.

Another important aspect of comb filter operation is that it requires two lines of information for cancellation. Chroma crosstalk is therefore not canceled on the first and last lines of a field on the TV screen. For example, notice that the leftmost arrows representing the summation output signal at point D in Fig. 6.35 indicate that crosstalk is still present for the first line of video output from the summation circuit. This occurs because signal path C has no signal to compare with for cancellation during the time of the first horizontal line. The signal from the first horizontal line following the path at point B has been delayed 63.5 microseconds and will not be arriving until one horizontal line time later. There are two reasons why residual crosstalk on the first and last lines of a field has little effect on the screen display. First, the horizontal lines are extremely fine on a TV screen since 525 lines make up the entire picture. Second, the first and last lines of video information are respectively above and below the normal viewing area of the screen.

There are also two reasons why comb filters do not completely double the color amplitude signal. First, for doubling to take effect, the amplitude of the chroma oscillation on adjacent lines must be identical in amplitude and precisely 180 degrees out of phase. These criteria are rarely met in a normal color program. Second, some minor distortion occurs as the chroma signal passes through the delay line. Both the lack of perfect doubling and the absence of complete crosstalk cancellation in the summation circuit allows some of the adjacent track crosstalk to pass through the comb filter. However, the residual crosstalk and the minor

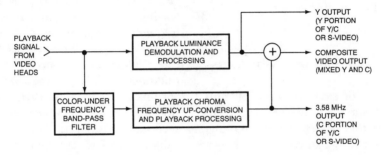

Fig. 6.35. Up-converted chroma output.

chroma distortions imposed upon the wanted chroma signal represent a very small percentage of the color quantity in the picture. Typically, the distortions present in the chroma signal and the remaining crosstalk information itself are not detectable on the TV. On the other hand, if a comb filter is not used in the playback processing, the crosstalk component would be very visible on the screen and the distracting colored beat pattern it overlaid on top of the video scene would be very distracting.

8mm and VHS Chroma Signal Output

After the signal has been up-converted and comb filtered, the playback 3.58 MHz chroma information is ready for mixing with the demodulated luminance signal. Figure 6.35 shows how this is done in the 8mm and VHS VCRs. The composite output is the combined 3.58 MHz chroma and luminance signals. This information is sent directly to the video line output connector and to the radio frequency (RF) modulator for conversion to channel 3 or channel 4 and then transferred to the tuner on the television set.

There is another output available on many VCR models called *Y/C* or *S-Video*. The S-Video (separate video) output provides separate Y/C (where Y represents luminance and C represents chrominance) signals. S-Video output is preferred over the RF modulated channel 3 or 4 and composite video outputs because it provides better playback picture quality.

Chroma Playback Failure Symptoms

Fortunately, the color circuitry is reliable and normally provides many hours of correct operation. Failures that occur in these circuits are normally caused by bad connections or a component external to the color processing

ICs. Integrated circuit failures are rare. When a problem occurs, failures in the 8mm and VHS color playback circuits exhibit similar symptoms.

To diagnose playback color failures, you must have a clear understanding of the circuit operation. Don't be discouraged if you cannot find the failed component right away. Experienced technicians have spent many frustrating hours locating the cause of a chroma circuit problem. This portion of the VCR is one of the most complex. In troubleshooting problems in this section, carefully observe the symptoms, then review the theory explained in this chapter. Finally, associate the symptoms you observe with the various functions of the individual blocks in the playback circuit. Identify the circuit that would cause the symptoms that you are observing if it failed.

The following list provides some basic playback circuit functions with the symptoms revealed when these circuits fail. Assume that the tape used for these examples plays fine on another unit and that the TV is working properly.

1. *No color.* This symptom may be caused by the absence of the color signal anywhere in the playback color processing path. For example, an open 743 kHz band-pass filter in the upper left-hand corner of Fig. 6.17 will stop color from being produced.

Another cause for no playback color may relate to the activation of the ACK because the up-conversion frequency generation circuitry failed. The ACK section prevents color from leaving the circuit if the up-converted color frequency is not phase- or frequency-locked. Therefore, a lack of color playback may indicate a failure in such circuits as the 3.58 MHz reference oscillator, the APC section, the chroma phase rotation (VHS) or phase inversion (8mm), the AFC (VHS) or APC ID (8mm) section, or a Burst ID failure. Each of these circuits requires specific pulses to operate such as burst flag, horizontal sync, and video head switching. An absence of these pulses may therefore cause a lack of color.

If it is accessible from the outside of the color IC, check the control voltage that activates the ACK to see if this circuit is activated. If it is, there is a strong possibility that the problem is in the phase or frequency correction section.

2. *Incorrect color.* This problem shows up as diagonal color bands moving across the screen, rainbow colors in the picture, and incorrect hues. Any or all of these symptoms may be the result of a malfunctioning APC or AFC (VHS), or APC ID (8mm) section. In this case, the ACK detector is not catching the error. The ACK detector may allow the chroma to pass through the system if the phase error is close enough to the reference standard. Missing video head switching pulses at the chroma phase rotation (VHS) section or the phase inverter (8mm), or a lack of horizontal sync or burst flag pulses are possible causes of incorrect playback color.

If the ACK circuit is defeated, it will allow the bad color signal to exit this circuit and be seen on the TV screen. Recall that the AFC in VHS and the APC ID in 8mm are used to correct frequency errors extending beyond the $\pm\frac{1}{2}$ fh (±7.5 kHz) range in the 3.58 MHz up-conversion process. When frequency errors occur, they create diagonal colored bands across the screen. The larger the frequency error, the more bands present. Each diagonal band represents 1 fh or 15.7 kHz. Two bands equal an error of 31.4 kHz, etcetera. Thus, colored diagonal bands can suggest the coarse frequency is off and that the AFC in VHS or the APC ID in 8mm is failing. When the coarse frequency is off the colored bands are the only color that appears on the screen. In other words, the bands are not superimposed on a correctly colored picture. Don't confuse the coarse frequency error colored banding with coarse banding that appears if there is an unwanted beat pattern gener-ated on top of an otherwise normally colored scene. (Some potential

sources for color beating have been previously described.) In the case of beat present in the video, the colored bands are superimposed over the correct colors.

Taking this one step further, assume that the coarse frequency correction in the VCR is working, but the APC is defective. (Remember, the ACK is defeated so it is still letting all color pass—correct or not.) In this case, there will be no horizontal bands, but the colors reproduced on the TV will not be correct. Red objects may appear as some other color, for example. Solid-color objects may change in color as the phase wanders. At times the colors may appear correct, but as time passes the colors may drift to another hue. There could be two or three vertical colored bands that occupy the entire screen. These vertical bands indicate that the coarse frequency is being maintained within the required range, but that the APC is not locking on.

When coarse frequency correction and phase errors like these occur, we don't normally see the symptoms because the ACK steps in and blocks the incorrect chroma from the TV set.

3. *Noisy color.* Noisy color may include multiple colored speckles in the picture, color flashes, or bands of incorrect colors streaking horizontally across the screen. Excessive color noise can also cause incorrect color to appear, so be sure to review the section above for incorrect color as well. Noise can come from a defective VCO, a bad amplifier, a malfunctioning APC or AFC (VHS) or APC ID (8mm) section, a defective chroma phase rotation (VHS) or phase inversion circuit (8mm), or a failure in one of the heterodyne mixers. Playback color noise can be caused by missing or intermittent video head switching pulses, horizontal sync, or burst flag pulses.

One reason the defective color is not blocked by the ACK may be that the noise is presented within an otherwise good signal. It is possible that a correct color signal may be mixed with short bursts of the wrong signal (noise) deceiving the ACK detector into allowing the noise to pass through the system.

4. *Color level fluctuations and flashes.* The ACC is responsible for correcting unwanted chroma level changes. Color level fluctuations in the reproduced picture are normally associated with this circuit. Occasionally a video head that is approaching the end of its useful life will play back with an acceptable luminance signal but will produce considerable chroma level differences between heads that are too large for ACC correction. The result is a chroma level that flickers on the screen at a 30 Hz rate.

Changes in chroma level that appear in horizontal bands or in thin streaks are not typically caused by worn video heads. They are

the result of a component failure or a defective connection.

COLOR-UNDER RECORD

8mm Chroma Record

8mm record automatic frequency control (AFC)

Figure 6.36 shows the 8mm chroma record process. The largest difference between this circuit and the 8mm playback is in the coarse frequency correction process. The 8mm record block uses AFC and the playback circuit uses APC ID for the coarse frequency control.

The record AFC is kept locked on to the frequency by monitoring the incoming sepa-

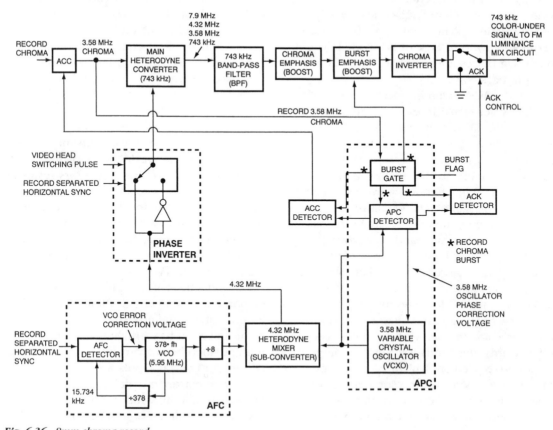

Fig. 6.36. 8mm chroma record.

rated horizontal sync from the record video source. The AFC produces a 378 fh (5.95 MHz) frequency that is phase- and frequency-locked to the input horizontal signal by sampling the VCO's output coming from the divide-by-378 block. Phase and frequency differences produce a correction voltage that brings the 378 fh VCO back into conformity with the horizontal sync reference.

The output of the 378 fh VCO is divided by eight (743 kHz) and applied to the 4.32 MHz oscillator where it beats against the 3.58 APC signal to generate the 4.32 MHz signal.

Shifts in the input video frequency are tracked by the AFC VCO enabling the VCR to record the signal precisely as it is received.

8mm record automatic phase control (APC)

The record APC operation is similar to the playback process except for some minor differences. First, the burst sample used by the APC burst gate comes from a point before the main heterodyne converter rather than from the main converter's output point used in the playback process. Second, there is no APC ID in the record APC because the AFC maintains the frequency of the input source signal. Third, the 3.58 MHz oscillator used in the record APC is a voltage controlled crystal oscillator (VCXO).

Recall that the playback APC circuit uses a 3.58 MHz VCO referenced to a fixed crystal standard. The playback APC requires a VCO because of the wide frequency correction range $\pm\frac{1}{2}$ fh ($\pm$7.5 kHz) necessary for timebase error tracking. An APC with a VCXO has a tighter operation window because the VCXO correction range is much narrower than that of a VCO. During record, the input signal is assumed to be relatively stable, so the record APC does not need the wide frequency correction range available from a VCO. VCXOs offer accurate error tracking, but are limited in the amount of error correction they can provide. They are not able to track the larger time-base instabilities produced during playback.

The record APC circuit monitors the phase of the input video color burst signal and compares it with the 3.58 MHz VCXO frequency. Differences between the two frequencies generate an error voltage that brings the VCXO output back into line with the input burst. During record, the source input burst serves as the oscillation reference signal for the APC detector and the VCXO frequency sample is the feedback signal.

The output frequency from the APC VCXO is applied to the 4.32 MHz heterodyne submixer where it is beat with the AFC's 743 kHz signal to produce the 4.32 MHz frequency.

8mm record sub and main heterodyne converters

The 4.32 MHz mixer's heterodyne output frequency does not contain the same jitter that was present in the playback mode. The subconverter's output does alter as the input video signal subcarrier frequency and phase varies, however. The input chroma signal is normally quite stable because the broadcast television signal being recorded is very reliable.

The output of the 4.32 MHz subconverter is fed to the phase inverter where the phase of the signal is altered for crosstalk elimination during playback. Here the carrier signal of every other horizontal line recorded by the channel 2 head is inverted. The inverter logic does not alter the phase of the record signal when the channel 1 head is in contact with the tape.

The output of the phase inverter is passed to the main heterodyne frequency downconverter where it beats with the incoming 3.58 MHz chroma signal. The frequencies output from the down-converter are the main input frequencies (3.58 MHz and 4.32 MHz), the sum frequency (7.9 MHz), and the difference frequency (743 kHz). Only the 743 kHz frequency is desired so the heterodyned output is sent to a 743 kHz band-pass filter where all other frequencies are stripped out. After band-pass filtering, the color-under signal is passed to the chroma and burst emphasis amplifiers.

8mm burst and chroma emphasis

Chroma emphasis enhances the chroma signal sidebands during record to improve the playback chroma signal-to-noise ratio. The characteristics of chroma emphasis are shown in Fig. 6.37. During playback, the improved chroma signal-to-noise ratio is achieved when the chroma signal is de-emphasized. By returning the boosted chroma levels to their original lower levels, noise picked up by the video heads during playback and noise generated in the playback up-conversion process are lowered simultaneously in the de-emphasis process. This places the visible noise amplitude at

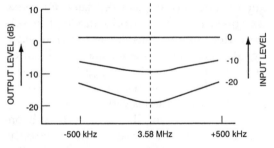

Fig. 6.37. *8mm chroma record emphasis characteristics.*

a lower level than it would be if the signal did not require de-emphasis.

During record, burst emphasis doubles (6 dB) the burst level. The burst emphasis circuit can identify the burst signal by the burst gate pulse applied to the amplifier.

Burst is emphasized during record to increase the burst signal-to-noise ratio in the playback process. The improved signal-to-noise ratio is achieved by reducing (de-emphasizing) the level during playback. The level of noise detected by the heads during playback is also reduced as the burst is de-emphasized. An improved burst signal-to-noise ratio enables the

playback circuits to accurately process the chroma signal.

8mm record automatic color control (ACC) and automatic color killer (ACK)

The record ACK circuit prevents a color signal from being recorded on the tape if the AFC or APC circuits lose lock. When the ACK is active, the signal recorded will be black and white.

ACK operates by comparing the APC error with a sample of the source input burst. The input record signal serves as the comparison reference. If the phase error exceeds predetermined bounds, the ACK detection voltage toggles the ACK switch to the ground position eliminating the color output to the record heads.

If the record AFC fails to achieve lock, the APC cannot lock on either. Thus, the ACK detector can monitor the status of both the AFC and APC by just observing the APC locking condition.

The ACC section works to stabilize unwanted chroma level changes present in the input video line. This is done by monitoring the input burst level and adjusting the amplification of the ACC section to keep burst constant. Unwanted chroma level shifts will alter both the burst and the chroma subcarrier signal proportionally. By applying corrections to restore the proper burst amplitude, the subcarrier representing the chroma saturation level in the picture will also be corrected.

The record ACC has a much slower response time than the playback ACC. Playback ACC reacts to rapid changes occurring from head-to-tape contact variables caused by tape surface flaws, head response differences, etcetera. The record ACC does not have the rapid change to overcome because the input signal normally comes from a stable source. Changes in record chroma level errors are typically gradual and are thus easily handled by the slower ACC response.

VHS Chroma Record Circuits

Figure 6.38 is a block diagram of the basic VHS color-under record circuitry. This drawing shows one circuit that has been popular. Here the combined luminance (Y) and chrominance signal enters a low-pass filter as shown in the upper left of the diagram. The low pass filter allows only Y information to pass so it can be frequency modulated and sent to the video head for recording. The combined Y and chroma signal is also passed to a band-pass filter where only the higher chroma frequencies are passed. The 3.58 MHz chroma subcarrier signal is heterodyned with a 4.21 MHz oscillation signal and converted to the 629 kHz difference signal. This color-under subcarrier information is recorded directly onto the tape using the FM Y signal as bias.

VHS record chroma automatic frequency control (AFC)

The 4.21 MHz oscillation signal is created from the sum of two signals—the 629 kHz from the AFC loop and the 3.58 MHz from the APC loop. In this circuit, a 2.517 MHz voltage controlled oscillator produces a signal that is 160 times the horizontal sync (160 fh). It is divided by four to yield 629 kHz. The 629 kHz output of the divide-by-four circuit splits to provide inputs to the phase rotation circuit generating the 629 kHz output for the 4.21 MHz signal generator, and to a divide-by-40 circuit that generates a 15.734 kHz input to a phase comparator. The AFC's phase comparator matches the separated horizontal sync coming from the source input video signal with the 15.734 kHz signal fed back from the VCO output to generate a DC correction voltage. The correction voltage phase-locks the 2.517 MHz VCO with the separated horizontal sync pulses from the incoming video.

The output of the 160 fh VCO is divided by four to produce a 40 fh (629 kHz) signal that is applied to the record chroma phase rotation circuit. (Record chroma phase rotation was previously described.) Shifts in the input video frequency are tracked by the AFC VCO enabling the VCR to record the signal exactly as it is received. This is possible because changes in the input video's chroma frequency generate changes in the input horizontal sig-

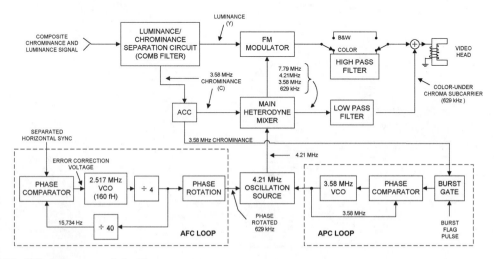

Fig. 6.38. VHS chroma record circuitry.

nal. Thus, the record AFC is able to track the incoming signal and cause the VCR to record the color signal precisely as it is received.

VHS chroma record automatic phase control (APC)

The other input to the 4.21 MHz subconverter is 3.58 MHz from the automatic phase control (APC) loop. A 3.58 MHz VCO produces this oscillation signal. One input to the APC circuit is burst. It provides a sample of the incoming record chroma signal. The APC identifies chroma burst with the assistance of another input pulse—burst flag. The APC detector compares the input video burst phase with the phase of the 3.58 MHz VCO and makes any corrections needed by causing a DC control voltage change at the 3.58 MHz VCO frequency and phase-locking the color conversion. The color 4.21 MHz conversion signal is thus frequency- and phase-locked to the incoming video signal.

VHS color record sub and main heterodyne converters

The 3.58 MHz oscillation output of the APC circuit is heterodyned with the 629 kHz oscillation from the AFC circuit to produce the 4.21 MHz oscillation. In this sub-converter, the 4.21 MHz signal is frequency- and phase-locked to the incoming color signal.

In record, the 4.21 MHz heterodyne oscillation signal doesn't contain the same jitter that was present in the playback mode. However, the heterodyned output does alter according to the input color video signal's subcarrier frequency and phase variances. The input chroma signal is normally quite stable because the broadcast television signal being recorded is very reliable.

Current VHS color record circuits

Figure 6.39 describes a current color record circuit. Digital circuitry is used in this processor to provide AFC and chroma phase rotation.

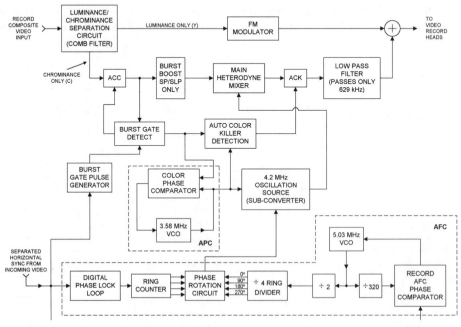

Fig. 6.39. *VHS digital chroma record processing is similar to earlier methods.*

The main difference in this AFC circuit is in the operating frequency of the AFC VCO. It runs at twice the 2.517 MHz frequency (5.03 MHz or 320 fh). The similarities between this circuit and previous circuits becomes evident once you realize that each of the frequencies used in these designs is a multiple of the horizontal line rate (15.734 kHz) and the fundamental 629 kHz AFC value.

The APC circuit in this design is the same one found in earlier models. The color phase comparator samples the input burst and compares that signal with the 3.58 MHz VCO frequency.

Figure 6.39 also shows other blocks used in the VHS chroma record section. These include the burst boost, ACC, ACK, low-pass filter, and the band-pass filter.

Other VHS record circuits

629 kHz low-pass filter

The output of the 4.21 MHz oscillation generator is passed to the main heterodyne frequency down-converter where it beats with the incoming record 3.58 MHz chroma signal. The outputs from the down-converter are the main input frequencies (3.58 MHz and 4.21 MHz), the sum frequency (7.79 MHz), and the difference frequency (629 kHz). Only the 629 kHz frequency is desired so the heterodyned output is sent to a 629 kHz low-pass filter where all other frequencies are blocked.

After low-pass filtering, the color-under signal is input to the burst emphasis amplifier. Here the burst signal is identified by a burst gate pulse and boosted to twice its amplitude (6dB).

Burst boost

Burst is amplified during select record speeds to increase the burst signal-to-noise ratio in the playback process. During playback, an improved signal-to-noise ratio is achieved by reducing (de-emphasizing) the signal level back to its original strength. Noise occurring around the color-under burst frequency and detected by the heads during playback is also reduced in level as the playback burst is de-emphasized. An improved burst signal-to-noise ratio helps the playback circuits produce accurate chroma signal processing.

As explained earlier, only the SP and SLP record mode burst is boosted.

Record automatic color control (ACC) and automatic color killer (ACK)

The ACC section works to stabilize unwanted chroma level changes in the input video. It accomplishes this by monitoring the input burst level and adjusting the amplification of the ACC section to keep burst constant. Unwanted chroma level shifts will alter both the burst and the chroma subcarrier signal proportionally. When the correct burst amplitude is restored, the subcarrier representing the chroma saturation level in the picture is also corrected.

The record ACC has a much slower response time than the playback ACC. Playback ACC is required to correct for rapid changes in head-to-tape contact variables caused by tape surface flaws, head response differences, etcetera. The record ACC doesn't have the same challenges to overcome because the input signal is normally from a stable source. Record chroma level errors are more gradual and are easily handled by the slower response record ACC circuitry.

The record ACK circuit prevents any color signal from being recorded on the tape if the AFC or APC circuits lose lock. When the ACK activates, the recorded signal is black and white.

ACK operates by comparing the APC error with a sample of the source input burst. The input record signal serves as the comparison reference. If the phase error exceeds predetermined bounds, the ACK detection voltage causes the ACK to block the chroma, eliminating the color output to the record heads.

If the record AFC fails to achieve lock, the APC cannot lock on either. Thus, the ACK detector can monitor the status of both the AFC and APC by just observing the locking condition of the APC.

After passing through the record ACK, the color-under signal is mixed with the frequency modulated luminance signal and passed to the video record heads.

Symptoms of 8mm and VHS Chroma Record Failure

Circuits in the color record section are reliable and will normally provide many hours of proper operation. Failures that occur in these circuits are often caused by bad connections or a component external to the color processing ICs. Integrated circuit failures are rare. However, when color record failures occur in the 8mm and VHS color playback circuits, they exhibit similar symptoms.

Failures in the record chroma circuitry can create confusion in determining if the problem is in the record or playback color processing sections. Confusion occurs for two reasons. First, consumer VCRs bypass the color-under and color-up conversion processes when viewing a television program through the VCR. This means that any time the VCR is in record, or just passing a television signal through it without recording the program, the color circuits are bypassed. Thus the VCR can have a color record problem and yet display a perfect picture until the operator decides to play back the recording just made. Second, record chroma failures generally produce abnormal color reproduction when playing back a tape known to have good color on it. This is because many of the circuits used in the color section are common to both record and playback processing modes. Obviously, a failure in one of these common components will affect both modes of operation. Often, you can differentiate between record and playback color circuit failures by recording a program and playing back that recording on another machine. If the color program plays back on the second VCR correctly, the record circuitry can be assumed to be functioning properly. Playback related failures can be isolated by playing back a previously recorded tape known to have good color in the suspect VCR.

To successfully diagnose record color circuit problems, you must have a knowledge of the circuit operation. Don't be discouraged if you cannot find the failed component right away. Experienced technicians have spent many frustrating hours locating the cause of a chroma circuit defect. The VCR record color circuitry is complex. When troubleshooting problems in this section, carefully observe the symptoms, review the theory, and then associate the symptoms you observe with the various responsibilities of the individual blocks in the playback circuit. Determine what circuit would cause the symptoms that you are observing if that circuit failed.

The following are some basic record circuit failure symptoms. In each of the examples, assume that the playback circuit operation may also be affected by the failure, so you will be playing back the tape on a machine known to be functioning properly.

1. *No color.* This symptom may be caused by the absence of the color signal anywhere in the record color processing path. For example, an open 3.58 MHz band-pass filter in the upper left-hand corner of Fig. 6.39 will stop color from being produced.

The absence of record color may also relate to the activation of the ACK due to a failure in the down-conversion frequency generation circuitry. The ACK section prevents color from leaving the circuit if the down-converted color frequency is not frequency- or phase-locked. Therefore, a lack of record color may indicate a failure in such circuits as the

3.58 MHz reference oscillator, the APC section, the chroma phase rotation (VHS) or phase inversion (8mm), the AFC, or the APC section. To operate properly, each of these circuits requires specific pulses such as burst flag, horizontal sync, and video head switching. An absence of one or more of these pulses may therefore cause a lack of color.

If it is available at one of the chroma IC pins, check the control voltage that activates the ACK to see if this circuit is activated. If it is, there is a strong possibility that the problem is in the phase or frequency correction section.

2. *Incorrect color.* This problem includes such symptoms as diagonal color bands moving across the screen, rainbow colors in the picture and incorrect hues. Any one or all of these symptoms may be the result of a malfunctioning AFC section. In this case the ACK detector is not catching the error and the erroneous signal is being recorded. The ACK detector may allow the chroma to pass through the system if the phase error is close enough to the reference standard. Missing video head switching pulses at the chroma phase rotation (VHS) section or the phase inverter (8mm), or a lack of horizontal sync or burst flag pulses can cause incorrect playback color.

If the ACK circuit is defeated, the bad color signal can exit this circuit and be recorded on the tape. Remember that the AFC is used in both the VHS and 8mm formats to correct frequency errors extending beyond the range of the APC correction window. This is helpful when troubleshooting defects in the 3.58 MHz down-conversion process. When frequency errors occur, they create diagonal colored bands across the screen. The larger the frequency error, the more colored bands present in the recorded video. Each diagonal band represents 1 fh or an additional 15.7 kHz error in frequency. Two bands equal an error of 31.4 kHz, etcetera. Thus, colored diagonal bands can suggest that the converted subcar-

rier frequency is off. This is the responsibility of the AFC section. When the coarse frequency is off, the colored bands are the only color that appears on the screen. In other words, the bands are not superimposed on a correctly colored picture. Don't confuse the AFC frequency error colored banding with banding that may also appear if there is a colored beat pattern generated on top of an otherwise normally recorded color scene. (Some potential sources of color beating were described previously.) In the case of colored beat present in the video, the colored bands are superimposed over the correct colors.

Take this one step further and assume that the coarse frequency correction in the VCR is working, but the record APC is defective. (Remember, the ACK is defeated, so it is still letting all color pass to the record heads—correct or not.) In this case, there will be no horizontal bands but the colors reproduced on the TV during playback will not be correct (e.g., red objects will appear in some other color). Solid color objects may change in color as the phase wanders. At times the colors may be correct, but as time passes, the colors drift to another hue. There could also be two or three vertical colored bands that occupy the entire screen. These vertical bands indicate that the AFC is holding the coarse frequency within the required range of the APC, but that the APC is not locking on.

When coarse frequency correction and phase errors occur, you don't normally see the symptoms because the ACK steps in and blocks the incorrect chroma from being recorded on the tape.

3. *Noisy color.* Noisy color may include multiple colored speckles in the picture, color flashes, or bands of incorrect colors streaking horizontally across the screen. Excessive color noise may cause the colors to appear incorrect. Color noise can come from a defective VCO, a bad amplifier, a malfunctioning APC or

AFC section, a defective chroma phase rotation (VHS) or phase inversion circuit (8mm), a failure in one of the heterodyne mixers, or a defective video head. Missing or intermittent video head switching pulses, horizontal sync, or burst flag pulses are possible causes of record color noise.

One reason the noisy color is not blocked by the record ACK may be that it is superimposed on an otherwise good signal. It is possible that a correct color signal may be mixed with short bursts of the wrong signal (noise) deceiving the ACK detector into allowing the noise to pass out to the heads for recording.

4. *Color level fluctuations and flashes.* The record ACC corrects for unwanted input chroma level changes. Color level fluctuations in the recorded picture are normally associated with a failure in the ACC circuit. It is possible that worn heads may cause record chroma level fluctuations. Occasionally a video head approaching the end of its useful life will record an acceptable luminance signal but will produce considerable playback chroma level differences between heads. In this case the playback ACC will not be able to correct for the large record level differences and will generate a 30 Hz color flicker in the playback picture.

Changes in recorded chroma level that appear in horizontal bands or in thin streaks in the playback picture are not typically caused by worn video heads. They are frequently the result of a component failure or a defective connection in the circuit.

CHAPTER REVIEW QUESTIONS

1. Why must the input color sub-carrier signal be down-converted to a lower frequency?

2. What is the function of the record and playback APC?

3. The VHS format uses AFC during playback to perform what two functions?

4. The 8mm format does not use AFC during playback. What operation corrects for gross frequency errors during playback?

5. Why isn't the chroma record signal emphasized in the LP mode for VHS machines?

6. Why is chroma burst emphasized during record?

7. Why is ACC important during playback?

8. Why is color burst used as the ACC reference?

9. ACK action is enabled by monitoring the status of what circuit?

10. The NTSC color video signal causes the 3.58 MHz color sub-carrier signal to alternate phase by 180 degrees on each adjacent line. Why?

11. Why must the color-under crosstalk signal be canceled electronically?

12. List at least four signals used by the color-up conversion process to stabilize the up-converted color signal during playback.

Chapter 7

VCR Luminance Processing Theory

The last chapter covered the theory and operation of VCR color processing. Chapter 7 will take you through VCR luminance. In this chapter you'll learn about recording and playing back the FM luminance signal and recent designs that improve resolution of VHS and 8mm video tape images.

RECORDING THE FM LUMINANCE SIGNAL

VCRs use frequency modulation to record the luminance portion of the signal as described in Chapter 5. In VHS units, the modulating frequency varies between 3.4 MHz (corresponding to "sync tip") and 4.4 MHz (corresponding to 100 percent white). Standard 8mm units use an FM deviation of 1.2 MHz, ranging from 4.2 MHz for sync tip to 5.4 MHz for peak white. The RF (radio frequency) in the VCR's FM deviation range contains the basic picture information. Most of the luminance signal's fine detail is contained within the sidebands and

must be retained to reclaim the high-frequency components that make up the video picture detail. Most of the upper sideband is lost in the record and playback process due to the effects of writing speed and frequency response losses within the video head itself.

Since the ability to retrieve the upper sideband of the recorded signal is limited, the lower sideband is used to recover the fine detail in the signal. Reproduction of the high video frequencies, which contain the finer detail in the picture, is thus dependent upon the machine's ability to recover the lower sideband. Before the prerecorded signal is frequency modulated, the fine detail of the video signal is boosted to emphasize the high-frequency components and to minimize the loss of resolution due to the limited bandwidth of the lower sideband. During playback the signal is de-emphasized to restore normal video.

While recording a color video program, the extreme lower portion of the FM sideband is chopped off at about 1.3 MHz to make room for the color-under signal and its sidebands. In some VCRs, the lower sideband extending below 1.3 MHz is allowed to pass when a black-and-white-only signal is detected. Making the full lower FM sideband available in the black and white mode provides higher resolution (more picture detail).

Figure 7.1 is a block diagram describing the processing steps for the Y or luminance portion of the video signal during record. This signal first encounters a Y-C filter that separates the Y and chroma signal. In Fig. 7.1A the chroma information consists of the high-frequency (3.58 MHz) portion of the video signal. Picture brightness and detail information are contained in the lower frequency portion (below 3.58 MHz) of the video band. The lower frequency video information is called *luminance* and designated with Y.

Figure 7.1B shows a comb filter design being used as a Y-C separator. This is typical in many VCR models. Comb filters provide excellent chroma and Y signal separation while preserving much of the high-frequency portion of the Y signal. The comb filter requires an automatic gain control corrected signal with a clamped voltage level input to function correctly so circuits using a comb filter as a Y-C

separator have these correction circuits placed in the front end of the Y record path.

Y Automatic Gain Control (AGC)

The levels of separated Y signal must be controlled before it can be modulated. Video levels vary substantially between recorded programs. To provide proper signal production, the FM modulator must have a video signal applied that varies within limitations regardless of the level of signal coming into the VCR. AGC (automatic gain control) ensures that the video signal supplied to the modulator is within required limits. Incoming video must be stabilized to 1 volt peak-to-peak between sync tip and 100 percent peak white.

The Y AGC circuitry performs complicated tasks to ensure that the incoming video signal is within the limits specified by the modulator. However, AGC must not react to normal variations in reproducing scene brightness. If the AGC were simply to detect and limit peak video levels, normal peak brightness would be reduced while average dark scenes would be brighter. This would produce normal scenes with brightness levels much darker than they

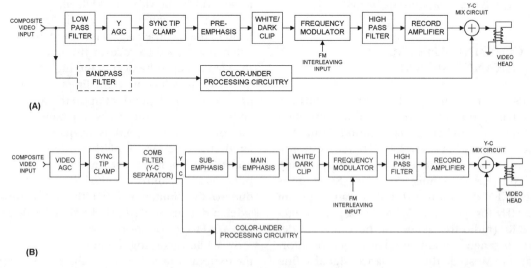

(A)

(B)

Fig. 7.1. *(A) The Y record circuit. (B) A Y record circuit with the AGC process occurring before comb filter Y-C separation.*

should be while dark scenes would have unusually high levels of brightness.

AGC is designed to correct for errors in overall amplitude without interfering with normal brightness variations. To accomplish its task, AGC monitors and responds to variations in the horizontal synchronization pulse amplitude. In a 1 volt peak-to-peak standard video signal, horizontal sync is fixed at a constant 0.3 V amplitude. This leaves the remaining 0.7 V for picture brightness levels. Picture brightness may vary with scene content but the sync amplitude levels should remain constant. Thus sync amplitude detection is a reliable way to monitor video level error in the incoming video signal.

Sync detect AGC is achieved by separating the incoming video horizontal sync, inverting, amplifying, and delaying the sync pulse, and reinserting the pulse in the horizontal blanking interval. This inverted and amplified horizontal sync pulse is applied during the horizontal blanking interval so it won't interfere with video information presentation. This inverted pulse is only used in the video AGC circuit for level detection. It's not recorded on the tape. This inverted and delayed horizontal sync pulse is amplified to provide 1 V peak-to-peak between sync tip level and the positive going peak of the inverted sync pulse.

Figure 7.2 shows the relationship between the incoming video signal containing both Y and chroma information; the Y-only signal after color has been filtered out; the separated horizontal sync; the horizontally delayed, amplified, and inverted horizontal sync; and

the combined signal applied to the AGC detector. The AGC detector monitors the relationship between the horizontal sync tip level and the internally generated 100 percent peak white reference level. It also adjusts the AGC amplifier to maintain a 1 V peak-to-peak detection level. As the video signal levels from the tuner vary between stations, or as the incoming recorded video signals differ between other external video input sources, the AGC makes the appropriate corrections to maintain the 1 V peak-to-peak reference.

The AGC detector is a peak detection circuit that is not influenced by normal variations in video scene brightness. If the video level exceeds 0.7 V peak-to-peak, the AGC detection circuit recognizes this as having exceeded the reference 1 V peak-to-peak signal internally generated in its circuitry. As the peak level is exceeded, the AGC amplifier begins to compress the sync level signal applied to the modulator. When the video portion of the input signal is more than the 100 percent peak white reference level, the AGC circuit acts to keep the video component at a constant amplitude. To do this, the AGC must compress the sync level slightly and add this compressed pulse level to the 100 percent peak white reference level to keep the total video information within the 1 V peak-to-peak signal required by the modulator. Figure 7.3A describes the video level before sync detect AGC correction is applied. Figure 7.3B shows the video level after AGC correction.

Sync Tip Clamp

The functional block to the right of the Y AGC in Fig. 7.1 is the sync tip clamp circuitry. Once the incoming video signal has been level-corrected by the AGC circuit, the negative peak of the video sync tip voltage is held to a predetermined level in the clamping circuit. Clamping is required to eliminate DC voltage shifts that occur in a video picture every time the brightness level changes. Sync tip must not vary from

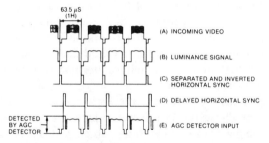

Fig. 6.38. AGC circuit waveforms.

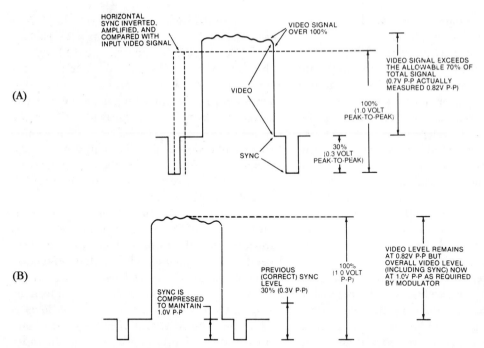

Fig. 7.3. Automatic gain control (AGC) compresses sync. (A) Video level before sync detect AGC correction is applied. (B) Video level after sync detect AGC correction.

a prefixed (clamped) voltage level. If it does, the modulator will generate a deviation frequency that is different from the desired frequency each time sync tip, offset by a new DC voltage, arrives at the modulator. Clamping the sync tip level also holds the incoming video signal within a predetermined voltage range for the FM circuit.

Some circuits use the sync tip clamp in the Y record process circuitry prior to AGC detection.

Comb Filter

The comb filter in Fig. 7.1B separates the Y from the C by recognizing the 180 degree out-of-phase relationship of the chroma subcarrier between adjacent lines. Comb filter Y-C separation is typically performed inside an IC. The operation of the comb filter was previously described. The primary difference between the standard comb filter and the Y-C separation comb filter used here is that the Y-C separation comb filter circuit is designed to preserve both Y and C signals. The circuit acts on the incoming composite waveform to separate the two signals and apply each to its own distinct path.

Sub and Main Emphasis

In the processing circuitry of Fig. 7.1, the record video signal passes from the Y-C separator comb filter into the sub and main emphasis circuits. The sub-emphasis section improves the signal-to-noise ratio of the low-level and high-frequency portions of the video signal. The main emphasis circuit (commonly called pre-emphasis for pre-frequency modulation emphasis) maintains good high-frequency response of the video signal during the record/playback process. Retention of the high-frequency content allows the VCR to reproduce fine detail in the playback picture. Since sub-emphasis acts to improve upon main emphasis, it will be described following main emphasis.

Main Emphasis and Pre-Emphasis

The main emphasis circuitry keeps the high-frequency content of the picture present by forcing the modulator to shift rapidly to a new modulating frequency every time there is a change in contrast. A video scene containing a black telephone pole against a white background is an example of a significant change in contrast. Contrast changes may also be more subtle such as varying shades of gray.

The high frequency (picture detail) is preserved because the main emphasis circuit emphasizes the area of sudden voltage changes in the video level that occur with the change in contrast. These emphasized excursions cause the modulator to rapidly change to the new FM frequency representing the new video level. If proper amounts of pre-emphasis are not applied, the modulator will change to the new frequency at a slower rate, generating smeared video at contrast level transition points. For example, when viewing a playback reproduction of the color bar test pattern, the playback Y signal would have rounded corners instead of well-defined corners on the video stair step. These square corners represent the high-frequency portions of the video signal. Main emphasis plays an important part in the recovery of picture detail because it aids the VCR in reproducing high frequencies, which would otherwise be lost in the record/playback process due to upper and lower FM sideband limitations.

Pre-emphasis levels must be carefully selected to maintain a suitable signal-to-noise ratio and to minimize the effects of spurious noise caused by too much emphasis of the video signal prior to frequency modulation (FM pre-emphasis). If too much pre-emphasis is applied, the FM modulator will be overdriven creating demodulation noise (black and white horizontal streaks) in the playback signal. Crosstalk from adjacent video tracks is also increased when too much pre-emphasis is applied. The trick is to emphasize the high-frequency portions of the signal just enough to provide good playback frequency response and yet keep the signal-to-noise ratio as high

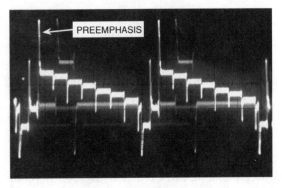

Fig. 7.4. *Video waveform with main emphasis.*

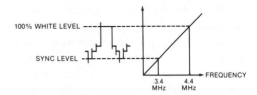

Fig. 7.5. *Relationship between VHS format main emphasis and FM modulation.*

as possible. Figures 7.4 and 7.5 show the effects of the main emphasis circuit and its relationship to the FM deviation frequencies.

Notice in these figures that the pre-emphasis spikes cause the FM modulator to exceed the 3.4 MHz and 4.4 MHz frequencies used for the video waveform. The VCR system must be capable of reproducing this increased FM bandwidth so it can recover the emphasis portion of the signal.

The relationship of main emphasis to the video signal is shown in Fig. 7.6. From left to right, notice the horizontal sync pulse that rises to the video reference black level. This is known as the video setup or pedestal level.

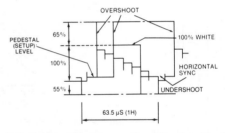

Fig. 7.6. *Relationship between the video signal and main emphasis.*

The setup level corresponds to pure black as viewed on the TV screen. There is an upward spike at the point where the video level rises from sync tip to the setup level. This spike is the applied main emphasis signal. Notice in the video information portion of the waveform that each time the signal goes from a dark level to a white level, the main emphasis pulse goes in the same direction. It overshoots the level of the original video signal. After the signal has reached the 100 percent white level, it begins to gradually step back down to the video setup level. This waveform is called a video stair step. The main emphasis circuit causes negative-going spikes to occur each time the video signal goes from a white signal to a darker signal.

The amount of main emphasis applied is determined by the rate of change between levels in a signal (time constant) and the amount of change (level) present in the same signal. Thus main emphasis levels are predetermined and injected into a signal stream in preset amounts according to the rate of signal change and the level of that same change.

Sub-Emphasis

The sub-emphasis circuit, included in the preemphasis block, operates differently than the main emphasis circuit. Sub-emphasis improves the signal-to-noise ratio of the playback picture. Unlike the main emphasis circuit, which reacts linearly to all changes in luminance, sub-emphasis operates in a nonlinear fashion. It concentrates on video signals with very low amplitude changes (minimal contrast level changes) in the higher video frequency range. These high-frequency signals are the fine detail and low-level contrast changes in a video scene. The finely pitched subtle changes in luminance visible on a painted and textured dry wall finish in your home is an example of this low-level and high-frequency signal. The small size and close proximity of the bumps in the textured finish represent the high-frequency detail, while the slight changes in brightness between the bumps and the rest of the wall surface represent small differences in the luminance level.

By definition, noise is an unwanted high-frequency and low-contrast level signal that can be present in a video signal. Unfortunately, noise and the desired low-level fine detail in the picture are similar in content. When video noise becomes too prevalent, the low-level fine detail gets lost in the reproduced scene. The greatest potential for noise generation occurs in the playback process, especially in the RF preamplification stages where the signal from the playback heads is first amplified.

The visibility of the noise originating in playback can be reduced by compensating for it in the record process. Figure 7.7 shows the nonlinear response characteristics for the 8mm and VHS systems. When the record signal is applied to the nonlinear sub-emphasis circuit, the desired low-level signal is increased according to the nonlinear response curves.

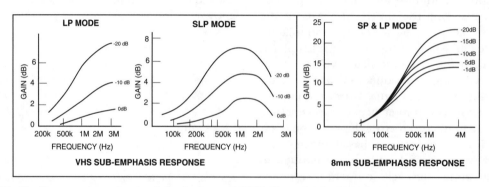

Fig. 7.7. Nonlinear response characteristics for 8mm and VHS sub-emphasis circuits.

The dB scale on the left side of the graphs shows the gain applied to the input signal, while the horizontal axis represents the frequency of the input signal. The numbers on the right side of the graphs indicate the levels of the input video signal. Notice that lower video input levels are boosted more than higher video input levels. Also note that the higher frequency signals receive a different amplification. There is sub-emphasis in the VHS SP mode, but the response level is much lower than for the LP and SLP (EP) modes.

During playback, the sub-emphasized video must be returned to its original values. This is done in the sub-de-emphasis section after demodulation. During playback, the sub-de-emphasis circuit not only receives the playback video signal, but it also receives the noise produced in the head amplifier circuitry. The sub-de-emphasis section restores the video signal by subjecting both the playback video and the introduced noise to an amplification stage with a designed response characteristic that is a mirror image of the graphs shown in Fig. 7.7. This is where the signal-to-noise improvement is made. While the video levels are being reduced back to normal, noise present on the same signal path is also reduced by the circuit. If the video signal has to be de-emphasized by –15 dB then the noise with similar amplitude and frequency characteristics will also be reduced by that amount. This action places the normal noise level below our range of visual perception.

White Clip/Dark Clip

The sub and main pre-emphasized signal is then passed to the white and dark clip sections as shown in Fig. 7.1B. White and dark clipping is used to control the amount of pre-emphasized overshoot. Pre-emphasizing the signal creates an overshoot in both the white and dark directions. If these excessive overshoots were allowed to enter the modulator, it could be overdriven, resulting in a streaking video display where transitions occur between black

and white. This streaking (also known as *demodulation noise* or *bearding*) would be most evident at points of large transitions between white and black. White clip and dark clip adjustments limit the pre-emphasis pulse excursions as shown in Fig. 7.8.

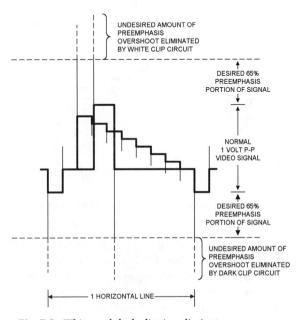

Fig. 7.8. White and dark clipping eliminates excess overshoots.

The FM Modulator

The FM modulator receives the processed signal and converts it into a signal whose frequency deviates in accordance with the incoming video level voltage. A frequency modulator is a form of *voltage controlled oscillator* (VCO) as shown in Fig. 7.9. A VCO converts an incoming voltage to a frequency. Each individual voltage level detected at the VCO's input produces a specific frequency. In the case of a frequency modulator, the VCO operates at a *free run frequency* without an input signal. In a VCR, this free run frequency is called the *carrier frequency* or *sync tip frequency*. The frequency output from the frequency modulator increases

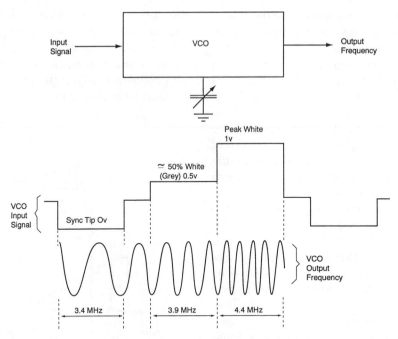

***Fig. 7.9.** A frequency modulator is a special purpose VCO.*

linearly in proportion to the increase of the input voltage. This change in frequency is called the *deviation frequency*. In a VCR, the maximum deviation frequency corresponds to the peak white video levels in the picture. Thus, the total FM deviation range extends from the sync tip to the peak white levels. In a conventional VHS machine, the sync tip frequency is 3.4 MHz and the peak white frequency is 4.4 MHz. Every video level occurring in the picture will fall within the sync tip and peak white levels. This means that, excluding the FM frequencies generated by the pre-emphasis spikes, the FM deviation frequencies will only range between 3.4 and 4.4 MHz.

Let's analyze how a video signal is frequency modulated. In our example, assume that the video signal voltage input to the frequency modulator ranges between 0 volts for sync tip and 1 volt for peak white. Also assume that the VCO sync tip frequency is 3.4 MHz and the total deviation output for peak white (1 volt) is 4.4 MHz. (These are the conventional VHS modulation specifications.)

When the sync tip voltage (0 volts) is sensed at its input, the VCO in Fig. 7.9 outputs a 3.4 MHz sync tip (carrier) signal. When the input voltage rises to 0.5 volts, the device produces a frequency of 3.9 MHz (500 kHz is one-half of the 1 MHz total deviation and 3.4 MHz plus 500 kHz equals 3.9 MHz). Finally, when the VCO input reaches 1 volt, 4.4 MHz is produced at its output. Each input level between 0 and 1 volts applied to the FM modulator (VCO) produces a corresponding and specific frequency output within the 3.4 MHz to 4.4 MHz range.

The color video signal is measured in units of voltage amplitude and consists of many complex signals. These signals include the chroma subcarrier, brightness or luminance (Y), horizontal and vertical synchronization pulses (sync), and burst. In consumer VCRs, the Y (including the horizontal and vertical sync) is processed and frequency modulated by itself, without the chroma subcarrier and chroma burst. Recall that the color information is separated for processing in the color-under circuitry.

Picture detail (resolution) is a result of rapid video signal changes from one brightness level to another. For example, consider a video scene containing a dark telephone pole standing in front of a brightly lit building. The changes in video level relate to a change in voltage level within the video signal. To reproduce the sharp transition between the telephone pole and the building, the frequency modulator must make a rapid change in its output frequency when the different voltages corresponding to the white and black video are detected at the VCO input. In a frequency modulation system, detail reproduction is thus dependent upon the modulator's ability to suddenly switch the output frequency to the rapidly changing input video signal levels. (Remember that the main pre-emphasis circuit in a VCR exists to help shift the FM VCO into this new frequency quickly.) If the VCO in the FM circuit changes frequencies slowly as the transitions between the dark telephone pole and the white building are reached, the video reproduced by the system would not contain clear and sharp edges at the luminance transition points. Instead, the video would appear to transcend through various shades of gray until the new video level is finally achieved. Our eyes would observe these gradual transitions as a smearing or an object with soft edges. The detail (resolution) would be lost.

In a conventional 8mm system, the Y portion of the TV signal is frequency modulated with a 4.2 MHz to 5.4 MHz signal. The horizontal and vertical sync tip pulses correspond to the 4.2 MHz portion of the frequency spectrum. Maximum brightness in the TV image corresponds to 5.4 MHz. The FM sync tip and deviation frequencies vary depending on whether or not the unit uses the VHS, SVHS, 8mm, or Hi 8mm format.

As the picture brightness varies from one part of the TV image to another, the signal frequency changes within the specified deviation range. Frequency modulation produces complex sidebands, which extend above and below the deviation frequency. These sidebands are recorded on the tape. Most of the sideband signal energy is contained in the frequencies closest to the main signal in the deviation passband. These sidebands contain the fine detail (high-frequency information) in the picture. During playback, enough sideband information is recovered to produce a close replica of the original video signal. For example, in VHS machines the sideband information extends from about 1 MHz to 5.1 MHz as shown in Fig. 7.10.

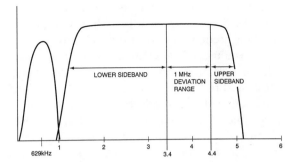

Fig. 7.10. The frequency spectrum for FM VHS video includes a narrow 3.4–4.4 MHz deviation band and upper and lower sidebands.

The upper sideband would extend higher than 5.1 MHz, but the inability of video tape to retain the higher frequencies and the high-frequency losses created by the head cause the upper sideband information to be lost during the playback process. In consumer VCRs, only the lower sidebands are recovered.

The amount of lower sideband signal recovered is important in reproducing the fine detail in the original signal. The higher resolution capabilities of the SVHS and Hi 8mm formats are a direct result of the extended bandwidths of the lower sidebands. Table 7.1 lists the upper and lower deviation frequencies for several formats. Figure 7.11 compares the VHS and SVHS sidebands. Notice that some of the higher resolution formats also have extended deviation ranges (as compared with standard conventional models). When the lower sideband is extended, the deviation range is typically increased to maintain good

Table 7.1. Frequency and Resolution Specifications
for Various Consumer Formats.

Format	Sync Tip Frequency (MHz)	Peak White Frequency (MHz)	Total FM Deviation (MHz)	Luminance Resolution (TV Lines)
VHS	3.4	4.4	1.0	240
SVHS	5.4	7.0	1.6	400
8mm	4.2	5.4	1.2	270
Hi 8mm	5.7	7.7	2.0	430

video signal-to-noise characteristics. All of the higher resolution formats have extended lower sideband bandwidths when compared to their predecessor conventional format.

The use of FM overcomes several obstacles associated with video recording that were discussed in Chapter 5 (see Chapter 5 section heading "FM Record/Playback"). During playback, the VCR monitors variations (deviations) from a set modulation frequency. Upon detection, the machine converts these signals into

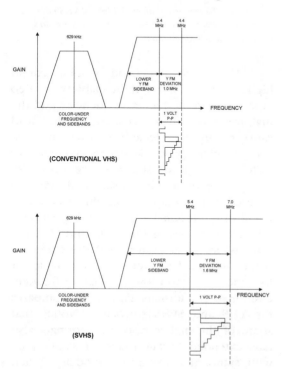

Fig. 7.11. *Frequency spectrum comparison between VHS and SVHS formats.*

the original received signal. Because the system detects variations in frequency and not amplitude, it isn't sensitive to minor amplitude variations in playback signal strength caused by head-to-tape contact flaws.

The FM process permits the recorded signal to saturate the video tape without causing adverse playback video signal distortion. This permits FM signal recording without using bias. During playback, the FM signal recovered off the tape is distorted, but because the FM demodulator is only looking for variations in frequency, the original signal can be adequately restored. The ability to saturate the tape during recording produces stronger playback signals and decreases the chance of tape or VCR mechanical imperfections affecting the playback signal.

In consumer VCRs, the free running frequency of the modulator is typically the same frequency as that of the sync tip level. For conventional VHS machines this frequency is 3.4 MHz; the conventional 8mm free running frequency is 4.2 MHz.

Typically this frequency is adjusted by eliminating all incoming video signals, placing the unit in record, and electronically adjusting for the required frequency. The DC control voltage entering the modulator is now the sync tip clamp voltage. Incoming horizontal and vertical sync pulses will be clamped to this same reference voltage when they are applied to the modulator.

The deviation frequency is also important in the alignment of the modulator. This alignment establishes the peak white frequency of the FM signal. Procedures vary with each manufacturer, so it's important to follow the alignment instructions carefully.

Both the sync tip and the peak white deviation alignments are very important to establishing good contrast and proper video levels, and therefore must be performed precisely. An incorrectly aligned FM modulator will produce several symptoms when playing back a self-recording. These symptoms include demodulation noise (brief periods of horizontal streaking following high contrast level changes), incorrect video playback luminance levels, beating, and loss of resolution.

To reduce production costs, some VCR models no longer provide FM modulator alignment capability. In these VCRs, the modulator frequencies are set by a combination of fixed resistors and capacitors which are hardwired into place to achieve the correct values. Fortunately, these factory-preset alignments normally don't change significantly during the life of the unit.

Symmetry

It's important for the output waveforms of an FM modulator to be symmetrical. Each half cycle of an individual oscillation must have equal amplitude, and the rising and descending excursions of the waveform should be the same. When the FM oscillation is symmetrical, the beat patterns, distortions, and noise generated in the record modulation and playback demodulation processes are minimized. This produces a demodulated playback video signal that is as free of these unwanted artifacts as possible.

The demodulated playback video signal normally contains some carrier ripple. This is interference that occurs because the FM modulation frequencies and the upper portion of the video passband frequencies are close to one another. These two signals beat together to form a carrier ripple frequency.

Carrier ripple appears on the screen as a very fine beat pattern. To eliminate this pattern, the FM demodulator doubles the FM playback frequency prior to demodulation and produces a carrier ripple frequency that is high enough to allow low-pass filters to remove this ripple without affecting the frequency response of the playback video signal.

For this double frequency scheme to work, the FM carrier waveform must be symmetrical. If not, the residual carrier energy will cause a fine herringbone beat pattern on the screen. This form of beat interference is called *carrier leak*. A few VCR designs provide adjust-ments in the record modulation process to optimize the FM modulator's waveform symmetry. When provided, these adjustments make the FM modulator VCO waveforms as symmetrical as possible.

Some high-end VCR models have a playback carrier leak adjustment independent from the record FM symmetry control. The addition of a carrier leak adjustment allows the technician to optimize the VCR playback video and further reduce the presence of this interference in the demodulated playback picture.

FM interleaving/half H shift

In VHS LP and SLP speeds, and in all 8mm record speeds, the FM modulator frequency is shifted by 7.867 kHz ($\frac{1}{2}$ fh) during the interval that a preselected video head is recording the FM signal. This process is called *half H shift* or *FM interleaving*. FM interleaving reduces the visibility of the beat pattern generated by the residual Y FM signal crosstalk not eliminated in the playback head gap by the azimuth offset. The use of head gap azimuth offset for elimination of the Y playback FM signal crosstalk was described in Chapter 5. Offsetting the head gap azimuth reduces the adjacent track crosstalk by approximately 38 dB. This is a significant reduction but not a complete elimination of the unwanted crosstalk signal. FM interleaving is an electronic processing method used to further reduce the visibility of the beat pattern coming from the remaining adjacent track FM crosstalk component.

FM interleaving makes the remaining FM carrier beat patterns produced from the close proximity of the tape RF tracks interleave on the television screen. The FM frequency difference between the two RF tracks alters the beat patterns generated on the TV screen so that they interleave as shown in Fig. 7.12. Since we see a composite of the interleaving effect, our eyes act as filters and visually cancel out the appearance of these unwanted beat patterns. Thus FM interleaving does not reduce the level of the adjacent track crosstalk below the −38 dB level achieved by the head gap azimuth

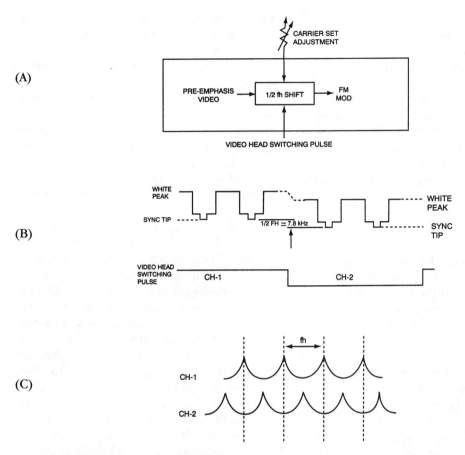

(A)

(B)

(C)

Fig. 7.12. *FM interleaving. (A) The ¹/₂ fh shift circuit identifies the recording head and the corresponding field. (B) The ¹/₂ fh shift circuit applies a DC offset voltage to one of the incoming vertical fields. (C) The frequency of the FM modulator is shifted up by 7.8 kHz to create an interleaved FM carrier between vertical fields recorded on adjacent tracks.*

angle offset. It just makes it harder to see the crosstalk that remains.

The 8mm machines take advantage of the interleaving pattern and use a comb filter to electronically cancel out the interleaved beat signal inside the machine instead of relying singly upon the visual cancellation ability of our eyes. The 8mm machines are able to take advantage of a Y comb filter because they FM interleave the record signal for all record speeds. VHS machines don't interleave the SP speed FM signal because guard bands exist in the tape between tracks at that speed. The use of FM interleaving in only two of the three VHS record speeds makes the use of a Y comb filter

impractical during playback so the VHS units depend entirely upon the ability of our eyes to visually cancel the interleaved beat.

Interleaving the beat is accomplished by shifting the frequency output from the frequency modulator upward ¹/₂ fh (7.8 kHz). This is accomplished by adding a DC voltage to the incoming vertical field corresponding to a predetermined video head writing the FM signal onto the tape. The video head switching pulse is used by the FM modulator to identify the vertical field and the recording video head that will write the shifted FM signal.

Since an increase in frequency produces an increase in luminance level, this shift in frequency will generate a slight unwanted difference in brightness between fields in the demodulated playback signal unless compensation is applied. See Fig. 7.12B. This change in brightness would show up as a 30 Hz flicker in brightness causing the playback to look like an early 1900s old time movie. The change in playback brightness levels between fields resulting from the shifted FM frequency is compensated for by removing the DC voltage offset during the playback demodulation processing.

High-Pass Filter

The Y FM signal is passed to the high-pass filter shown in Fig. 7.1. This filter is enabled during the recording of a color program to limit the lower sideband frequencies, thus making room in the frequency spectrum for mixing in the color-under signal. In some models, this filter is bypassed if a black and white program is being recorded, allowing maximum picture detail on the output of the VCR.

FM Record Amplifier

The frequency modulated signal in the record amplifier circuit shown in Fig. 7.1 is boosted in a record amplifier so it can magnetically saturate the video tape.

In some models, another function of the record amplifier is to provide additional record current during insert edit modes. Insert edit places a new recording in an already existing recorded program. The new video is inserted into the previously recorded signal without disturbing other information on the tape, such as audio.

The full track erase head cannot be turned on during an insert edit, or the previously recorded audio and servo control track pulses would also be erased. Since the full track erase head is de-energized, increased Y FM current is applied to the video heads to overcome previously stored video and thus help reduce the

color rainbow screen effect that appears when an incomplete erasure of a previously recorded program has occurred. The video heads do not completely erase the previous video information during record so a full track erase head is used during normal record. Once an insert edit has been performed, a second insert edit cannot successfully be performed on the same section because extra record current that is applied during the second insert operation will not properly overwrite the previous video signal because it was also recorded with extra record current.

When performing an edit, additional record current is applied under system control of the record amplifier. This extra record current doesn't eliminate all of the rainbow pattern, but it does remove enough so the rainbow problem is negligible. Not all VCRs have this feature. It is generally used on models without flying erase heads.

Some frequency equalization is provided in the record amplifier by slightly boosting the low end frequency portion of the signal. This equalization is provided to compensate for the frequency response of the video head during playback, as described back in Chapter 5.

Most consumer VCRs have preset record equalization adjustments established at the factory by a series of fixed resistors and capacitors. When record current equalization adjustments are provided, they are performed in the record amplifier using an *RF sweep generator*. The RF sweep generator is a test instrument that rapidly sweeps through the frequency ranges selected. For VCRs, the sweep generator is required to sweep through the lower sideband and FM deviation frequency range. The sweep generator also provides identification markers at frequencies selected by the technician to enable accurate equalization alignments.

Typically, RF equalization specifications call for equal frequency response across the FM deviation range. As video heads approach the end of their useful life, record equalization characteristics change causing a loss of detail in the recorded picture. In the case of severely worn video heads, record equalization qualities will diminish to the point that excessive noise and beat patterns are recorded in the program material.

After the frequency modulated signal has been voltage and current amplified, the color-under chroma signal is mixed with the FM. The amplitude varying color-under signal requires bias to be recorded onto the tape. The Y FM signal provides this bias.

ROTARY TRANSFORMER

To be recorded, the FM Y and chroma signal must pass through the high-speed rotating video heads. This signal is applied to the video heads through a rotary transformer comprised of two sections. One section is stationary and is attached to the lower video head drum assembly. The second portion is connected directly to the rotating video head drum. The rotating transformer is a ferrite core with small copper wires wound in such a way that the stationary part of the transformer is able to conduct energy to the rotating portion. Figure 7.13 shows a cutaway of a rotary transformer and a disassembled video head drum assembly with the rotary transformer exposed. The rotary transformer is not a serviceable item. Failure of the rotary transformer necessitates replacement of the entire drum assembly. Mechanical precision is required for proper placement of the upper and lower portion of the rotary transformer to provide maximum energy transfer during high-speed rotation.

The rotating video head drum assembly is connected to the upper portion of the rotary transformer with very tiny wires. When the video head assembly is replaced, care must be used so these tiny leads don't get damaged. Breaking a wire would result in replacement of the entire drum assembly.

Units with flying erase heads also conduct erase current to the erase heads through the rotary transformer.

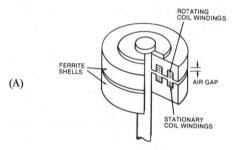

(A)

Cutaway of rotary transformer.

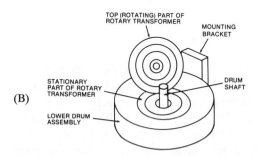

(B)

Rotating portion of the rotary transformer lifted off the drum shaft.

Fig. 7.13. *The rotary transformer. (A) Cutaway of rotary transformer. (B) Rotating portion of the rotary transformer lifted off the drum shaft.*

Rotary transformers seldom fail. Failures are typically caused by mishandling the device—especially when removing and reinstalling the upper video head drum during service. Symptoms of a rotary transformer problem appear as a noisy picture caused by low playback RF signal strength. A defective rotary transformer may give symptoms similar to having one or both of the video heads clogged. The appearance of clogged heads occurs because one (or both) of the heads is (are) unable to transmit the playback RF signal to the RF amplification stage connected to the other end of the rotary transformer.

Low playback RF signal strength can be caused by improper rotary transformer spacing. Spacing between the stationary and rotating sections of the rotary transformer is critical for good RF signal transfer from the heads. This is especially important in the playback mode because of the weak signal coming from the playback head. If the gap between the transformer sections is too wide, the weak signal will not transfer across to the other side. This problem produces a weak RF signal, which in turn generates increased noise in the reproduced picture. As the gap widens, the signal becomes weaker until it eventually disappears. At this point no playback picture is produced—only noise. The rotary transformer gap width is not an adjustable item on most consumer models. This is a factory setting. A problem in this area often requires replacement of the entire drum assembly. Great care must be exercised when servicing any portion of the video head assembly to prevent disturbing the rotary transformer part of the drum mechanism.

A defective rotary transformer may also cause a symptom similar to head clogging. If both playback head signals are missing, no video will be present on the screen. This may result from bad connections in multiple areas of the rotary transformer, or too large of a gap between the rotary sections. An excessive gap will affect the strength of both head signals. Connection failures typically occur in just one head signal path. Rotary transformer continuity can be checked with an ohmmeter. If the bad connection exists within reach of a soldering iron, a repair can be made.

PLAYING BACK THE FM LUMINANCE SIGNAL

Video Playback Preamplifier

Because the playback signal coming off the video head is low voltage (less than one millivolt), amplification must be provided to increase the signal to a usable level for processing. This is the function of the preamplifier circuit. The preamplifier must provide gain while introducing as little noise as possible into the playback signal. Components are carefully selected to meet these high gain and low noise requirements.

The preamplifier is a tuned tank circuit. A tank circuit has frequency response characteristics that are determined by the values of its capacitor and inductor components. Transformers such as the rotary transformer and the windings and core material of the video head are inductors and are part of the tuned preamplifier circuitry. Adding a capacitor to a transformer produces a tuned tank circuit. The tank circuit is tuned by selecting the capacitance that will make the circuit resonate at a desired frequency. The preamplifier tank circuit is tuned to peak at approximately 5 MHz so that maximum output is recovered from the video heads during playback (especially at the upper end of the FM spectrum).

Most current VCR models have nonadjustable preamplifier peaking circuits. These circuits have been preset at the factory. Peaking adjustments were common on older units. Whether the preamplifiers are preset or adjustable, the peaking circuits in the preamplifier are used to optimize the frequency response in the playback picture.

If the peaking circuit in your VCR is adjustable you make the adjustments by changing the setting on a variable capacitor, a variable coil, a variable resistor or any combination of these devices. If the peaking circuit is adjusted with only a capacitor and a coil, the playback response characteristic of the tuned preamplifier circuit is typically set so those signals near the 5 MHz frequency point are peaked.

A damping resistor is placed in the tuned circuit to dampen or reduce the amplitude of the 5 MHz peak and broaden the response characteristics of the tank circuit so all video frequencies are retrieved and amplified equally. It is important for the preamplifier circuit to equalize (flatten out) the amplification characteristics of all frequencies coming from the playback video head because of the 6 dB per octave response of the head during playback. If the levels weren't equalized, the higher playback FM frequencies corresponding to the brighter video signals would play back with amplitudes higher than those for the darker portions of the picture (the lower FM frequencies). A nonequalized signal would produce an amplitude modulation in the playback FM signal and induce beat patterns in the playback video. In severe cases this could generate demodulation noise (short duration horizontal black and white streaks).

When provided, adjustments of the preamplifier tuned circuit are made using a factory alignment tape that has an RF sweep waveform recorded on it. The RF sweep section of the alignment tape sweeps the frequency range of the entire video spectrum picked up by the video heads in 1/60 of a second. Figure 7.14 shows the playback waveform pattern from a factory RF sweep alignment tape. Frequency markers are placed to identify important frequencies as the sweep range varies from 0 to 5.1 MHz. Playback alignment of the tuned circuit is performed by adjusting the variable resistor control so it has no effect on the circuit. While watching the sweep waveform on the oscilloscope, the technician adjusts the variable capacitor so the peak of the tuned frequency is outlined in the alignment instructions. This is normally between 4.5 and 5.1 MHz. The variable resistor is then adjusted to produce an RF sweep envelope with amplitude characteristics that are as equal as possible over all frequencies.

Proper video RF playback amplifier alignment is important to ensure that the playback characteristics of the video signal are equal for both video heads. Each video head has its own preamplifier and tuned circuit. If the preamplifier tuned circuit for one head is misadjusted, the FM portion of the signal, which corresponds to peak white video in the deviation range, would not be played back correctly. The result could be demodulation noise or streaks at peak white levels flickering at 30 Hz on the TV screen. This occurs when the video head preamplifier doesn't properly play back the FM frequency associated with peak white. Streaks or noise could flicker on the screen each time that particular video head attempted to play back peak white.

Many preamplifier circuits also permit an adjustment which balances the RF envelope levels picked up and amplified by the preamplifier circuit, providing equal playback RF signals from each preamplifier. Figure 7.15 shows an unequal RF envelope. This VCR has no preamplifier balance adjustment. The imbalance is caused by a worn video head.

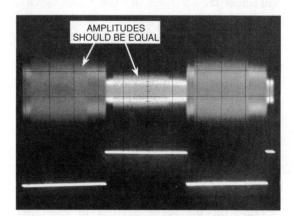

Fig. 7.15. Unequal video head output after preamplification.

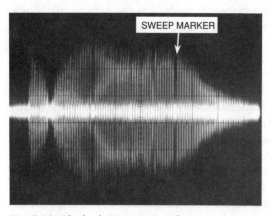

Fig. 7.14. Playback RF sweep waveform.

The video head preamplifiers must be electronically turned on and off at appropriate times, as shown in Figure 7.16. When head A is playing back RF track A, the video head B preamplifier must be switched off. Video head B is

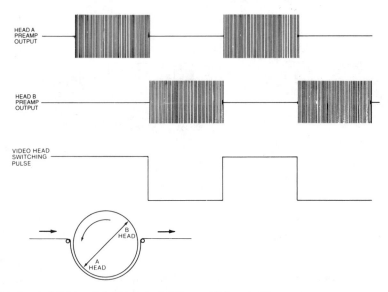

Fig. 7.16. *The video head switching pulse turns on one preamplifier at a time.*

not in contact with the tape most of the time that video head A is playing back a signal. If the preamplifier for video head B was on during head A playback, the B head preamplifier would have no signal to amplify. Instead, the preamplifier would pick up external noise signals and send these to be demodulated along with the video head A FM signal, which would produce an extremely snowy picture.

Figure 7.17 is a photograph of an oscilloscope presentation showing a preamp output and switching pulse.

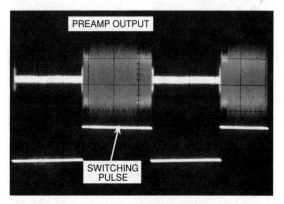

Fig. 7.17. *Preamp output with switching pulse.*

A video head switching pulse is generated by the rotating video head drum. This signal is processed in the servo circuitry and will be described later. The important point to remember is that this head switching pulse is responsible for turning on the correct preamplifier at the appropriate time. Figure 7.16 also shows the relationship between RF signal output from head A, head B, and the video head switching pulse.

Y FM Limiting

Figure 7.18 is a block diagram of Y playback processing. The pre-amplifier FM signal is applied to a limiting circuit where the top and bottom portions of the FM playback signal are chopped off. Signal limiting eliminates noise, playback FM level fluctuations, and waveform distortion, which can all occur at the positive and negative peaks of the FM signal. Any waveform distortion that occurs as a result of saturating the tape during record is also chopped off in the limiting process. Limiting provides the cleanest demodulated output possible. In some designs, the limited signal is passed

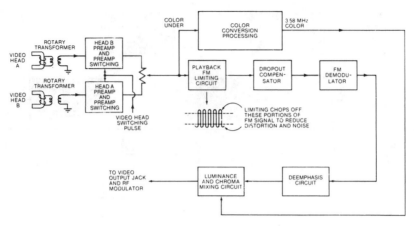

Fig. 7.18. *Luminance playback block diagram.*

through a dropout compensator to reduce the unwanted visual effects that occur as a result of missing pieces of tape oxide. Other models apply dropout compensation later, after the signal has been demodulated.

Dropout Compensation

A possibility exists that a few lines of video information could be missing during playback because of imperfections in the video tape. If the magnetic coating on the video tape surface is missing because it was scratched or damaged, or if the tape binder itself is missing or not strong enough to hold the magnetic coating on the tape, a phenomenon known as *dropout* will occur. Dropout occurs when the playback RF signal is absent for short periods of time. Dropout is seen as tiny lines of snow that travel rapidly across the screen from left to right. The actual amount of dropout varies greatly, from less than a line to one or more lines. Generally, most video tapes have a limited number of dropouts. Typically when dropouts do occur, only one or two horizontal lines are affected.

The use of FM helps to reduce the sensitivity of the video playback processing circuits to amplitude variations in the playback FM signal. FM permits magnetic saturation of the

video tape during recording. This allows the maximum amount of magnetic information to be retrieved from the tape in playback. Minor variations in playback amplitude levels are eliminated by the limiter circuit and do not affect the FM demodulator. A limiting circuit is used to literally chop off the upper and lower portions of the playback FM signal. Minor variations in the level of the playback signal are literally ignored because the limiting circuit chops off and discards those portions of the signal anyway.

If total dropout of the FM signal occurs, however, all video during the dropout is lost. A dropout compensator minimizes the visual effects of dropout on the TV screen. Dropout compensation (DOC) is a special circuit that identifies missing sections of RF information and inserts a good video signal from a previous line in its place.

Figure 7.19 shows a block diagram of how one type of DOC operates. This type of dropout compensator is known as an *RF DOC* because it acts to replace the missing RF signal when dropout is detected. Another type of DOC operates on the demodulated video signal by filling in the missing demodulated video. This type of DOC is known as a video DOC.

Let's explore the operation of the RF DOC. Playback RF from the limiting circuitry enters the DOC and is split in two directions. The

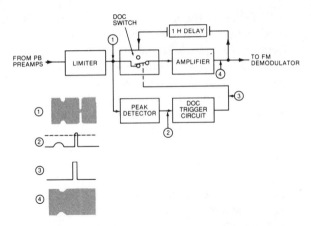

Fig. 7.19. *Basic RF type dropout compensation (DOC).*

first path for the RF information is through the main amplifier. For discussion purposes, this will be considered a dropout-free RF envelope. After amplification, the signal is sent to the FM demodulator. It's also passed to a horizontal delay line, which delays the RF signal by a period equal to one horizontal line (1H or 63.5 msec). Keep in mind that this good RF field contains 262½ horizontal lines. Each of these 1H delayed horizontal lines is supplied to one of the inputs of the DOC switch waiting to be used. If it is not needed by the time the next line of information comes along, it is replaced by the following good line of RF video information.

Now go back to the point where the limiter outputs the RF envelope. Imagine that the next vertical field of RF information coming along contains dropout. The peak detector circuit identifies the exact point where RF energy is missing and sends a pulse to the DOC trigger circuit. The sensitivity level of the DOC trigger circuit has been set to ignore minor fluctuations in RF energy levels and to generate a trigger pulse when RF information drops below a threshold. The DOC trigger pulse is output to the DOC switch while dropout is present. The DOC trigger switch toggles into the delayed RF position. The switch will remain in that position as long as the DOC trigger pulse is present. The DOC operates by

injecting the previous good line of RF information from the second field and sending that on to the demodulator.

We then see the same horizontal line twice—our eyes are not sensitive enough to observe this repeat action. Keep in mind that this single line of repeated video information is only 1 out of 525 lines on the screen. That's hardly enough for us to notice. If DOC were not to occur in the playback process, the result would be tiny visible lines of black and white snow rapidly moving across the screen and detracting from the overall picture. Figures 7.20 and 7.21 show video playback before and after dropout compensation.

Fig. 7.20. *Video presentation before dropout compensation.*

Fig. 7.21. *Video presentation after dropout compensation.*

It is interesting to note that dropout compensation works very well when there is only one line of horizontal information missing. Dropouts for longer than one horizontal line are also handled fairly well by the DOC. It's the last line of good horizontal video information that continues to be repeated over and over again, however. In this case, DOC forms a loop continually repeating the last good line of RF information.

Careful examination of Fig. 7.21 will reveal some image distortion occurring near the top of the boy's head because the same line of good video is being repeated to compensate for several missing lines. In reality, dropout compensation is effective for only a few lines of missing video information. DOC occurs only for the luminance portion of the signal. Consumer VCRs don't provide DOC for the chroma signal. The dropout-compensated, limited-RF signal is fed into the FM demodulator circuitry.

Newer VCR designs no longer use delayed RF dropout. Instead, the circuitry delays the demodulated video in a CCD and makes the demodulated signal available for reinsertion into the playback video when a dropout occurs. Detection of the RF level is still required to produce a trigger pulse that will identify missing RF from the off-tape signal. However, rather than injecting delayed RF into the point of the missing signal, the VCR electronics will inject the last good line of demodulated video to mask the missing picture information.

Delaying the RF was popular in early VCR models because it was easy to delay the high-frequency playback RF signal. Until CCDs were designed into VCR equipment, it was too costly to delay the broad frequency spectrum associated with the baseband video signal using anything other than delayed RF dropout.

FM Demodulation

The FM demodulator is based on the electronic principle demonstrated in Fig. 7.22. This principle recognizes that if a square wave signal of equal amplitude and time duration is rectified and filtered, the average DC level will be exactly 50 percent of the original square wave. If the positive-going portion of that square wave maintains the same width or time duration and yet the pulses are placed closer together, the rectified and filtered DC voltage will increase because the negative going portion of the square wave is diminished.

Figure 7.23 shows the basic operation for one FM demodulator. This is only one example of how FM demodulation is accomplished. FM demodulation techniques vary not only between the different manufacturers but also from one model to the next within the same manufacturer's product line.

In our example in Fig. 7.23, we see that limited FM enters a differentiator circuit that removes the DC component of the incoming

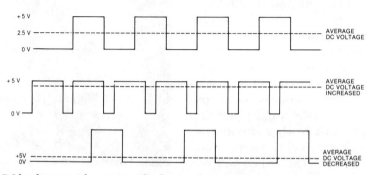

Fig. 7.22. Average DC levels vary with respect to the distance between square wave pulses.

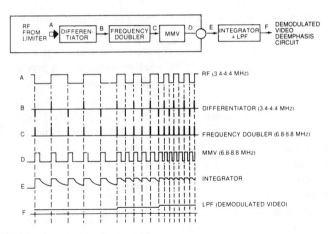

Fig. 7.23. *FM demodulator.*

square wave by converting the negative- and positive-going transitions of the square wave into sharp spikes. The positive-going spikes correspond to the positive-going portion of the incoming square wave; the negative-going spikes correspond to the negative-going portion of the original square wave signal. The frequency of the differentiated square wave is doubled in a doubling circuit. Diodes connected like those found in a common full-wave rectifier circuit can be used to perform this function. The doubling circuit inverts and adds the negative-going spikes to the positive-going spikes.

The positive-going spikes from the frequency doubler enter a monostable multivibrator (MMV) whose output is controlled by pulses from its incoming signal (in this case, the frequency doubler). The duration of the period of the MMV output pulse is controlled by external components such as resistors and capacitors. The positive-going portion of the MMV output is relatively short. As the incoming FM square waves bunch closer together, representing higher frequencies, the changes in the output voltage level of the MMV are also compressed. The output frequency has, in effect, been doubled. The output of the MMV is integrated and passed through a low-pass filter to produce a DC voltage whose level directly

corresponds to the frequency and period of the MMV.

Current FM demodulators are contained within ICs. VCR FM demodulators vary from one model to the next; however, they all share two main characteristics. First, they act as pulse count or pulse density detectors. The higher the density of the FM signal, the higher the detected voltage output from the demodulator. Second, they all double the frequency of the carrier in order to place the carrier ripple well above the video passband frequency. Carrier ripple occurs in the FM modulation/demodulation process because the operating frequency of the FM oscillator and the upper frequencies of the input record video signal are close together. These two high frequencies beat together and produce a difference frequency known as *carrier ripple*. Doubling the RF frequency input to the FM demodulator places this carrier ripple frequency well above the passband of the VCR system and eliminates the unwanted oscillation.

The FM demodulator must also remove the DC offset voltage produced by the FM interleave circuit in the record mode. Since the FM interleave circuit shifted the carrier frequency of the record FM signal by 7.8 kHz ($\frac{1}{2}$ fh) for one of the fields, the slight increase in brightness produced by this shift must be eliminated.

De-Emphasis

The demodulated Y signal now begins to closely resemble the original luminance portion of the recorded signal. Only two things need to be done to restore the signal to its original form. The signal needs to be de-emphasized and the chroma needs to be re-inserted back into the luminance. During record, sub-emphasis (also known as non-linear emphasis) and main emphasis were applied to the luminance signal in the record pre-emphasis circuit to preserve high-frequency picture detail and to improve the SNR. These emphasis characteristics must be removed.

De-emphasis is performed on the video signal by first passing the demodulated signal through a low-pass filter to eliminate unwanted high frequencies not removed from the playback video by the demodulation process. The band-pass of the low-pass filter that is used depends upon the VCR format. Conventional VHS VCRs use a low-pass filter in the 3 MHz range while the SVHS units use a 5 MHz low-pass filter.

Sub-emphasis is removed by a circuit that performs the opposite function as the record sub-emphasis circuit. Signals that were boosted in the record mode must be returned to their original levels. This action also reduces the noise picked up by the playback heads, making noise less apparent on the screen. Remember, sub-emphasis exists to improve the video playback signal-to-noise ratio.

The main de-emphasis is applied to the sync tip portion of the signal by clamping the negative-going excursions. Main de-emphasis of the video portion of the signal is accomplished by placing capacitors in the circuit amplifiers within the main de-emphasis block. The capacitors act to reduce the high-frequency compo-

nent of the signal a slight amount. The values of these capacitors are carefully selected to attenuate the unwanted pre-emphasized video spikes while maintaining proper high-frequency video information.

If the system uses a video DOC circuit, the de-emphasized signal is sent to the CCD for delay and possible use when dropout occurs.

8mm Y Comb Filter

In the 8mm format, the demodulated Y is passed through a Y comb filter before being sent to the final video output amplification stage. This comb filter removes the FM interleaving signal described previously in this chapter. As was shown in Fig. 7.12C the FM signal crosstalk from an adjacent track is interleaved making it possible for a comb filter to cancel out any of the remaining crosstalk not eliminated by the head gap azimuth offset.

VHS units do not use a Y comb filter, so the de-emphasized Y signal is sent directly to the video output amplification stage.

Playback Video Outputs

After de-emphasis, the demodulated video signal is applied to the output amplifiers for amplification and then made available at the TV output connector. VCRs often provide more than just a TV output choice. They can have a Y/C (S-Video) output, a composite (LINE) video output, and an RF (channel 3/4) output.

The Y/C output offers separate playback luminance and chrominance connections between the VCR and the TV. This connection gives the viewer the best video reproduction possible because it requires the least amount of signal processing in the VCR output circuitry and in the receiving television's electronics. In this connection, the demodulated video is maintained separate from the playback up-converted chroma signal.

When the composite output is used, the chrominance and luminance signals are applied to a mixing circuit inside the VCR before being sent to the composite video line output jacks. This mixed output choice provides the second best playback video quality. When a composite signal is sent to the television, the luminance and chrominance information are identical to the S-Video signals in every way except for one thing; they are combined into a single waveform to form a composite signal. The television electronics must separate the composite waveform into Y and C components for processing inside the receiver. The additional combination and separation of the VCR output adds additional noise and some unwanted artifacts into the viewed picture making this connection the second preference to the S-Video output.

The final choice in VCR output connections is the radio frequency (RF) output. This connector is the female type screw-on connector that looks like the wall connector used by a cable company to hook up a 75 ohm coaxial cable to your TV set. This connector is called an *F* connector and is often labeled as the "to TV" connector on the back of a VCR.

The RF output is a composite chrominance and luminance signal that also contains the playback audio. The audio and composite video signals are RF modulated to around 70 MHz to produce the channel 3 or channel 4 television tuner frequency. The VCR's RF modulator obtains the playback composite video and audio signals and modulates these sources to a frequency that the television receiver can use. The RF modulated signal is received by the television's tuner and RF demodulation circuitry where it is demodulated back into a composite video waveform for processing. The modulated audio signal is separated from the video RF signal inside the TV's RF demodulation circuitry and sent to the TV's audio circuitry.

While the VCR's RF output provides a convenient way to hook up both playback audio and video on a single wire, it is considered the less desirable method of output connection to the TV set because added noise and unwanted artifacts are placed on the playback signal through the additional processing required in both the VCR and receiving TV set. More information on tuners and RF modulators is provided in a later chapter.

RESOLUTION IMPROVEMENTS

Manufacturers have discovered ways to improve the video picture reproduction by modifying and improving the video record and playback processing circuitry. The first manufacturer to introduce improved picture quality was Sony. They modified their conventional Betamax format by changing the FM modulation and demodulation circuitry to provide a way to play back a picture with 20 percent greater detail. They called this new procedure *SuperBeta*.

SuperBeta circuitry changed the Beta FM deviation specifications from the original 3.6 to 4.8 MHz to a new 4.4 to 5.6 MHz deviation. (The 4.4 MHz corresponds to sync tip black, while 5.6 MHz corresponds to 100 percent peak white.) This moderate shift in the FM deviation increased the amount of room available for the lower sideband portion of the signal. Recall that much of the picture detail is contained within the FM sideband frequencies. With increased space now available for the lower FM sideband, a greater portion of the lower sideband frequencies can be recorded onto the tape without interfering with the color-under signal. Sony claims a 20 percent increase in picture detail over that obtained by the conventional Betamax system.

One advantage of making only a slight shift in the FM deviation frequency specifications is that the resolution can be improved without requiring changes in the video tape's magnetic coating. Another advantage is that tapes recorded in the SuperBeta mode can be played back on conventional Betamax machines with minimal increase in demodulation noise. Both of these advantages provided downward compatibility while offering some increase in picture detail.

Although Sony still manufacturers some Betamax and SuperBeta models, these formats are less popular than the VHS and 8mm designs.

VHS High Quality (HQ)

After the introduction of the SuperBeta format, it didn't take long for the VHS manufacturers to also find a way to make slight picture quality improvements in their units. VHS manufacturers designated the first of these improvements HQ for *high quality*. VHS HQ does not modify the FM deviation frequency.

With VHS HQ, three areas of the video process were improved. First, the white clip level adjustment was modified to provide a 20 percent increase in the emphasis spike level before signal modulation. Second, the luminance noise reduction circuitry was improved. Third, the chroma noise reduction circuitry was improved.

Improvements in picture quality for VHS HQ are noticeable when comparing previous VHS units with those incorporating HQ technology. Video resolution is not increased, but overall picture quality appears cleaner (due to improved noise reduction), and sharper (due to picture detail enhancement brought about by the increased white clip levels). HQ improvements are not as dramatic as those in the SuperBeta machines because SuperBeta improved both the video resolution and the noise reduction.

SVHS and Hi 8mm Formats

Advances in video head manufacturing and in magnetic media technology have resulted in head and tape designs that opened the door for manufacturers to produce consumer formats with greatly improved resolution. In fact,

a "resolution revolution" is well under way.

In recent years three higher resolution formats have appeared in the consumer market. These are the Extended Definition Beta (ED Beta), Super VHS (SVHS), and Hi 8mm. While Sony still manufacturers some Betamax and SuperBeta models, the ED Beta format has now been discontinued.

Higher resolution is possible because of improved video head core designs and improved magnetic coatings on video tapes. These improvements enabled higher FM frequencies to be recorded on the tape (higher writing speeds) while the tape is moved at the same longitudinal speed as it is in conventional models.

The introduction of the SVHS and Hi 8mm formats gave us approximately 50 percent increase in resolution over the conventional design. While these formats provide a substantial resolution improvement compared to the increase offered by the SuperBeta units, the SVHS and Hi 8mm signals cannot be played back on their conventional predecessors like the SuperBeta recordings can. This is because the large change in FM recording frequencies in these higher resolution machines is beyond the operating range of the conventional VCR's FM demodulator. Hi 8mm and SVHS machines may be considered extensions of their respective predecessors, but since tapes recorded in these higher resolution modes cannot be played back on the original format machines they are considered to be separate formats.

This does not mean that a VHS tape cannot be played in an SVHS machine or that a conventional 8mm tape will not play back in a Hi 8mm unit. They will. To provide downward compatibility, manufacturers have designed additional circuitry into the higher resolution equipment to permit playback of both conventional and high-resolution recordings. SVHS and Hi 8mm units are also capable of recording in their respective conventional modes to allow playback in the standard format models. This requires some duplication in circuitry so the signal processing requirements unique to both conventional and higher resolution formats could be handled. Different FM modulation alignment and white and dark clip circuitry is required for conventional and high-resolution recording so two similar circuits are installed. Individual playback RF equalization circuits are also incorporated for each mode.

But compatibility between the higher resolution and the standard formats ends when attempting to play back a higher resolution recording in a conventional machine. In this case, the display is typically unrecognizable because the deviation range and FM carrier frequencies occur beyond the design parameters of the VHS or standard 8mm circuitry, so the earlier design systems cannot demodulate the signals.

To achieve higher resolution recordings, SVHS and Hi 8mm designs had to change the FM modulator and record amplifier circuitry. The modulator carrier and deviation frequencies are both increased and new pre-emphasis and white and dark clip levels were established. The record amplifier design was also modified and different equalization parameters were incorporated. Even the video input filtering circuitry was changed to accept the higher baseband video frequencies associated with the increased resolution signals.

The playback electronics were also redesigned to handle the higher RF frequencies being detected off the tape and the higher demodulated video frequencies passed by the system. Preamp equalization was set to equalize higher frequencies, and the FM demodulator was redesigned for a wider deviation range at higher frequencies. Even the de-emphasis was redesigned.

With higher FM record frequencies came increased lower sideband bandwidth. This is the basis for the increased playback resolution. The secret to higher resolution recording is to move the FM deviation range up significantly, increasing sideband recovery and thus vastly improving resolution. This means that the finer detail, once lost in conventional designs, can now be easily reproduced.

Figure 7.24 shows that along with the sync tip and peak white frequencies that are designed much higher for the Hi 8mm and SVHS units, the deviation ranges have also been extended. (A review of Table 7.1 may be helpful when studying Fig. 7.24.) The FM deviation was expanded along with using higher FM frequencies to optimize the signal-to-noise ratio (SNR) in the luminance playback. An important relationship exists between the deviation range and the lower sideband that is recovered. Recovering a wider, lower sideband requires a greater FM deviation range to maintain an acceptable SNR. For example, if the SVHS FM deviation was held at around 1.0 MHz (the value used in the VHS format), and only the carrier frequency was moved in the new design, the SNR would suffer.

It is interesting to note that SuperBeta does not change the total deviation frequency. Both the sync tip (carrier) and peak white frequencies were shifted up by 800 kHz. But the deviation range between the carrier and peak white values was left untouched. It was possible to leave the deviation range the same because the carrier frequency was only moved 800 kHz. An expanded deviation is only necessary when the carrier frequency is substantially increased. This slight shift in the carrier frequency with no change in the deviation range keeps the SuperBeta design within the operational limits of non-SuperBeta machines. This makes SuperBeta completely compatible with earlier Betamax models. One disadvantage of maintaining compatibility between Betamax and SuperBeta is that only limited resolution improvements are possible.

The theory of operation doesn't change just because higher resolution is engaged. Only the frequencies change, so the circuitry is modified to handle these different frequencies. The fundamental operation of the circuitry remains the same.

Higher resolution requires increased writing (recording) speeds, and higher writing speeds require densely packed magnetic parti-

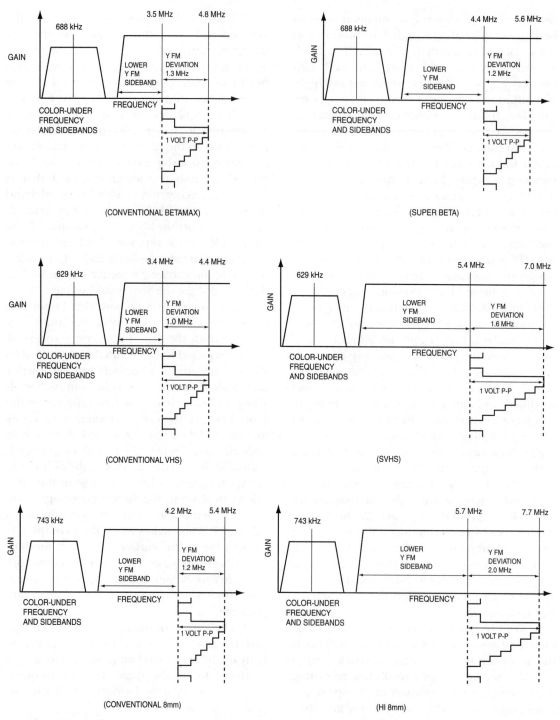

Fig. 7.24. *Comparing FM sync tip (carrier), peak white, and deviation frequencies in conventional and higher resolution formats.*

cles on the tape surface. These demands also require magnetic particles that exhibit increased retention characteristics. Finer particle size, improved size uniformity, and increased retentivity means that new tape formulations are necessary. In Chapter 5 you learned that the SVHS format uses an improved oxide tape, which is an improved version of the conventional oxides and that Hi 8mm machines require metal particle tape, which meets more stringent requirements than the standard 8mm metal tape. Both of these tapes are more expensive than the standard VCR tape. Metal tape is the most expensive.

Noise Reduction and Edit Switches

When dubbing or editing tapes on a VCR, carefully follow the manufacturer's suggestions concerning the placement of switches for noise reduction and editing. Correct placement of the noise reduction switch gives you the best picture quality during playback but will offer poor results when dubbing or editing. Noise reduction features provided in a VCR are specifically designed for the best reproduction quality of playback recorded signals. The noise reduction switch activates circuits that modify the noise reduction activity and change the characteristic of the frequency response during normal playback. The picture tube containing the TV screen responds best to video frequencies around the 1 MHz range. The signal output from the VCR noise reduction circuit is frequency-emphasized at the 1 MHz range in specific quantities to improve the image quality on the screen.

However, this emphasis causes picture quality deterioration when a tape is being dubbed from one machine to another. The problem occurs because the FM modulator assumes the emphasized 1 MHz portion of the video signal is actual video information. The result is reduced quality of a dubbed or edited tape. Some machines have an edit switch that

configures the playback machine circuitry during the dubbing for the best picture quality. If you are using a VCR that offers noise reduction and editing switches, de-energize the noise reduction circuitry and activate the edit switch while you are editing. Otherwise, you'll lose some quality on edited tape.

E-to-E

During the recording process, VCRs allow the video program being recorded to be displayed on the TV screen. This feature is known as *electronic-to-electronic* or E-to-E. E-to-E means that the incoming video signal is accepted by the machine and sent directly out to the video output jacks and RF modulator. This lets the operator simultaneously view and check the recorded signal.

The E-to-E signal does not show all of the signal processing applied to the input video waveform. In consumer VCRs, E-to-E normally means that the signal coming into the VCR from the tuner or line input jacks is available at the RF modulator output or video and audio line output jacks after having gone through minimum signal processing. For example, an E-to-E signal is not FM modulated and demodulated before being sent to the output connectors. E-to-E bypasses this stage.

During the record mode, E-to-E is selected when the TV/VCR button is placed in the VCR position. This causes the audio and video signals from the machine to be E-to-E signals.

SYMPTOMS OF LUMINANCE RECORD AND PLAYBACK FAILURE

To diagnose a failure in the VCR's luminance section, the trouble must be isolated to the record or playback processing. For example, horizontal streaking in the video can result from a playback or a record problem. When attempting to isolate the problem, use a tape with a known good recording on it to check

Table 7.2. Luminance Troubleshooting Table

Luminance Failure Symptom	Luminance Record Failure (If Symptom Is Present on All Self Recordings)	Luminance Playback Failure (If Symptom Is Present on Played Back Tapes)
Dark Picture	1. Y-C comb filter trouble 2. AGC circuit wrong 3. FM modulator deviation frequency wrong 4. white clip/dark clip alignment 5. main and sub-emphasis 6. defective amplifier; power supply voltage trouble 7. defective connection/cable 8. improper termination	1. Main and sub-de-emphasis 2. defective amplifier 3. demodulator failure 4. defective video cable 5. defective RF modulator (only if the dark video is present on channel 3 or 4 and the video out of the output connectors is O.K.) 6. power supply voltage trouble 7. improper termination 8. bad connection
Picture Too Bright (overexposed)	1. Y-C comb filter trouble 2. AGC circuit 3. FM modulator deviation range 4. main and sub-emphasis 5. defective video amplifier 6. power supply voltage trouble 7. poor termination 8. defective connection	1. Main and sub-de-emphasis 2. demodulator failure 3. defective video cable 4. defective RF modulator (only if the excessive video is present on channel 3 or 4 and the video out of the output connectors is O.K.) 5. power supply voltage trouble 6. poor video termination 7. defective connection
Horizontal Streaks in Picture (demodulation noise)	1. FM modulator carrier frequency error 2. FM modulator deviation frequency error 3. record amplifier equalization problem 4. incorrect record current 5. defective rotary transformer 6. worn video heads 7. incorrect white/dark clip	1. Preamplifier RF amplification circuit 2. RF equalization alignment problem 3. limited balance alignment error 4. demodulator balance alignment error 5. worn video heads 6. defective rotary transformer 7. demodulator failure
Poor Resolution (soft picture)	1. Y-C comb filter trouble 2. FM record amplifier equalization alignment 3. worn video heads 4. incorrect white/dark clip levels 5. main emphasis or sub-emphasis error 6. FM modulator carrier frequency error	1. Preamplifier RF amplification circuit 2. RF equalization alignment 3. worn video heads 4. demodulator low-pass filter failure 5. Y comb filter failure 6. defective RF modulator (only if the problem is present on the channel 3/4 output)
Noisy Video (snow)	1. No FM modulator output 2. weak or defective record amplifier 3. bad video head 4. clogged video head 5. defective rotary transformer 6. bad connection	1. Clogged video head 2. defective video head 3. bad connection 4. defective preamplifier 5. defective rotary transformer

the playback circuit operation. If the good tape shows the same problem, then the failure is with the playback circuits.

You should also make a recording on the troublesome VCR and play back that recording on a VCR that you know is in good working condition. The playback picture from that machine will help you discern if the record circuits in the defective VCR are functioning properly.

Once you have isolated the problem to the record or playback circuits, use Table 7.2 to assist you in further troubleshooting. Table 7.2 contains some common failure symptoms and a list of associated circuits that could cause that failure. This table assumes that the failure symptom shows up in only one of the modes. If possible, isolate the failure to either a record or playback problem before using this chart. In the event that your failure symptom shows up in both record and playback modes, you need to pick one mode and troubleshoot that symptom. In this event, you can still use this table but limit yourself to troubleshooting the symptoms for just one mode of operation. We generally like to repair playback problems first so that we know that we have a good playback reference when troubleshooting record failures. Often you will find that repairing a problem in one mode will also lead you to (if not solve) the cause of the failure in the other mode.

Don't become confused if E-to-E looks fine when you feel you have a record failure. Remember that E-to-E bypasses most of the record and playback processing circuits; so record problems will only show up when playing back the recordings you make.

CHAPTER REVIEW QUESTIONS

1. What is another name for the FM carrier frequency?

2. At what frequency do most consumer VCR FM circuits operate when no video signal is applied?

3. Why must the FM signal be equalized during playback?

4. What do changes in the FM deviation frequency represent in the video signal?

5. What is the function of the white and dark clip circuits?

6. What is the purpose for interleaving the FM video signal?

7. Why is it important to recover the lower sideband of the FM luminance recording?

8. Why must the signal input to the FM modulator be clamped?

9. What is the function of sub-emphasis?

10. Briefly explain the function of a dropout compensator.

11. List some possible causes of horizontal streaking during playback of a known good tape.

Chapter 8

VCR Servo Control

*T*his chapter deals with an electromechanical operation of a VCR called the servo. In a VCR, servo circuits control the speed of the tape as it is pulled through the machine. They also control the rotational speed and position of the video head disk as it rotates over the moving tape. Precise tape speed and video head rotation must be maintained during both record and playback operations to enable recovery of the information recorded on the tape. In Chapter 5 you learned that mechanical tolerances in the tape path must be maintained to ensure tape interchangeability between different VCRs. Correct servo operation is another important factor in achieving good tape interchangeability.

SERVO FUNDAMENTALS

To properly play back recorded information, the video head must travel over the same RF track area on the tape as it did during record. This action is called *tracking*. Head A must travel over the head A RF track and head B must travel precisely up the recorded B track. If this doesn't occur, the video head will pick up noise from the adjacent RF track. With azimuth recording, head A doesn't recognize RF track B so the TV displays a screen full of snow or random video noise when head A is traveling up the B track and vise versa.

Several factors ensure that a video head travels over the correct RF track. First, the tape must be traveling at the same speed in playback as it did in record. Second, the video head drum assembly must rotate at the same speed during record and playback. Third, video head rotation must have the same phase relationship with vertical sync during record and playback.

In order for the VCR to play back the video RF tracks correctly, RF information must be recorded in precise locations on the tape and the VCR must be able to identify the location of this information. Therefore, RF tracks are marked to enable the machine to relate track A to video head A, and track B to video head B. In addition, the speed and exact position of the video heads on the rotating video head drum must be carefully controlled. The video tape must be drawn through the machine at a precise speed to ensure proper playback of sound and picture. Servo control circuitry maintains the correct head drum rotation and video tape speed.

Servo control is performed by electronic circuits that govern the speed and phase of the drum and capstan motors. Figure 8.1 shows the basic operational principles used in a typical analog servo circuit of some early VCR models.

All servo circuits require a reference signal and a sampling signal to provide feedback speed and phase information. The servo circuit compares the feedback signal to the reference in order to monitor motor speed and phase and to make any required corrections. Successful operation of this analog circuit lies with its ability to sample the DC voltage near the midpoint of the ramp coming from the trapezoid generator and then to hold that sampled voltage for motor control. Sampling is accomplished inside the sample and hold block shown in Fig. 8.1.

The ramp-shaped signal is produced by a 30 Hz trapezoid generator. The trapezoid signal has a peak amplitude of 5 volts. In this circuit the trapezoid is the comparison reference signal.

The sampling signal is used as the motor feedback information. This waveform is derived from the 30 Hz square wave output of a divide-by-24 circuit. The divider's source is a 720 Hz signal created by a series of magnets mounted in a circle on the base of the motor. This series of circular magnets is called an *FG (frequency generator) ring magnet*. It rotates as the motor turns. The rising edge of the counted-down FG 30 Hz square wave acts as a gate for the sampling action in the sample and hold back.

As the motor rotates, the 720 Hz signal is produced. Notice the phase relationship between waveforms A and B in the lower part of Fig. 8.1. If the motor speed varies, the divider circuit outputs a different frequency and phase referenced to the steady ramp waveform. When the motor is running at the correct speed and phase, the sampling occurs at the midway point of the ramp. This halfway position is the 2.5 volt level on the ramp slope.

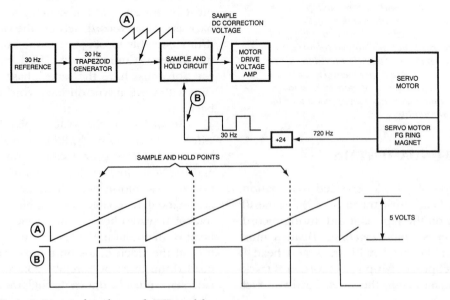

Fig. 8.1. Basic servo control used on early VCR models.

Here's what happens to this circuit when the motor changes phase (rotational position). As the motor drifts out of phase, the positive and negative transitions of the counted-down 30 Hz square wave change phase relative to the steady ramp waveform. Since sampling takes place at the rising edges of the counted-down square wave, the difference between waveforms causes the sampling point to occur at a different position (a different DC voltage point) on the ramp's slope. When the sample is taken at a different DC potential, a voltage other than 2.5 volts will be output from the sample and hold circuit. The different ramp sampling voltage has become the new sample and hold voltage supplied to the servo motor. This new motor drive voltage acts as a correction voltage forcing the motor to reposition (correct) its phase so that the sample is again made at the 2.5 volt point of the ramp.

Today, digital circuits are used to produce the speed and phase correction voltages for VCR servo motors. Digital equivalent signals have replaced many of the analog waveforms previously used for the reference to feedback comparisons and sample and hold operations. The operations of these digital circuits will be covered later in this chapter.

The servo circuitry in a VCR is comprised of two main sections: the video head drum servo section and the capstan servo section. These sections each have two subsections. One subsection governs the coarse (free-run) speed of the motor while the other subsection controls the phase of the motor (fine tunes the speed).

For example, the video head drum can rotate at a coarse speed of 1798 RPM, but the exact placement or position of the video head at any given instant in time must be accurately controlled. The free-run speed of the head drum is controlled by the free running speed section, while exact placement of the video head is controlled by the phase control unit.

Table 8.1 lists the basic responsibilities of the drum and capstan servo sections for the record and playback operation modes. It also lists the reference and feedback signal sources used by the servo to maintain control.

The functions described in Table 8.1 have been simplified to introduce you to the responsibilities of the individual servo sections. Some interaction occurs between these sections, however. For example, the table indicates that the drum motor is solely responsible for maintaining precise speed control of the head disk in the playback mode and that the capstan does all of the correction to place the RF tracks precisely beneath the constant speed of the video head. In actual operation, the head drum will shift phase slightly as the capstan works to maintain precise tracking control. While the capstan is responsible for most of the tracking control, the drum servo does respond slightly to corrections made by the capstan servo.

CAPSTAN AND DRUM RECORD SERVO OPERATION

Capstan Record Servo

As shown in Table 8.1, the capstan's responsibility during record is to pull the tape through the machine at a steady speed. The capstan does not require a reference or feedback associated with the incoming record video signal to perform its task. Instead, it's referenced to a crystal oscillator and its own self-generated pulses. The capstan servo does not need to be tied to the incoming record video to produce smooth tape movement.

A review of Figs. 5.33 and 5.34 will help you understand the importance of precise tape movement by the capstan servo. Notice the accurate placement of the RF tracks on the tape relative to the adjacent tracks. If we assume that the video head is rotating at a consistent speed and that the tape is being moved through the tape path with a steady motion, the tracks will be positioned as shown in those illustrations. If, however, the capstan servo allows the record tape speed to vary, the posi-

Table 8.1.
Drum and Capstan Servo Functions

	Drum		Capstan	
	Record	*Playback*	*Record*	*Playback*
FUNCTION	SPEED: Control rotation speed (maintain constant 1798 RPM). PHASE: Rotational position control. (Place the recording video head precisely on the tape precisely 10½H before the start of vertical sync.)	SPEED: Coarse rotational speed control (maintain a constant 1798 RPM). PHASE: 1) Precise rotational speed control. 2) Rotational position control. (Assure that the video heads first contact the tape at precise intervals as the drum rotates.)	SPEED: Maintain a consistent linear tape speed (coarse tape speed control). PHASE: Refine the coarse tape speed (precise tape speed control).	SPEED: Maintain a constant linear tape speed (coarse tape speed control). PHASE: Position the recorded RF track precisely under the scanning video head so that the playback head can scan up the center of the track.
REFERENCE SIGNAL	SPEED: 3.58 MHz crystal oscillator counted down to 30 Hz. PHASE: Incoming video vertical sync.	SPEED: 3.58 MHz crystal oscillator counted down to 30 Hz. PHASE: 3.58 MHz crystal oscillator counted down to 30 Hz.	SPEED: 3.58 MHz crystal oscillator counted down to 30 Hz. PHASE: 3.58 MHz crystal oscillator counted down to 30 Hz.	SPEED: 3.58 MHz crystal oscillator counted down to 30 Hz. PHASE: 3.58 MHz crystal oscillator counted down to 30 Hz.
FEEDBACK SIGNAL	SPEED: Pulses coming from the drum motor frequency generator (FG). PHASE: Pulses coming from video head position magnets mounted on the video head drum motor (called PG for pulse generator).	SPEED: Pulses coming from the drum motor frequency generator (FG). PHASE: Pulses coming from video head position magnets mounted on the video head drum motor (called PG for pulse generator).	SPEED: Pulses coming from the capstan motor's frequency generator (FG). PHASE: Counted down capstan pulses coming from the capstan motor's frequency generator magnets (PG Pulses).	SPEED: Pulses coming from the capstan motor's frequency generator (FG). PHASE: Control track (VHS) or ATF (8mm) signal recorded on the tape.

tion of each recorded track relative to the next track would not be consistent. Therefore, the location of the recorded RF tracks relative to each other depends on the precise speed control of the capstan servo. Accurate placement of these tracks during record is critical because it assures good playback tracking and determines the playback picture quality.

The capstan servo section has one other responsibility not indicated in the chart. The coarse speed servo circuit controls the SP/LP or SLP record and playback speeds selected by the user.

Figures 8.2 and 8.3 are block diagrams of the basic capstan record and playback servo circuits used in many units. The capstan motor

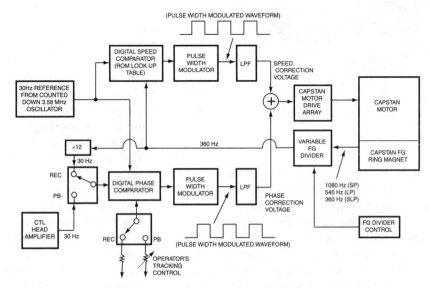

Fig. 8.2. *Basic capstan record and playback servo operation used in current model VHS units.*

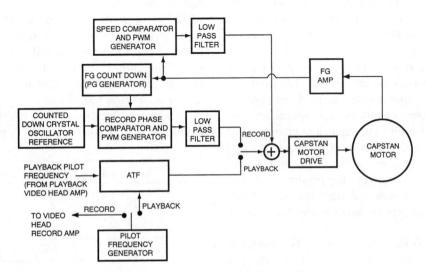

Fig. 8.3. *Basic capstan record and playback servo operation used in 8mm units.*

has a permanent magnet attached to the fly-wheel that produces an FG signal. Two direct drive capstan motors and one belt drive capstan motor are shown in Fig. 8.4.

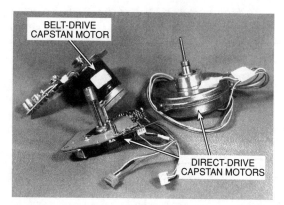

Fig. 8.4. Capstan servo motors.

Figure 8.2 shows incoming FG pulses passing through a divider circuit that produces a 360 Hz signal. (FG frequencies vary according to manufacturing designs. The 360 Hz frequency is used for discussion only.) The divider circuit is key to correct capstan speed servo operation as the record and playback tape speeds (SP, LP, SLP/EP) are changed by the operator. It divides its input by one, two, or three as directed by the incoming divider control circuit (to be outlined shortly) and always provides a 360 Hz output for the speed comparison block.

During record, the capstan servo pulls the tape through the machine at the speed selected by the operator. In playback the capstan servo circuitry monitors the speed of the recorded tape and automatically changes the video tape speed for proper playback. If, during the recording mode the video tape speed select button is changed, the playback capstan servo circuit senses the new speed and adjusts the tape speed accordingly for proper playback.

The 360 Hz signal from the FG divider is passed to the digital speed comparator where it is digitally sampled and then verified against a

30 Hz reference pulse. The result of the comparison is checked against a digital read only memory (ROM) look up table in the same block where a decision is made regarding correction to the motor speed. If a speed correction is necessary, the ROM contains information as to how much of a change is necessary.

The digital speed servo data is applied to the *pulse width modulation (PWM)* block. This circuit generates a pulse that will eventually become the speed correction and motor drive voltage.

Within Figs. 8.2 and 8.3 are two blocks each labeled "pulse width modulator (PWM)." These are square wave oscillators whose period is controlled by the digital correction signal from the ROM. Figure 8.5 shows the output of the PWM during correct and incorrect rotation speeds of the motor being controlled. When the motor is rotating at the correct speed, the PWM output is a 50/50 duty cycle. If the motor is rotating faster than normal, the PWM signal outputs a square wave whose positive-going period has a longer duration. If the motor is rotating slower than it should, the positive period of the square wave has a shorter duration. Figure 8.6 is a PWM waveform as viewed on an oscilloscope. These square wave PWM signals are filtered and converted into a DC control voltage. The amount of DC voltage produced by the filtered PWM signal corresponds to the length of time the positive portion of the signal is present. The faster the motor rotates, the higher the DC voltage produced. The opposite is true if the motor is rotating too slowly.

When evaluating a servo PWM output, it is helpful to note the waveform appearance of

CYLINDER MOTOR SPEED	SLOWER ROTATON SPEED	PROPER ROTATION SPEED	FASTER ROTATION SPEED
PWM CIRCUIT OUTPUT	HIGH FOR SHORTER DURATION	50% DUTY CYCLE	HIGH FOR LONGER DURATION
DC VOLTAGE	VOLTAGE DECREASES	CENTER VOLTAGE	VOLTAGE INCREASES

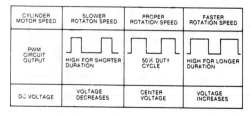

Fig. 8.5. Servo motor speed control output.

Fig. 8.6. *Oscilloscope presentation of PWM.*

the normal and correction (increased and decreased) pulse duty cycles shown in Fig. 8.5. When the motor's speed and phase are correct, a 50/50 duty cycle exists. Note that when the motor speed needs to be increased (motor is running too slow) the duty cycle has a shorter positive duration. This is caused by the lower FG frequency coming from the motor.

Notice what happens to the DC voltage output from the low-pass filter in Fig. 8.2 as the capstan speed PWM changes. When the PWM signal has a 50/50 duty cycle, the output will be 50% of the peak amplitude of the square wave. This is the center voltage described in Fig. 8.5. In other words, if the waveform is 5 volts in amplitude, a 50/50 duty cycle will produce a 2.5 volt motor drive voltage. When the motor is running slow, the voltage out of the LPF will be reduced because of the shorter positive cycle time. (Also see Fig. 8.5.) A low voltage at the output of the LPF causes the motor to speed up (more torque).

When the motor runs too fast, Fig. 8.5 shows the PWM having a longer positive duration time output. This is due to the increased FG pulses coming from the motor. This in turn increases the LPF output voltage in Fig. 8.2, which directs the motor to slow down.

In our example, when the motor is running at a correct speed, the speed PWM circuit outputs a motor drive voltage of 2.5 volts. This is the nominal motor operation voltage. Any change in this voltage will cause the motor speed to change. Compare this motor drive

voltage with the analog sample and hold voltage produced in Fig. 8.1. The digital circuits use many of the common voltages and principles applied to the earlier analog model designs. Both circuit designs produce an analog DC correction voltage for the motor.

In normal operation, the PWM waveform rapidly shifts between a 50/50 duty cycle and some other slightly different duty cycle. This is because the servo constantly acts on the signal to make corrections to the motor speed and phase. When observing a PWM waveform on an oscilloscope, the rising and falling edges may not appear well defined because of this continued rapid duty cycle shifting. You should be able to see that the mean value of the waveform duty cycle is near 50/50, however.

Now evaluate the capstan record phase operation shown in Fig. 8.2. The 2.5 volt signal used to turn and correct the motor's rotation is not the product of the speed PWM and speed LPF alone. It is derived from a combination of the speed and phase paths shown in the figure. The PWM blocks in both sections have their products added together in the summation circuit to create the complete 2.5 volt motor drive and correction voltage. The speed section is responsible for generating a major portion of the 2.5 volts. The phase correction voltage is used to modify the coarse voltage from the speed section to fine tune the result. The end product is a nominal 2.5 volt motor drive and correction potential.

During record, the 360 Hz capstan FG signal is divided by 12 to produce a 30 Hz signal for the digital phase comparator. The record/playback selection switch in Fig. 8.2 routes the counted down FG signal to the digital phase comparator in playback. This generated 30 Hz is labeled "PG" from this point on in the phase processing path. PG is the motor feedback frequency used by the comparator to indicate the motor phase status. This pulse is labeled PG so it won't be confused with the FG signal used in the speed control path. PG signals are used in a servo for phase correction, whereas FG signals are used for speed control.

Inside the digital phase comparator, the PG is digitally sampled and compared to the applied reference 30 Hz signal. If the comparison of both the reference and feedback 30 Hz signals indicates that the phase of the motor is correct and stable, the digital phase comparator sends a digital code to the phase PWM block telling it to produce a 50/50 duty cycle. If the motor phase requires correction, the digital phase comparator is able to sense this and causes the PWM generator to output a different duty cycle that modifies the 2.5 volt motor drive voltage and restores the correct phase.

Note the tracking control selection switch that supplies a signal to the bottom of the phase comparator in Fig. 8.2. The variable resistor represents the playback tracking control used by the operator during playback. During record, this input is switched to a fixed adjustment setting. In some designs, during record, this switch connects to an internal control device that has been adjusted by a technician during record servo alignment.

Also note in Fig. 8.3 that there is no playback tracking control available to the operator. The 8mm tracking is completely automatic and cannot be overridden by the operator.

It's common for a VCR servo circuit like the one in Fig. 8.2 to have both speed and phase PWM generators operate with a 5 volt peak-to-peak square wave output. Both of these PWM waveforms are passed to their respective low-pass filters where they each produce a 2.5 volt correction voltage. The speed and phase voltages are then combined in an algebraic summation circuit to generate a composite 2.5 volt signal that represents the motor drive and speed and phase correction voltages.

Drum Record Servo

Table 8.1 indicates that the drum servo must adjust the head position to the incoming record signal vertical sync phase. Therefore, the head's rotational position (phase) must be able to track the incoming video signal vertical sync pulse timing. This requires the drum servo to have a video input (vertical sync) as one of its reference signals.

Drum record speed
During record the drum rotation must be steady. Figure 8.7 shows why. Notice the loca-

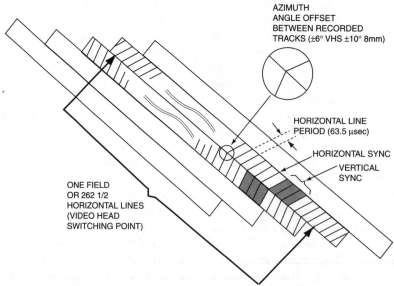

Fig. 8.7. An RF track showing horizontal sync and vertical sync pulse positions.

tions of the FM modulated horizontal sync pulses indicated in the figure. There are 262.5 horizontal pulses per field. Each RF track contains one field. When the capstan is moving the tape through the tape path smoothly, the RF tracks containing these sync pulses are evenly spaced on the tape. It is the steady rotational speed of the head that places the consistent distance between the modulated horizontal sync pulses on each track. In other words, the capstan servo determines the even placement of the RF tracks on the tape and the drum servo that defines the placement of the sync pulses within each track.

If the head speed is allowed to vary as the head writes the information up the track, the distance between the horizontal sync pulses on the tape will change. Higher rotation speeds will cause the pulses to be farther apart and slower speeds will cause them to bunch up. Also observe the position of vertical sync in Fig. 8.7. This too will change position if the recording head speed is not precisely maintained.

Figure 8.8 shows the basic drum record/ playback speed and phase servo operation. The digital drum record servo operates much like the digital capstan record servo. The speed

and phase paths each contain their own PWM blocks with their corresponding digital comparison circuits. Each of these PWM circuits produces a 50/50 duty cycle when the motor speed and phase are correct. Like the capstan servo, the drum motor drive and correction voltages center around 2.5 volts.

The drum speed operation is identical to the capstan speed servo function. The 360 Hz FG signal from the drum's magnetic ring is compared with a counted down crystal reference and the comparison data is evaluated in a ROM look up table to check for the need of speed PWM correction action.

Drum record phase

Design differences between the drum and capstan servo occur in the servo phase sections and relate to the generation of the drum PG signal and the source of the record phase reference pulse. Recall that the capstan servo used a divided down motor FG to generate its record PG signal. This was possible in that circuit because the capstan speed and phase are self-governing and do not require lock up to an external reference such as the record video signal. The capstan phase control is only responsible for fine tuning the motor's speed.

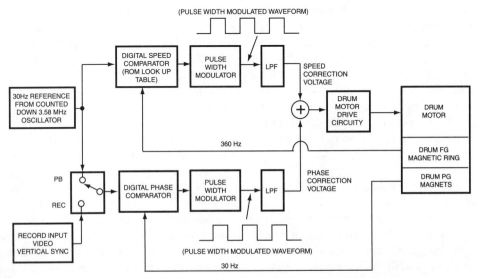

Fig. 8.8. Basic drum record and playback servo operation used in current model VHS and 8mm units.

On the other hand, the drum's phase control is required to have the video head contact the video tape 10.5 H (where H = horizontal lines) or 667 microseconds, before the start of the record video's vertical sync pulse. To establish this head-to-vertical sync pulse timing, the drum servo system must recognize the position (phase) of the record video head and the timing (phase) of the vertical sync. With this information, the drum phase servo can compare the vertical sync timing and the recording head's position and alter the rotational position of the head drum, if necessary, to maintain this advanced 10.5 H contact of the recording head. In other words, the drum position (phase) is adjusted to accommodate the arrival of the vertical sync. The record vertical sync is the drum phase servo reference while the PG signal provides the feedback drum position (phase) data.

Notice that the drum motor in Fig. 8.8 has separate PG and FG sections. Unlike the capstan motor, the drum motor must maintain a separate PG and FG magnet arrangement because of the video head position requirement of the design. The magnetic ring used to develop the FG and PG signals in the capstan servo is not capable of providing this information. An FG magnet is a continuous ring that only provides a constant oscillation, whereas the drum PG magnets are strategically mounted on the drum motor to provide pulses corresponding to the head position. Since the capstan motor only uses its PG data for speed refinement and not rotational position control, it is able to use a counted down FG as the PG signal. That is not true for the drum motor however, because the position of this motor is an important factor.

Video head switching pulse formation

On the video drum head assembly, two permanent magnets are used to identify the exact position of the heads. As these magnets rotate on the head drum assembly, they pass a pickup coil mounted on the stationary lower drum portion of the assembly. The pickup coil senses the magnetic lines of force cutting across its windings and produces the series of pulses that are used to form the 30 Hz drum PG. These pulses pass through an adjustable electronic delay circuit that compensates for small mechanical errors in the placement of magnets on the drum assembly. This adjustment is made for proper drum phase timing during servo alignment. The drum PG signal is also used to form the video head switching pulse required to turn on the individual video head playback preamplifiers. The system control monitors the video head switching pulse to be sure that the video head motor is rotating at the correct speed. If this circuit senses that the head is not rotating correctly, it will shut down the machine.

The drum PG signal in Fig. 8.8 is applied to the digital phase comparator where it is checked against the vertical sync reference. The source of the vertical sync reference is the input video signal being recorded. The drum phase servo functions to place the recording video head on the tape 10.5 H before the start of the vertical sync interval. If the drum phase is incorrect with respect to the vertical sync, the phase PWM is caused to alter the 50/50 duty cycle accordingly so that the motor drive voltage will correct the phase error.

The drum phase correction does not function until the speed servo has stabilized the drum speed to the correct velocity. If the drum speed is off by just one RPM (revolution per minute) the phase servo cannot perform its positional correction.

In summary, the capstan record servo is responsible for pulling the tape through the tape path at a constant speed. This action is responsible for the location of the recorded helical RF tracks on the tape. The capstan speed servo is responsible for governing the coarse speed and the capstan phase control refines the coarse correction. The drum record servo is responsible for positioning the vertical sync and horizontal sync pulses at precise locations within the RF track and for maintaining exact distances between the adjacent sync pulses throughout the track. The drum speed servo is responsible for all speed maintenance while the phase servo directs the video head rotational position. Keep Table 8.1 close at hand when reviewing VCR servo functions. If you consider each servo section's responsibility under various modes of operation while troubleshooting a servo failure, isolating a defect will be much easier.

CAPSTAN AND DRUM PLAYBACK SERVO OPERATION

Now use Table 8.1 to help explore the tasks of the playback servos. With the exception of the operation of the capstan's automatic playback speed selection circuit, the playback speed control for both the drum and capstan servo functions identically in the record mode. Since the fundamental servo speed control is the same for the two modes, and the record speed was already described, only the automatic playback speed selection section of the capstan speed servo will be presented as we explain how playback speed control works. Since the phase servos operate differently between record and playback, we will describe the playback phase operation in detail.

Drum Playback Phase

The operation of both the capstan and drum phase servos change somewhat in playback as compared to how they work in record. Recall that the record drum phase servo had to synchronize the position of the video head with an external signal while the capstan speed and phase record servo sections were only required to move the tape at a consistent speed. In playback, the drum phase servo no longer requires the external video sync reference because it only needs to move the video head at a precise rotational speed. In playback the drum must turn with precision so that the

capstan can accurately place the recorded RF tracks underneath the playback scanning head's path.

While the drum's rotation needs to be concise during playback, it is not phase locked to any external input. This is why Table 8.1 shows the drum playback phase reference changed from the separated vertical sync input to a counted down crystal reference.

During playback, the drum phase servo ensures that the video head first contacts the tape at exact intervals as it rotates. This is accurately done by comparing the VCR's counted down internal crystal reference to the video head switching pulses generated by the head PG magnets. Switching to an internal reference during playback allows the VCR's drum phase servo to play back correctly without the need for a constantly connected external video signal. When the video head drum's playback phase playback servo is performing properly, the vertical sync recorded on the tape will play back at precise intervals and at a stable position out of the viewing area on the playback monitor.

A review of Fig. 8.8 reveals how the playback drum servo functions in a VHS and 8mm digital servo circuit. Note that the operational principle of the speed servo does not change for the playback mode. The drawing shows that the drum phase control receives a different reference input during playback. With the exception of this one input, the drum phase servo operation is identical in the record mode. In playback the drum servo is only concerned about precise rotation speed—not phase locking onto an external video reference. Using an internal reference for playback phase operation allows the VCR to play back correctly without being hooked up to a video source.

Thus, the drum's playback phase servo becomes a speed refinement circuit during playback. The drum must rotate with precision so that the capstan's phase servo can predict where the video head is going to start scanning the tape. The capstan phase servo must be able to forecast where the video head will start its scan so that it can accurately position the RF tracks underneath the playback video heads.

Capstan Playback Phase

During playback, the capstan's phase servo can no longer be relegated to just moving the tape at an even speed. It must now manipulate the tape's longitudinal position to place each recorded track underneath one of the moving video heads. The capstan playback phase section performs this task. To do this, the capstan phase servo circuit needs to know the location of each of the RF tracks. It must also be able to identify which track was recorded by a particular video head because video heads use different head gap azimuth angles to record the individual tracks. If the wrong head is used to read a particular track, the azimuth cancellation within the head gap will prevent the RF signal from leaving the head core.

The RF track identification and position information is contained in a tracking signal recorded on the tape at the same time the video FM signal is written onto the tape. In VHS machines, this tracking signal is called the control track pulse—abbreviated CTL (for control track inductance—where L is the symbol for inductance). In the 8mm format, this tracking signal is called ATF for auto-track-find. The 8mm ATF signal is a series of frequencies recorded within the video RF track. The VHS CTL signal is comprised of 30 Hz (actually 29.97 Hz) pulses recorded on a separate longitudinal track at the tape's bottom edge. These tracking control signals will be described later in this chapter.

The VHS capstan servo uses the tracking control signal to identify the location of each RF track and the particular recorded azimuth angle used to record each track. As shown in Table 8.1 the capstan playback servo uses the tracking signal to position the RF track beneath the path of the scanning video head. The capstan servo is responsible for steady tape

speed and precise RF track positioning. When you move a VHS machine's tracking control, you are electronically altering the tracking signal information. Movement of this control alters the phase of the feedback CTL signal and causes the capstan servo to either advance or retard the position the RF tracks relative to the head's scanning path depending on which way the tracking control is moved. This customer adjustment is provided to allow recorded video tape to be interchangeable between different machines.

The 8mm format capstan playback servo does not allow the operator to override the automatic phase control. Optimum interchangeability is achieved by using the ATF signal. During playback the drum servo must maintain a precise video head rotation speed and ensure that the video head contacts the moving tape at the correct time. It is at this point of operation that the capstan and drum playback servos must work together to achieve the desired result. The drum speed control governs the coarse rotational speed while the phase control refines the drum speed. The drum playback phase servo makes certain that the head-to-tape contact occurs at precise intervals as the head rotates. In other words, in playback, the head servo maintains a precise drum speed and the capstan servo places the tracks underneath the moving head. The drum's speed and phase servo control work together to assure that the timing of each playback vertical sync pulse is 33 milliseconds from the next vertical pulse on the adjacent track (59.94 Hz) and that each of the horizontal sync pulses on a given RF track are a constant 63.5 microseconds apart (Fig. 8.7).

Now look back at Figs. 8.2 and 8.3 to see how the digital capstan playback servos operate in VHS and 8mm machines. In each of these units, the capstan servo control works the same in playback as it does in record mode with two exceptions. First, the FG divider circuit is no longer controlled by the record speed selection switch on the front of the VCR. It is now controlled automatically by an auto speed selection circuit. The second difference is the source of the 30 Hz square wave feedback signal. In a VHS machine, in the record mode, the source of this signal is from FG pulses that have been divided down. In playback the source of the 30 Hz signal is from the control track head. The 8mm format uses a correction voltage from the ATF servo section for phase control.

The VHS Control Track

Notice in Fig. 8.2 that the VHS digital phase feedback signal is switched to the amplified CTL signal in playback. This is necessary because the VCR needs to know the position of the RF tracks on the tape during playback. Control track pulses identify the location of the RF tracks on the tape. This is why the capstan servo must use the control track as a playback feedback source.

VHS VCRs use the CTL pulse to phaselock the capstan motor rather than the counted down 30 Hz FG feedback pulse. The 30 Hz CTL signal from the tape identifies the exact position of the vertical sync pulse as it is recorded. Earlier in this chapter, we described how the capstan servo made sure that each video head RF track reaches the video head drum at the exact time that the corresponding head begins to sweep across the tape in playback. The head drum and capstan servos must work closely to accomplish this task. The responsibility of the playback head drum servo is to make sure that the video head contacts the tape at precise 29.97 Hz (30 Hz) intervals and exactly 10.5 horizontal lines prior to vertical sync. The capstan servo must make sure that the recorded RF track is there at the same time that the head arrives. In other words, in playback, the head drum servo maintains a stable rotating speed and phase of the video head drum while the capstan servo advances or retards the speed of the tape slightly to match the appropriate RF track with the correct head.

The control track pulse from the tape is sampled by the digital phase comparator and verified against the counted down 30 Hz reference. The tracking control shown near the lower left corner of Fig. 8.2 is the variable control used by the operator to adjust the tracking (if necessary). This control allows the VCR user to alter the standard tracking phase by shifting the digital playback phase. Normally the tracking control is left in a center (preset) position unless a tape interchange problem exists during playback. External tracking control is only available during playback.

The capstan digital phase control circuitry compares the tracking control setting and the reference and feedback pulses, and then issues a PWM generation signal to the PWM block in the form of a digital pulse stream. The PWM output is filtered to become a DC phase correction voltage and part of the capstan motor drive input.

The VHS Tracking Control

Tracking control, an adjustment found on the outside of a VHS machine, applies a control voltage to the digital phase comparator circuit and fine tunes the capstan servo's phase control during playback. Mechanical tracking controls have a center detent position to physically indicate the setting for normal operation. Even VCRs using electronic pushbuttons for tracking control typically have a means of electronically centering the range of their adjustment. In either case, this adjustment enables fine tuning of the capstan servo control electronics to correct for tracking differences between different machines. Ideally, all machines will play back correctly with the tracking control in the center detent position. If for some reason yours doesn't, the tracking adjustment will help correct this error.

The tracking control advances or retards the phase of the capstan servo by altering the relationship of the reference and feedback pulses. The tracking control is moved from the standard position to slightly advance or retard the phase relationship of these two pulses and to obtain the best playback picture. Advancing the phase of the capstan servo with the tracking control causes the tape to kick forward a little so the RF tracks on the tape are positioned properly for detection by the video head. Retarding the phase of the capstan servo with the tracking control causes the video tape to slow down briefly and allows the video heads to catch up with RF tracks on the tape. Thus the tracking control adjusts the phase relationship between the rotating video head drum assembly and the RF information recorded on the tape by altering the phase of the capstan motor.

In 8mm VCRs, there isn't a control track signal for RF track location and head identification. These machines use an auto-track-find (ATF) signal that identifies the location of each RF track on the tape. The use of ATF supersedes the need for a separate CTL pulse on the tape and also negates the need for a tracking control on the outside of the machine.

VHS machines require that a tracking control be available to the user to enable playback tape interchangeability between machines. Interchangeability errors are caused by mechanical differences between units. These differences are possible because the video head recording the RF track and the CTL record/playback head that records the CTL pulse are at different points in the tape path. If the audio/control track head is not positioned correctly during mechanical alignment of the machine, the CTL pulses recorded on a tape by that unit will cause tracking errors when that tape is played on another VCR. The VHS track-

ing control provides a way to correct for this mechanical error.

Since 8mm units use the video heads to record the ATF signal, mechanical interchangeability errors resulting from an incorrectly positioned CTL head do not exist. Therefore, an external tracking control is not provided for the operator on 8mm machines.

Auto Speed Selection

For a VHS machine, playback automatic speed selection is performed using the capstan speed servo and the electronics outlined in Fig. 8.9. The FG divider in the capstan circuit acts on the incoming FG signal to produce a 360 Hz FG signal at its output regardless of capstan speed. To accomplish this task, the FG divider must know how to divide the incoming FG signal. This information comes from the audio speed select circuitry by measuring the distance between each of the CTL pulses. Two voltages from the auto speed select circuit directly correspond to the speed of the tape.

Assume that during record the first 5 minutes of tape are recorded in LP speed. While still recording the TV program, the record speed selection switch is accidentally changed to the SP position. The tape is recorded at SP speed until suddenly you realize that the program you are recording is longer than you anticipated so you quickly turn the record speed selection knob to the SLP position.

After completing the recording session the tape is rewound and played back. The part of the program recorded in LP is played back first and control track pulses occur at exactly

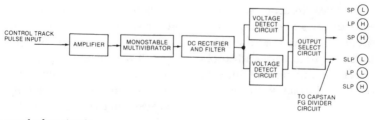

Fig. 8.9. *VHS auto speed select circuit.*

30 Hz as they should. The VCR playback continues at this speed until the point is reached where the recorded speed changes to SP. Since the change in record from LP to SP speed doubles the linear tape speed, the distance between control track pulses recorded on the tape is also doubled. Therefore, when the VCR, which is playing back in the LP speed, comes upon the section of tape recorded in SP, it detects the control track pulses as being twice as far apart or $\frac{1}{2}$ the normal frequency (15 Hz).

The auto speed select circuit forms square wave pulses using a *monostable multivibrator* (MMV). The MMV generates a short duration square wave every time it receives an incoming control track pulse. The farther apart the incoming control track pulses, the greater period of time that passes before the MMV generates another positive-going pulse. These pulses are filtered and converted into a DC voltage.

Because these pulses are farther apart, the average DC filter voltage is lower. This voltage level triggers two voltage detection circuits designed to change output voltage in proportion to the input voltages they sense. The output voltages toggle between a high logic state of 5 volts or more and a low logic state of 0 volts (logic high or logic low). The respective output pulses are passed to the capstan FG divider circuit which is preprogrammed to divide incoming FG signals by a predetermined amount as selected by the logic high and low levels from the auto speed select circuit.

Next, in this playback scenario, the point is reached where the recorded speed changes from SP to SLP. The control track pulses were recorded in SLP but are being played back in SP so now they occur at a 90 Hz rate rather than the normal 30 Hz. The MMV generates a 90 Hz signal which is filtered and converted into a higher than normal DC level. The auto speed select voltage detectors modify their outputs to a logic state which allows the FG divider circuit to function properly.

While these speed changes are taking place, the capstan speed servo circuitry must make dramatic voltage changes to adjust the capstan motor to the right speed. This is done by forcing the speed PWM block to generate a duty cycle other than the normal 50/50 which in turn generates the appropriate motor speed correction voltage. The capstan speed is changed to the new value because the FG divider circuit has shifted its countdown ratio to a new quantity. This in turn causes the FG feedback signal input to the speed comparison circuit to be incorrect (something other than 360 Hz). The speed comparator refers to the speed look up table in the ROM and issues the digital correction data to the PWM block. The speed PWM duty cycle is changed from the 50/50 duty cycle to an appropriate duty cycle and corrects the motor speed. Once the capstan FG feedback signal out of the FG divider reaches 360 Hz again, the speed PWM returns to the nominal 50/50 duty cycle. This speed correction process starts all over again when the auto speed selection circuit senses a recorded speed change and alters the capstan FG divider circuit parameters.

Although 8mm units don't use a CTL pulse for tracking control, they still have auto playback speed detection capability. This feature is possible through a combination of operations between the playback ATF circuitry and the microprocessors used in the system control and capstan servo circuits. Just as in VHS machines, it is the 8mm capstan servo that changes the linear tape speed between the SP and LP modes. One difference between the VHS and 8mm formats is in the number of record/playback tape speeds available. Many VHS models provide up to three tape record/playback speeds. The 8mm format is limited to two speeds. The automatic playback speed selection in the 8mm machines will be presented later in this chapter as part of the 8mm ATF servo theory.

VHS Auto Tracking

Some recent VHS models provide an auto tracking feature for the customer. This option can be switched off by the operator. The auto tracking circuit monitors the playback RF level coming from the video head playback RF amplifier section in the video processing stage. Since optimum tracking causes the video head to scan down the center of the recorded track, maximum RF level is obtained when tracking is optimized. Therefore, the auto tracking circuitry is able to adjust the capstan servo for best tracking by monitoring the playback RF level and automatically adjusting the tracking control circuit for peak RF level.

When auto tracking is active, the external tracking control available to the customer is defeated and the auto tracking circuitry takes over the tracking adjustment. The customer can regain manual tracking control by turning off the auto tracking feature and manually adjusting the external tracking control.

Basic ATF Operation

All 8mm units use ATF to identify when the video head is tracking down the center of the correct video track. ATF not only eliminates the need for the CTL head used in the VHS format, but it negates the need to have a customer tracking control.

VHS units require an external tracking control so the operator can adjust for tape playback interchangeability. External tracking control is required in VHS designs because of the potential for different mechanical distances between the initial video head-to-tape contact point on the video head drum and the placement of the CTL head in the tape path among individual machines. The tracking control lets the user electronically correct for these mechanical differences.

One advantage of using an ATF signal in 8mm units is that this signal is recorded along the entire RF track allowing the machine to adjust the tracking throughout the entire trace of the head up the track. In the VHS format, a single CTL pulse is only present at the beginning of each track. This single pulse allows the

VHS unit to make a tracking correction only once during each head scan. The ATF feature simplifies the 8mm tape path alignment and assures good tape playback interchangeability during the entire period that the head scans the track.

The ATF servo operation relies on four pilot frequencies that are injected into the FM modulated video signal and into the PCM audio portions of the 8mm RF tracks during record. Figure 8.3 shows that these pilot frequencies are produced by the reference pilot generator section of the capstan servo circuit and sent to the video RF and PCM record amplifier for writing onto the tape during the record mode. The low amplitude and low frequency of these pilot signals keeps them from interfering with the FM video or the FM audio signals in the video RF section of the tape or with the PCM audio information in the PCM section. Figure 8.10 shows that the first pilot signal, F1, is 102.54 kHz and is recorded on the first video track. The second pilot signal, F2, has a frequency of 118.95 kHz and is written on the corresponding video track.

Notice in Fig. 8.10 that each PCM audio segment has a different pilot frequency than the video portion of the same track. This is because the recorded pilot frequency is changed at the video head switching point on the track. Since video head switching takes place between the PCM audio and video RF sections of the track, the record ATF pilot frequency also changes between these two sections of the same track. The change in the pilot frequency is controlled by command lines supplied to the pilot generator from the system control section.

The next two video tracks in the sequence are the F3 and F4 tracks. These two video tracks accommodate the pilot frequencies of 165.21 kHz and 148.69 kHz respectively. Once the four field sequence has completed its cycle, the progression starts over beginning with the F1 pilot frequency. This process repeats throughout the recording. These pilot frequencies are recorded

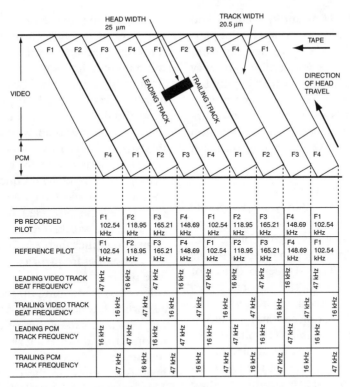

	F1	F2	F3	F4	F1	F2	F3	F4	F1
PB RECORDED PILOT	F1 102.54 kHz	F2 118.95 kHz	F3 165.21 kHz	F4 148.69 kHz	F1 102.54 kHz	F2 118.95 kHz	F3 165.21 kHz	F4 148.69 kHz	F1 102.54 kHz
REFERENCE PILOT	F1 102.54 kHz	F2 118.95 kHz	F3 165.21 kHz	F4 148.69 kHz	F1 102.54 kHz	F2 118.95 kHz	F3 165.21 kHz	F4 148.69 kHz	F1 102.54 kHz
LEADING VIDEO TRACK BEAT FREQUENCY	47 kHz	16 kHz	47 kHz	16 kHz	47 kHz	16 kHz	47 kHz	16 kHz	47 kHz
TRAILING VIDEO TRACK BEAT FREQUENCY		16 kHz	47 kHz	16 kHz	47 kHz	16 kHz	47 kHz	16 kHz	47 kHz	16 kHz
LEADING PCM TRACK FREQUENCY	16 kHz	47 kHz	16 kHz	47 kHz	16 kHz	47 kHz	16 kHz	47 kHz	16 kHz
TRAILING PCM TRACK FREQUENCY		47 kHz	16 kHz	47 kHz	16 kHz	47 kHz	16 kHz	47 kHz	16 kHz	47 kHz

Fig. 8.10. *The 8mm pilot frequencies and their relationship to the video and PCM tracks on the tape.*

on the tape at −14 dB below the color-under signal level.

Figure 8.10 shows that, in playback, the video head scans up the center of the F1 track and extends onto the two adjacent tracks. This occurs because the head is 25 microns wide and the recorded tracks are only 20.5 microns wide. (Chapter 5 explained how the playback tracks end up being narrower than the recording head width.) Since the pilot signals on each of the tracks have relatively low frequencies, the ±10 degree azimuth offset of each head gap has no effect on canceling the pilot signal crosstalk detected by the playback head from the adjacent tracks. Thus, the pilot frequencies of the F4 and F2 tracks are picked up by the video head as it scans up the center of the F1 track.

It is important to realize that as long as the head is scanning up the center of the desired track, the pilot signal crosstalk from both of the adjacent tracks will be equal in amplitude. It is also important to realize that if the head drifts slightly off the center of the desired F1 track towards an adjacent track (F4 for example), the crosstalk level of the pilot signal from the F2 track will decrease while the level of the F4 crosstalk will increase. These principles are fundamental to ATF servo control.

While in playback, the 8mm servo constantly monitors the levels of the pilot frequencies on the two adjacent tracks and adjusts the position of the tape track under the scanning head to maintain equal amplitudes of both adjacent track pilot crosstalk signals.

To correct for tracking errors, a method was devised for the ATF servo system to advance or retard the tape position relative to the scanning video heads. The ATF system was designed to recognize which way to shift the tape to correct for tracking errors during the entire time that the PCM audio sections and video segments are being scanned by the playback heads. The capstan servo accomplishes this by comparing the playback pilot frequency being detected from the tape with the pilot frequency from the reference pilot generator. If the compared signals do not match, the capstan servo jogs the tape position to reposition the track underneath the scanning playback video head.

ATF record operation

Since the capstan servo's only responsibility during record is to maintain a precise tape speed, only the capstan's digital phase and speed control portion function to perform this task during record. The ATF servo does not make any corrections to the capstan servo motor speed during record. In this mode, the ATF servo error correction voltage is held at about 2.5 volts. Holding the ATF voltage here during record allows the correction voltage from the digital phase section of the capstan's record digital servo to have complete phase control. During record, the pilot frequency generator in the ATF circuit produces the F1 through F4 frequencies for recording on the tape, but the ATF correction circuits sit idle until the playback mode is engaged.

ATF playback operation

During playback, the capstan's digital phase control voltage is held to approximately 2 volts letting the ATF servo stage make all the tracking phase corrections. For playback, the ATF error correction voltage has a dynamic correction range of 0.8 to 4.0 volts, which is sufficient to control the tracking phase without the assistance of the digital phase circuitry.

Tracking the video RF section

As the video head scans up a track during playback, the ATF frequencies detected by the video head are compared with the reference pilot frequency signal generated during that head scan period. The capstan servo's system control knows which of the four fields the video head should be scanning at a given time and causes the pilot frequency generator to produce the

corresponding reference pilot frequency for that field. If the video head is on the right RF track, the frequency read by the video head and the reference frequency produced by the pilot frequency generator will be the same. When the identical frequencies from the playback video head and the reference pilot generator are combined, the beat frequency produced from these two signals is zero Hz. (A beat frequency is an oscillation that results from mixing or heterodyning two dissimilar frequencies. As long as the two frequencies are the same, the beat frequency will be zero Hz.) If the video head is scanning the wrong track, the beat frequency is something other than zero.

The pilot frequencies were selected to enable the beat frequencies to indicate when the video head is scanning up the wrong track and also when it is not centered on the correct track.

To understand how the system detects incorrect head centering as it scans up the correct track, consider this case. Assume that the head is traveling up the F1 track as it should. The table in Fig. 8.10 indicates that the detected pilot frequency coming off the tape and the reference pilot generator frequency are the same (102.54 kHz). When the F1 RF track pilot frequency and the F1 reference pilot generator frequency are compared, they produce no beat or 0 Hz (102.54 kHz compared to 102.54 kHz = 0 Hz), indicating to the capstan servo that the head is on the right track.

Notice in Fig. 8.10 that the playback video head is slightly wider than the RF track that it is scanning. In this drawing the video head is not only detecting the F1 pilot frequency, but it is sampling some of the F2 and F4 frequencies from the adjacent tracks as well. The F2 and F4 frequencies from the adjacent tracks become crosstalk, which also beat with the reference generator's F1 pilot frequency. When the reference pilot generator's F1 frequency beats with the F2 crosstalk pilot frequency the result is 16 kHz (118.95 kHz − 102.54 kHz = 16 kHz). This result is called the *trailing track frequency*.

Note that the trailing track is located to the right of the center track. When the reference F1 pilot frequency beats with the F4 crosstalk pilot frequency, it produces a *leading track frequency* of 47 kHz (148.69 kHz − 102.54 kHz = 47 kHz). The leading track is located to the left of the center track. As long as the servo senses that the amplitudes of the leading and trailing track crosstalk frequencies are equal, and that the trailing track frequency is 16 kHz, and that the leading track frequency is 47 kHz, it recognizes that the video head is on the right track and is properly centered.

When the video head is positioned slightly off the center of the F1 track towards the F2 (trailing) track, the ATF servo senses that the tape is moving slightly ahead of the scanning head's path (too fast) and has pulled the trailing track too far underneath the playback video head. To correct this, the servo system slows the tape down to recenter the head on the F1 track. Conversely, when the head is slightly off center towards the F4 (leading) track, the tape is being pulled too slow relative to the scanning head, and the ATF servo causes the capstan motor to speed up the tape movement to establish proper scanning in the center of the F1 track.

As you examine the potential possibilities in Fig. 8.10, notice that the table associated with the drawing indicates that sometimes the leading and trailing track pilot frequencies don't beat with the reference to produce a 16 kHz trailing frequency or a 47 kHz leading frequency, even when the video head is on the right track. When the F2 and F4 tracks are scanned, the leading and trailing pilot beat frequencies are reversed compared to the F1 and F3 adjacent track crosstalk pilot signals. To simplify this introduction to ATF servo operation, consider just the operation of the pilot signals and the crosstalk beating while the F1 and F3 tracks are being scanned. Later, you'll read how the servo handles the inverted relationship of the pilot signal crosstalk when the F2 and F4 tracks are scanned by the playback head.

The capstan ATF servo makes phase adjustments to keep the head centered on the RF track throughout the entire scan. The system considers the video head centered on the RF

track when the levels of leading and trailing beat frequencies are equal in amplitude. As soon as the head-to-track relationship begins to deviate off center and favor one of the adjacent tracks, the amplitude of the corresponding beat frequency increases while the level of the other beat frequency from the other adjacent track decreases. The ATF servo uses this information to determine how much of a correction is necessary and which way to shift the tape to reestablish center tracking.

Now, suppose that the video head begins traveling up the F3 track when it is supposed to be tracking up the F1 track. In this case, the reference pilot frequency generator is outputting the F1 pilot frequency, but the video head is outputting the F3 pilot signal. At this time, the center track beat frequency becomes 63 kHz (165.21 kHz − 102.54 kHz = 63 kHz) instead of zero hertz because the reference pilot generator F1 frequency is beating with the F3 track pilot signal detected by the video head. The 63 kHz beat frequency is well above the servo system's frequency passband and is therefore ignored. Interestingly, the leading track crosstalk frequency is now 16 kHz instead of 47 kHz, and the trailing track beat frequency is 47 kHz instead of 16 kHz. This is opposite to the normal relationship of the leading and trailing track beat frequencies.

Continuing with this example, note what happens when the video head shifts off the center of the F3 track. (Remember that this is all happening while the video head is scanning the F3 track, when it should be moving up the F1 track.) Assume that the head is veering off the F3 track center towards the F4 track (the trailing track). The corrective tracking action of the capstan servo will cause the pilot beat signal crosstalk to increase (instead of diminish as it normally would) as it attempts to restore the correct tracking center position. This occurs because the head was initially centered on the wrong track reversing the relationship of the crosstalk beat frequency to the direction of the capstan motor's corrective movement. Since the capstan servo is designed to correct the tape's position towards the leading track when the system detects the increased 47 kHz beat signal level, the capstan motor will incorrectly jog the tape in the wrong direction as it attempts to reduce the 47 kHz beat level and recenter the track under the video head. This causes more of the adjacent track to appear under the video head increasing the level of the F4 pilot signal even more. Thus, it is this inverted relationship of the leading and trailing track beat frequencies that causes the 47 kHz beat to continually increase.

As the amplitude of the adjacent track 47 kHz beat signal continues to grow, the capstan servo causes the capstan motor to continue slowing the tape speed as the system attempts to achieve tracking center. This shifting of the tracks continues until the tape has moved to the point where the video head has crossed over the F4 track and begins to detect the F1 track. Now the capstan servo finally begins to sense that it is shifting the tape tracks in the right direction because the 47 kHz signal is beginning to decrease. The servo continues to jog the tape in the same direction until the F1 track becomes the center track and the amplitude of the 16 kHz crosstalk beat signal from the leading track matches the amplitude of the 47 kHz beat frequency from the trailing track.

Instabilities in VCR tape movement require that the capstan servo constantly correct for proper tracking. Therefore, if the VCR does start out scanning up the F3 track when it is supposed to be on the F1 video section, the servo will act to place the correct track under the video head. Although this method of wrong track correction is within the realm of the servo's capability, it takes a period of time for the servo to position the correct track under the playback head. The playback picture will display noise and unstable video as the ATF servo makes this two track correction. Therefore, another method of correction has been developed to shorten the response time of the capstan ATF servo and prevent unwanted picture disturbances when this condition is present. This method uses an electronic form of correction called *rear lock detection*. It will be described later in this chapter.

Tracking the PCM audio segment

Notice in Figure 8.10 that the pilot frequencies for the PCM segments of the RF track are different than the video sections. The PCM audio track under the F1 video track has an F4 pilot frequency for example. Recall that the reference pilot generator's frequency changes to the new value near the head switching point and that the head switching takes place between the PCM and video segments of the track. Therefore, when the PCM segment under the video F1 segment of the track is being played back, the pilot reference generator frequency is still at the F4 frequency. This occurs because the last frequency being generated by the reference circuit was the F4 value and the video head switching pulse has not yet occurred for the F1 video track.

As the PCM section containing the F4 pilot is scanned by the playback head, the F4 pilot frequency (148.69 kHz) is output from the head and beats against the 148.69 kHz frequency from the reference generator. The playback and reference pilot signals combine producing zero Hz (no beat). This result is not used by the ATF servo circuitry and is therefore ignored.

However, the leading (F3) and trailing (F1) PCM pilot crosstalk frequencies recorded on the tape also mix with the reference pilot generator's F4 frequency and produce leading and trailing pilot beat frequencies of 16 kHz and 47 kHz respectively. These crosstalk pilot beat frequencies maintain the same leading and trailing track beat frequency relationships as the video tracks, enabling the capstan servo to correct tracking errors in the PCM segment of the track as well. Figure 8.10 shows that all combinations of the PCM pilot frequencies will produce adjacent track beat frequencies of either 16 kHz or 47 kHz. Further examination of this figure reveals that the leading and trailing frequencies alternate between tracks just as the video tracks do. Notice that when the F1 and F3 PCM tracks are scanned, the leading track beat frequency is 47 kHz and the trailing track beat frequency is 16 kHz. Therefore, the PCM segment tracking can be processed by the

ATF servo in a manner similar to the video track sections.

ATF tracking servo theory

Figure 8.11 is a block diagram of a typical ATF servo playback circuit. In the lower right-hand corner of Fig. 8.11 is the pilot frequency generator. This generator produces the pilot frequencies written onto the tape in the record mode. It also acts as the reference pilot generator during playback.

Inputs to the pilot generator include the stable 5.9 MHz (378 fh) crystal oscillation and the Select 1 and Select 2 command lines from the microprocessor. The Select command lines cause the programmable divider inside the pilot generator to divide the 5.9 MHz reference oscillator signal into the F1 through F4 frequencies that are passed to the mixer in the upper left of the figure.

In record, the Select 1 and Select 2 lines shift at the video head switching point (between the PCM and video RF sections on the track) and the pilot generator divides the 5.9 MHz (378 fh) to supply the correct record pilot frequencies (F1 through F4). The record ATF pilot generator's timing and logic chart are shown in Fig. 8.12.

Notice in Fig. 8.11 that the pilot signal is applied to the video head amplifier where it is combined with the RF and PCM signals for recording.

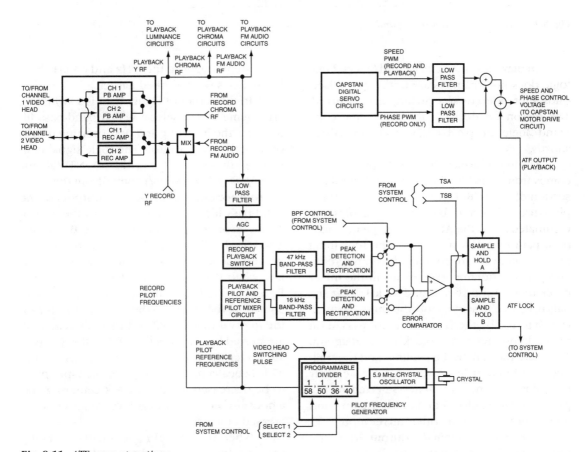

Fig. 8.11. *ATF servo operation.*

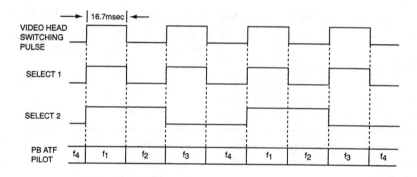

Fig. 8.12. ATF pilot generator timing and logic chart. (Record)

Pilot Signal		Control Signal	
Name	Frequency	Select 1	Select 2
f1	102.54 kHz (378/58 fH)	H	H
f2	118.95 kHz (378/50 fH)	L	H
f3	165.21 kHz (378/36 fH)	H	L
f4	148.69 kHz (378/40fH)	L	L

In playback, the Select 1 and Select 2 line logic does not change state until 2.0 milliseconds after video head switching has occurred. The reason for delaying the Select line command logic in playback will be explained shortly.

During playback the ATF pilot signal comes from the video head playback amplifier section of the VCR. Figure 8.11 shows that the playback video and color-under signals are eliminated from the ATF pilot frequency playback path by a low-pass filter. The remaining pilot frequencies are combined with the reference pilot generator frequencies in the mixer block to form the 16 kHz and 47 kHz beat signals used by the servo. Only the three frequencies detected by the video head are passed on to the frequency mixer block in the diagram. These are the center, the leading, and the trailing pilot frequencies. At this same time, only the pilot reference frequency selected by the pilot reference Select 1 and Select 2 command lines is provided to the other mixer input.

The beat frequencies output from the mixing circuit are passed to two band-pass filters as shown. These are the filters that cause the system to ignore the 0 Hz and 63 kHz beat frequencies we described earlier.

In the center of Figure 8.11 is a switching arrangement that selects which band-pass filter feeds the inverting and noninverting inputs of the error comparator. The logic signal controlling these switches (BPF CONTROL for *band-pass filter control*) comes from the servo microprocessor and causes the outputs of the band-pass filters to be directed to the appropriate error comparator input. The BPF CONTROL logic allows the ATF servo to properly process the inverted adjacent track pilot crosstalk when the F2 and F4 video tracks and the F2 and F4 PCM sections are being played back. Switching the inputs to the comparator during the interval that the F2 and F4 track segments are played back allows the ATF to always operate under the premise that an increase in the 47 kHz beat amplitude requires the tape speed to be decreased to obtain track centering and a boost in the 16 kHz beat level requires an increased tape speed.

When the F1 and F3 segments are being scanned by the head, the switches are configured so that the 47 kHz band-pass filter supplies the noninverting (+) input to the comparator and the 16 kHz band-pass filter is feeding the inverting (−) input. When the F2 and F4 sections are being played back, the switch conducts the 16 kHz signal to the non-inverting (+) input of the comparator and the inverting (−) input receives the 47 kHz signal.

VCR models that use the band-pass filtering selection switches record with the Select 1 and Select 2 timing and logic sequence pattern shown in Fig. 8.12. The playback mode timing and logic of the Select 1 and Select 2 signals in these units is shown in Fig. 8.13.

Some 8mm designs don't implement this band-pass filter switching arrangement when the F2 and F4 tracks are being played back. Instead, these units modify the Select 1 and Select 2 logic signals during the time period that the F2 and F4 sections are being played back. Altering the logic of the Select lines changes the frequency output from the playback pilot reference generator during the F2 and F4 video and PCM track playback intervals. Figure 8.14 shows the playback logic and timing of the ATF servo Select 1 and Select 2 lines using this method of ATF playback operation.

Compare the playback timing and relationships of Select 1 and Select 2 with the F1 through F4 tracks in Fig. 8.14 with the same data shown for the record mode (Fig. 8.12)

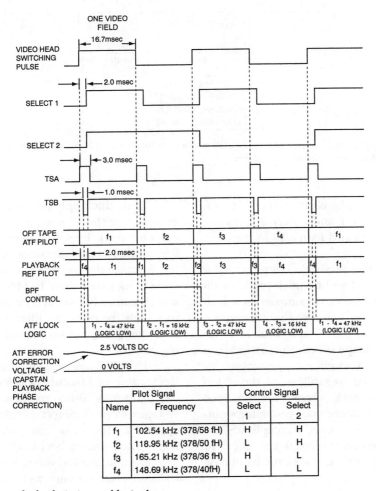

Pilot Signal		Control Signal	
Name	Frequency	Select 1	Select 2
f_1	102.54 kHz (378/58 fH)	H	H
f_2	118.95 kHz (378/50 fH)	L	H
f_3	165.21 kHz (378/36 fH)	H	L
f_4	148.69 kHz (378/40fH)	L	L

Fig. 8.13. ATF servo playback timing and logic chart.

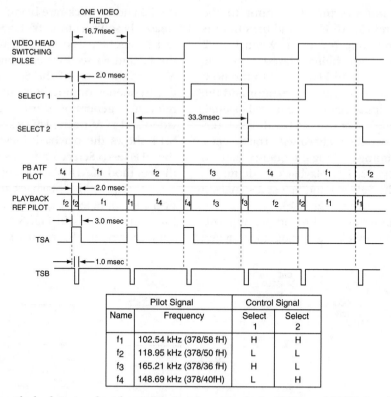

	Pilot Signal	Control Signal	
Name	Frequency	Select 1	Select 2
f₁	102.54 kHz (378/58 fH)	H	H
f₂	118.95 kHz (378/50 fH)	L	L
f₃	165.21 kHz (378/36 fH)	H	L
f₄	148.69 kHz (378/40fH)	L	H

Fig. 8.14. *ATF servo playback timing chart for models that do not toggle the 47 kHz and 16 kHz band-pass filter outputs when the F2 and F4 tracks are played back.*

and the playback timing chart in Fig. 8.13. Notice how the Select 1 and Select 2 lines are modified for the F2 and F4 tracks in the playback mode.

Modifying the playback reference pilot frequency for the F2 and F4 track playback intervals assures that the error comparator will always function correctly under the assumption that an increase in the 47 kHz beat level requires a decreased tape speed while an increase in the 16 kHz beat amplitude requires an increased tape speed, regardless of the video track being played back. Altering the sequence of the playback reference pilot generator's output frequency pattern thus accomplishes the same task as toggling the band-pass filters in other designs without the use of the band-pass filter switching.

Units that apply the modified pilot reference during the F2 and F4 tracks will develop a center track beat frequency of some value greater than 0 Hz. This does not affect the system because the center track frequency will be something other than 16 kHz or 47 kHz and will therefore be blocked by the band-pass filters. Only the resulting adjacent crosstalk pilot beat signal amplitudes are used for ATF tracking control.

For all 8mm designs, the amplitudes of the leading and trailing beat frequencies from the band-pass filters are converted into DC voltages in peak detect circuits before being applied to the error comparator. When the leading and trailing beat frequencies are equal in amplitude, the DC voltages output from the DC converter circuits will also be equal in value. These equal voltages are applied to the noninverting (+) and inverting (–) inputs of the error comparator. When the two inputs to the comparator are equal, the comparator's output will be zero indicating correct tracking.

ATF Lock and ATF Error Correction

As shown in Fig. 8.11, the output of the error comparator is applied to two sample and hold circuits (Sample and Hold A and Sample and Hold B). Sample and Hold A causes the ATF error circuit to ignore the pilot playback beat frequency data produced for the first 3.0 milliseconds following video head switching. During the first few milliseconds of comparison, wrong ATF error correction information is produced because the change in the playback reference frequency is delayed by 2.0 milliseconds from the point of video head switching and the generated beat frequency does not give the desired result for ATF error correction. The Sample and Hold A circuit causes the signal to be ignored during this first 3.0 millisecond period to prevent the ATF servo from issuing erroneous correction voltages to the capstan motor during that time.

During playback, the pilot reference generator's frequency change is delayed by 2.0 milliseconds to aid in detecting correct ATF Lock. The sample and hold section of the ATF Lock path is used by the servo microprocessor to determine when the video head is scanning up the wrong track and is also used in selecting the correct playback speed. Therefore, the unique beat signal generated during the 2.0 millisecond interval at the start of each video track segment is important for ATF Lock detection, but can confuse the ATF error correction circuit supplying the capstan motor.

The TSA and TSB pulse inputs from the servo system control circuit to the sample and hold boxes in Fig. 8.11 are the *tracking sample A* and *tracking sample B* pulse commands that activate the sample and hold circuits. Sampling of the comparator voltage is performed only when the respective TSA or TSB is at a logic low. The sampled voltages are held and output from the sample and hold circuits when TSA or TSB pulses are logic high. Logic timing of the TSA and TSB pulses is critical for correct ATF Lock and ATF Error correction signal operation.

Notice that the timing of the TSA pulse in Fig. 8.13 prevents the ATF error correction path from using the comparator's output for the first 3 milliseconds that the video section of the RF track is being scanned by the playback video head. This means that the ATF error correction voltage supplied to the capstan motor ignores the erroneous pilot frequency information output from the ATF comparator during the first 3 milliseconds that the video segment of the track is being scanned.

The outputs from the two sample and hold sections are individually amplified and then passed to the system control microprocessor as *ATF LOCK* and to the capstan motor drive circuit as *ATF ERROR*. ATF ERROR is a tracking voltage supplied to the capstan motor to phase control the rotation of the capstan shaft. The other path, ATF LOCK, is a tracking status signal that indicates whether or not the servo is centered on the correct track. ATF LOCK is also used for automatic playback speed selection.

The processing of ATF ERROR will be covered before ATF LOCK. If the signals from the 16 kHz and 47 kHz band-pass filters are not equal (indicating tracking error), the error comparator's output will be something other than its nominal voltage. The output of the comparator is sampled at the moment the TSA pulse arrives at Sample and Hold A producing ATF ERROR, the ATF error correction voltage that corrects the tracking. ATF ERROR is nominally 2.5 volts (error comparator output during correct tracking). Tiny changes from the nominal 2.5 volts cause the motor to run faster or slower. These small changes in the correction voltage reposition the tape tracks under the scanning playback head. The ATF ERROR voltage is combined with the output voltages of the capstan digital servo to form the capstan motor drive voltage.

During record, ATF ERROR is held constant to prevent the ATF servo from interfering with the capstan servo's digital phase and the speed PWM regulation of tape movement. Once the unit is placed into the playback mode, the ATF correction voltage will change as it works with the capstan's digital speed control to correct playback tracking errors. During playback the capstan's digital phase control PWM signal is held at 2.0 volts to give the ATF error correction voltage complete control over the capstan phase.

Sections A, B, C, and D in the lower right corner of Fig. 8.15 show various head-to-RF track conditions under which the capstan servo may need to operate to maintain proper tracking. When the video head is moving up the track, as shown in section A, the off-tape pilot frequencies and the pilot reference generator output combine to produce a low-amplitude 16 kHz trailing beat frequency and a high-amplitude 47 kHz leading beat frequency. After passing through the band-pass filters, these two signals enter the comparator producing a high output to the sample and hold circuits (only the + input of the comparator receives a signal from the band-pass filters). The ATF ERROR voltage to the capstan motor will become higher than the nominal 2.5 volt level causing the capstan drive circuitry to slow the motor down.

When the VCR tracking is off as shown in Fig. 8.15, section C, the off-tape pilot crosstalk will produce an increased 16 kHz trailing beat frequency and a low 47 kHz level. This in turn causes the output of the comparator to decrease the ATF ERROR voltage increasing the capstan motor speed.

Notice that when the ATF servo is producing ATF ERROR voltage corrections, the

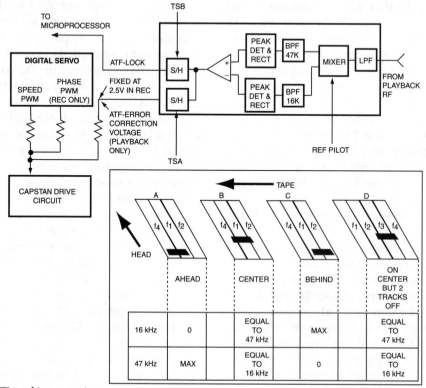

Fig. 8.15. ATF tracking operation.

changes in the 2.5 volt level are only made until the 16 kHz and 47 kHz amplitudes are once again equal. Once the two levels match, the ATF ERROR voltage returns to its nominal level until the next correction is required. In actual practice, this 2.5 volt signal is constantly fluctuating in tiny amounts as the ATF servo works to correct tracking errors that continually occur.

Now consider the situation when the head is on center but it is positioned two tracks off (Fig. 8.15, section D). In this situation, the head is scanning the F3 track when it should be over track F1. Recall that in this situation the relationship of the 47 kHz leading track pilot crosstalk and the 16 kHz trailing track pilot crosstalk has been inverted. This means that the leading track crosstalk pilot frequency is now 16 kHz instead of 47 kHz and that the trailing track pilot frequency is likewise reversed to 47 kHz instead of 16 kHz. As soon as the video head begins to move off the center of the F3 track, the ATF servo acts to make its first tracking correction. This correction will inevitably happen a very short time after the playback process is initiated.

Assume that the head-tape position starts to deviate off the F3 track center towards the F4 track. This shift is towards the trailing track, but the ATF servo believes that tracking is drifting towards the leading track since the 47 kHz beat level is growing. The ATF servo therefore causes the ATF ERROR correction voltage to increase slowing the motor down. As the motor slows, the tape is repositioned, placing more of the F4 track under the playback head. This causes more of the 47 kHz beat to appear in the ATF circuit forcing the ATF ERROR correction voltage to rise even more. The motor will be continually forced to slow down until the F1 track appears under the head and is centered. Once the F1 track is under the video head, the relationship of the pilot crosstalk beat frequencies to the leading and trailing tracks will be restored. Thus the ATF servo will cause the capstan motor to shift the position of the recorded tracks until the pilot frequency

generator output and the correct RF track are matched (amplitudes of 16 kHz and 47 kHz signals equal).

Next, study the operation of ATF LOCK. Figure 8.13 shows that the playback reference pilot generator signal is changed for each field (video track) and that the frequency change is delayed by 2.0 milliseconds after each video head switching pulse transition. Delaying the change of the playback reference frequency creates an error in the ATF comparator's output for the first 2.0 milliseconds of each video track segment. (This is the point where the head senses the portion of track just following the PCM audio segment.) The delay in the recorded pilot frequency is used by the ATF LOCK portion of the ATF servo to sense correct tracking of the system and to determine the correct playback speed.

The ATF servo samples the error voltage with the TSB pulse. This sample produces the ATF LOCK information used by the servo microprocessor control section to indicate correct tracking. A logic high on the ATF LOCK line indicates a tracking error.

Now study the TSB pulse timing in Fig. 8.13. TSB goes low for 1.0 millisecond just after the video track segment appears under the playback head and during the time that the playback reference pilot generator is still at the frequency of the previous track. When the F1 (or F3) video track is played back and the TSB pulse is low, the delayed playback pilot reference frequency will mix with the recorded center track pilot frequency of the F1 (or F3) track producing a 47 kHz beat frequency of significant amplitude. Since the band-pass filter switching logic (BPF CONTROL) maintains the band-pass filter outputs in the reversed positions, the 47 kHz signal is supplied to the inverting input of the error comparator, which generates a low output. This low level is sampled by the TSB pulse in Sample and Hold B (Fig. 8.11) producing a logic low on the ATF LOCK line. The system control microprocessor recognizes this low as an indication of correct tracking. As long as the system is locked onto the correct track, a logic low will be present on the ATF LOCK line.

When the F2 and F4 track segments are being played back, the mixing of the delayed change in the playback pilot generator output and the pilot frequency of the video track being scanned produces a 16 kHz signal. But, since the band-pass filter outputs are switched during this time, the 16 kHz beat is fed to the inverting input of the comparator. Therefore, the output from the error comparator will still be low when the F2 and F4 tracks are being correctly scanned indicating correct tracking to the microprocessor.

Since the TSB sample pulse goes low for only 1.0 millisecond shortly after each video head switching pulse occurs, Sample and Hold B must maintain (hold) its output logic level until the next TSB low pulse arrives. Samples are taken at the beginning of each video track to monitor the ATF error comparator's output for each new track being played back. As long as the error comparator output is low, ATF LOCK will be low, indicating to the capstan servo system control microprocessor that both the tape speed and phase are correct.

As described earlier, the ATF servo can correct within four fields (tracks) of error. If the VCR is tracking a tape two fields off from where it is supposed to be (for example, on F3 track rather than F1), the ATF error correction voltage supplied to the capstan motor will operate normally as long as the video head is traveling up the center of the track. As soon as the ATF servo has to make its first phase correction to recenter the track under the head (which will inevitably happen a very short time after playback is first initiated), the system will move the tape to place the correct track under the scanning head. In terms of video information time (milliseconds and microseconds), this process of moving the tape position so the correct track is under the video head can take a relatively long period of time. While this two-track correction is taking place, the picture will display disturbance—snow and other forms of interference. The snow will occur as the active channel 1 head, with the +10 de-

gree azimuth head gap, crosses over the track recorded with the –10 degree head gap. The other forms of instability will cause chroma errors and cause the picture to jump horizontally and vertically as the servo stabilizes on the correct track.

To speed up the process of locking the VCR onto the correct track, and to reduce the video disturbance associated with this process, the 8mm format uses a method of track detection and correction called *REAR LOCK DETECTION*. The term "REAR LOCK" indicates that the servo is tracking (locked on) two tracks off from where the microprocessor (and thus the playback reference pilot frequency generator) thinks it is supposed to be. REAR LOCK DETECTION and correction matches the playback reference pilot generator frequency to the recorded pilot frequency on the track being scanned rather than forcing the tape to shift by two tracks to align the tape's recorded pilot frequency to the pilot generator reference frequency. This form of correction allows the system to rapidly lock up and track correctly without creating the forms of interference associated with jogging the tape by two tracks.

The ATF LOCK signal is used by the servo microprocessor to detect a REAR LOCK condition. Figure 8.16 shows the logic and timing pulses that are used to indicate a REAR LOCK condition.

When REAR LOCK occurs, the recorded pilot frequency present on the tape beats with the delayed playback pilot reference frequency during the time that the TSB sample pulse is low. As shown in Fig. 8.16, when the VCR is two tracks off, the combination of the delayed playback pilot frequency generator and the recorded pilot frequency on the track generates 16 kHz when the F1 and F3 video tracks are being played back and 47 kHz when the F2 and F4 tracks are scanned.

During the time that the VCR is two tracks off, the servo microprocessor is generating the Select 1 and Select 2 commands and the BPF CONTROL switching logic as if the unit were on the correct track. Thus, when

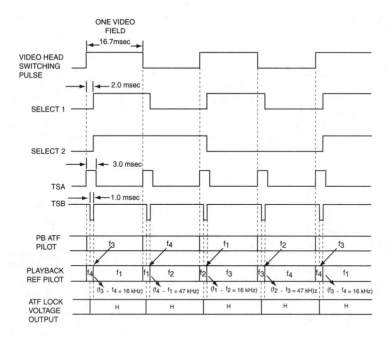

Fig. 8.16. REAR LOCK detection.

REAR LOCK occurs, ATF LOCK is a logic high because of the incorrect relationship between the BPF CONTROL and the beat frequencies generated during the time that the TSB signal is taking the sample.

The high ATF LOCK signal is then detected by the servo microprocessor causing it to change the phase of the Select 2 command line 180 degrees. This directs the programmable divider in the playback pilot reference generator to change its frequency output sequence so that it matches the pilot frequency of the track being played back. The sample and Hold B signal and ATF LOCK logic will be output as a high level ATF LOCK until the microprocessor forces the playback pilot generator to change and track the REAR LOCK track pilot frequency.

After the playback pilot reference generator has made the change, ATF LOCK will go low telling the microprocessor that all is well. Thus, REAR LOCK detection forces the pilot generator's playback reference to immediately match the track that the head is on rather than

force the ATF servo to reposition the tape by two tracks.

8mm Automatic Playback Speed Selection
The ATF LOCK signal is also used by the 8mm system to determine the correct tape playback speed. Models offering different speeds are limited to SP and LP speeds only. This format does not provide an SLP (EP) speed.

Tape speed changes are detected by the microprocessor as it monitors the ATF LOCK signal. In the case of playing back an SP recording where the recorded speed was changed to LP, the ATF LOCK signal will have the characteristics shown in Fig. 8.17. In this case, the VCR is playing back in SP speed when it encounters an LP recording. At the instant the LP section is played back, the playback pilot reference generator output interacts with the playback pilot frequencies shown in Fig. 8.17C to produce a pattern of highs, lows, and ignored voltages at the error comparator's output. These error comparator voltages are sampled by the TSB pulse causing the Sample and

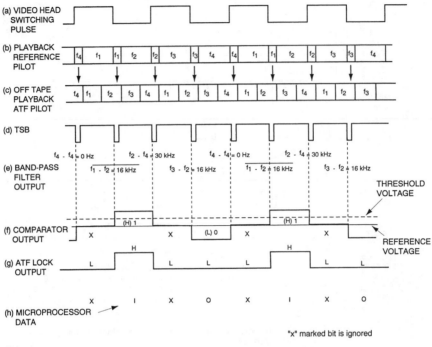

(a) VIDEO HEAD SWITCHING PULSE

(b) PLAYBACK REFERENCE PILOT

(c) OFF TAPE PLAYBACK ATF PILOT

(d) TSB

(e) BAND-PASS FILTER OUTPUT

THRESHOLD VOLTAGE

(f) COMPARATOR OUTPUT

REFERENCE VOLTAGE

(g) ATF LOCK OUTPUT

(h) MICROPROCESSOR DATA

"x" marked bit is ignored

Note:

1. Formulas in section (e) with a bar over them are redirected by the band-pass filter control switch. Therefore the 16kHz beat frequency is applied to the + input of the error comparator to produce a high at its output.

2. The 30kHz beat frequency produced by some comparisons in section (e) produce a 0 volt output at the error comparator just like the 0 Hz beat frequency does. This is because the 30kHz beat frequency is blocked by the 16 kHz and 47kHz band-pass filters.

Fig. 8.17. *ATF LOCK signal logic when the 8mm VCR is playing back an LP recording in the SP mode.*

Hold B to output the ATF LOCK pattern as shown in Fig. 8.17F.

The comparator output in Fig. 8.17F shows the high (1), low (0), and reference (X) voltage outputs. The reference voltage (X) occurs when the reference pilot and the recorded pilot frequencies are the same. This output is 0 volts DC and is not used by the microprocessor. Only the high (values above the relative voltage line) and low (voltages below the reference 0 volt level) are used by the microprocessor.

Figure 8.17G shows the ATF LOCK data used by the microprocessor to recognize that the tape is moving too fast relative to the recording. This pattern (X 1 X 0 X 1 X 0) will occur every time the VCR is in the SP speed and attempts to play back an LP recording. By recognizing this pattern on the ATF LOCK line, the system control knows that it must force the

capstan digital speed control into the LP mode. It takes a total of 12 playback RF video tracks for the system to recognize that this speed error has occurred.

Figure 8.18 shows the other wrong playback speed possibility. In this case, the VCR is playing back in LP speed and comes upon a section recorded in the SP mode. Notice that more than one TSB sample pulse occurs during the period that a single video track is being played back. The patterns of highs (1), lows (0), and unused (X) quantities are shown in Fig. 8.18F.

The pattern of microprocessor data ATF LOCK input shown in Fig. 8.18F (1 1 X X 0 0 X X 1 1 X X 0 0) will always occur when the VCR is in the LP speed and encounters a section of tape recorded in the SP speed. This causes the microprocessor to direct the capstan digital speed control to speed up the tape.

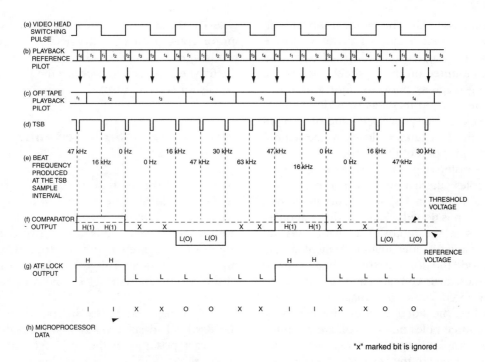

Fig. 8.18. *ATF LOCK signal logic when the 8mm VCR is playing back an SP recording in the LP mode.*

Whenever a tape speed change is required, the digital speed section of the capstan servo is the circuit that makes the speed change. The ATF servo is only capable of making phase (fine) speed adjustments to the coarse speed control.

DIRECT DRIVE SERVO MOTORS

The direct drive servo motor is a DC motor whose speed and phase are controlled by the incoming servo control voltage. Since control is so important, VCRs have a DC brushless motor. The brushless motor is ideal because of its durability, and ability to be accurately controlled.

A typical video head drum motor consists of three main coils with three poles for each coil, as shown in Fig. 8.19. The coils are mounted on the stationary portion of the video

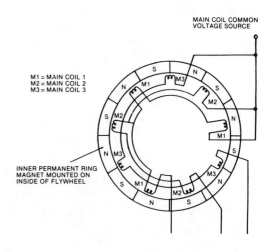

Fig. 8.19. *Direct drive drum motor has three main coils with three poles each.*

head drum assembly. The motor flywheel is attached to the rotating shaft of the motor. Three permanently magnetized individual ring magnets are mounted on the flywheel. Each of the three ring magnets performs individual functions. The two top magnets provide feedback to the servo circuitry for speed and phase control. The inside magnet is closest to the stationary coils and is part of the motor.

The inside ring magnet consists of 12 individual poles alternating between north and south, as shown in Fig. 8.20. This inner ring magnet is positioned close to the stationary main coils in the motor. The main coils are turned on and off individually by switching transistors inside the drum motor integrated circuit (IC). As these main coils are energized, magnetic lines of force cause the magnetic flywheel mounted on the video head drum shaft to rotate. The three poles on each coil are mounted 120 degrees apart so that by energizing the three poles of each coil in the correct sequence, each set of poles will take its turn in forcing the magnetic flywheel to rotate. As the motor rotates, each of the three main coils is energized 120 degrees after the previous main coil.

The direct drive capstan motor is constructed very similar to the direct drive drum motor. One exception is the number of ring magnets present on the capstan flywheel. The capstan has only two ring magnets—not three.

One of these ring magnets is used by the motor coils to turn the flywheel. The second ring magnet, located on the circumference of the flywheel, is used to produce the FG and PG signals for servo control.

Direct Drive Motor Speed and Phase Control

Switching of the main drive coil transistors is performed inside the drum motor IC. In order for the transistors to turn off each of the respective main coils at the correct time, the position of the video head drum flywheel must be identified. This is a function of one of the two magnetic rings on the rotating flywheel.

The magnetic poles are positioned on the flywheel in such a way that they identify the exact position of the flywheel. Fig. 8.20 shows the magnetic polarity of these magnetic rings. As the flywheel rotates it passes across two Hall effect ICs (called *Hall IC's*) which are permanently mounted on the stationary part of the lower drum assembly.

A Hall IC is an integrated circuit that is sensitive to the lines of force from a magnetic field. It acts like a switch and changes state each time the field of a magnetic pole is reversed. Figure 8.21 shows the nine poles and a Hall IC after the flywheel has been removed.

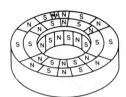

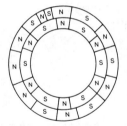

Fig. 8.20. Head drum flywheel has three ring magnets.

Fig. 8.21. A dismantled head drum motor.

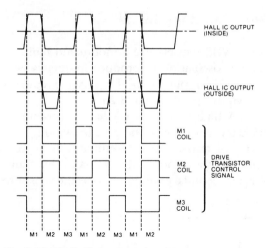

HALL IC OUTPUT (INSIDE)

HALL IC OUTPUT (OUTSIDE)

M1 COIL

M2 COIL — DRIVE TRANSISTOR CONTROL SIGNAL

M3 COIL

M1 M2 M3 M1 M2 M3 M1 M2

Fig. 8.22. Hall effect output signals.

Two Hall ICs are mounted on the video head drum motor. One of these ICs detects the magnetic information from the outside upper ring on the flywheel. The second IC detects the magnetic information on the inner ring of the upper portion of the flywheel. The two Hall ICs produce pulses that correspond to the exact position of the rotating head drum motor at any given moment. This information is returned to the drum motor drive IC where

it is processed and used for switching the main coil drive transistors. Figure 8.22 shows the output of the inside and outside ring Hall ICs, as well as the derivation of the electronic switching transistor pulses.

Recall that the outside upper ring shown in Fig. 8.20 has two magnetic poles that are much narrower than the other poles on the magnet. These two poles are used in one drum motor design to identify the position of the video head on the rotating drum assembly. This information is used to generate video head switching pulses.

Information from the Hall ICs on the video head drum motor is processed inside the IC, as shown in Fig. 8.23. In this model, the Hall IC signals not only switch the motor main coil drive transistors, they also serve as the source of the *PG (pulse generator) pulses* and *FG (frequency generator) pulses.* (The source of the PG and FG signals is not shown in Fig. 8.23.) PG pulses identify the position of the video head drum and are used to form the 30 Hz video head switching pulse. FG pulses represent the speed at which the motor is rotating. For this motor design, the FG and PG information exits the drum motor drive IC and

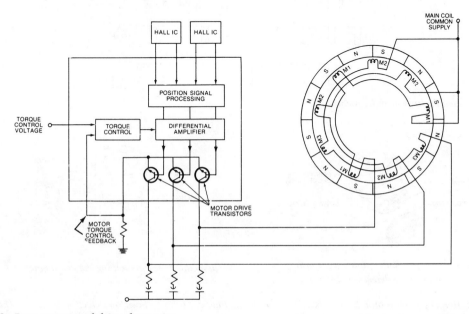

Fig. 8.23. Drum motor and drive electronics.

is processed by another IC to produce the video head switching pulses.

While the motor is rotating at 1800 RPM, the FG pulses occur at a rate of 180 Hz. The relationship between the two Hall IC outputs, the 30 Hz PG signal, the 180 Hz FG signal, and the video head switching pulse is shown in Fig. 8.24.

FG and PG signals are used by the video head drum servo to control both phase and speed of the motor.

Modulated FG Signal

Some VHS models use a modulated high-frequency oscillation to produce frequency generated (FG) pulses in the servo circuit, as shown in Figs. 8.25 and 8.26.

A high-frequency oscillation (nominally 65 kHz) is applied to the primary winding of a series of transformers mounted 120 degrees apart in the motor (Fig. 8.26A). The transformer secondary winding acts as the sensor to

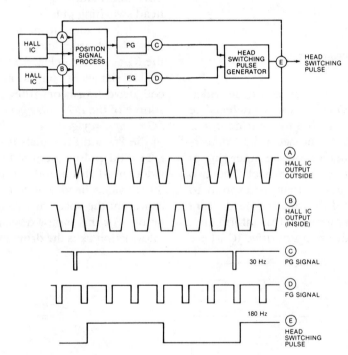

Fig. 8.24. *Head switching pulse formation.*

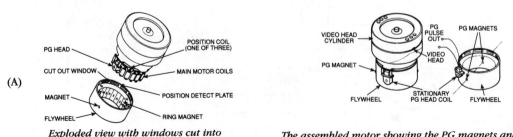

(A)

Exploded view with windows cut into the position detection plate.

The assembled motor showing the PG magnets and sensor.

Fig. 8.25. *Head cylinder motor. (A) Exploded view with windows cut into the position detection plate. (B) The assembled motor showing the PG magnets and sensor.*

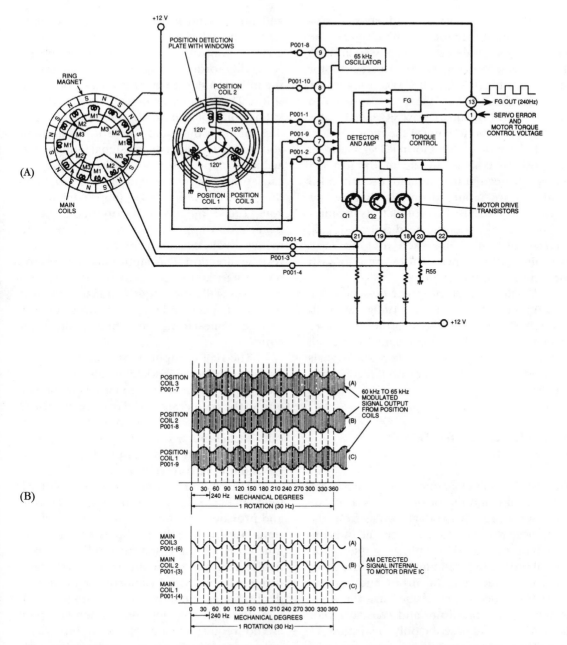

Fig. 8.26. *The cylinder motor drive. (A) Block diagram showing the motor drive coils, the FG position coils, and the position ring. (B) AM modulated 65 kHz head cylinder position coil outputs and the detected AM signals.*

detect the motor's rotational position. Under normal conditions, the 65 kHz oscillation would be conducted to the secondary winding without amplitude distortion, but window holes cut into a position detection plate mounted on

the rotating ring magnet cause the secondary winding's oscillation output to modulate.

As each window hole passes across a position coil, the oscillation conduction efficiency between the primary and secondary windings is altered. This change in conduction acts to amplitude modulate the incoming signal. The AM modulation is detected and passed into the drive IC where it triggers the correct motor drive transistor (Q1, Q2, or Q3) causing motor rotation. The detected signal is also used to produce a 240 Hz square wave FG signal that is used by the servo to control the motor speed.

These motors don't form PG signals from the position coils as in the motor just described. Instead, a magnet and a separate PG head coil are mounted on the flywheel, as shown in Fig. 8.25B. This coil detects the magnet as it passes under the head coil producing 30 Hz pulses that are used to form head switching pulses. These pulses are also used for servo motor phase control.

Whether the drum motor design uses a separate permanent magnet configuration to develop PG pulses, or creates the PG pulses from a customized FG ring magnet design, all drum motors must develop separate signals for head switching (head position) feedback to the drum phase servo.

Motor Drive ICs for Direct Drive Motors

Figure 8.23 shows that one of the inputs to the video head drum motor drive IC is a torque control voltage. This voltage comes from the video head drum motor servo circuit and controls both the speed and phase of the motor. Digital processing techniques derive the servo control voltages for the drum and capstan drive ICs. Since many designs use the same technology for the drum and capstan motor drive ICs, we will describe only the drum IC.

The motor servo circuit develops the speed and phase voltage as previously described. This control voltage is the motor torque control voltage input into the motor drive IC. This torque voltage is typically 2.5 volts, but may vary between manufacturers. This 2.5 volts is the product of combining the speed

and phase control PWM signals in the servo circuit and will change slightly as the servo operates to control the motor's speed and phase.

This torque voltage is not what causes the motor to start rotating from a stopped position. The motor start-up command is the responsibility of the motor direction control line input to the motor drive IC. This input voltage is a tri-state control signal. A high voltage level (approximately 12 volts) at this input causes the motor to turn in the forward direction, a medium voltage (approximately 5 volts) causes the motor to stop rotating (electronic breaking), and low voltage (approximately 1 volt) causes the motor to turn in the reverse direction. This input thus controls the motor start and stop action and the rotation direction while the torque command controls the motor speed and phase. The tri-state level voltage comes from the system control circuitry.

The Hall IC inputs to the motor drive IC provide the motor flywheel magnetic position data. This information is required by the system so the correct set of coils can be energized and force the flywheel to turn. The Hall IC outputs let the motor drive IC know which transistor to turn on so it can provide a path to ground for the correct motor coil.

The motor torque control feedback loop in Fig. 8.23 senses the motor current and provides a means of self-protection if the transistors in the IC begin to conduct too much current. Increased current flow is a particular concern when the motor first starts turning from a dead standstill or must suddenly change rotation direction or stop. Sudden direction changes and sudden stopping are not required in a drum motor, but are requirements of the capstan motor when special tape functions are engaged by the VCR operator.

The direction and torque control inputs to the motor drive IC are one place where a technician may choose to "break the servo loop" when troubleshooting a servo problem.

By disconnecting the torque control and motor direction control inputs to the motor drive IC, a technician can have complete control over the motor's direction and speed. This action prevents the servo and microprocessor control sections from controlling the motor and allows the servicer to divide (break) the circuit (loop) into two sections. By applying the correct voltages to the motor drive IC direction and torque control voltage input leads, the motor will operate if the motor and drive IC are functioning.

Even when the external voltage is very well regulated, accurate phase control of the motor will be difficult when injecting an external voltage into the torque control line. This is because the servo circuit shifts the torque control voltage in millivolts when controlling motor phase. Nevertheless, applying an external voltage to the isolated pins of the motor drive IC may help isolate a servo control problem.

SPECIAL EFFECTS

VCRs featuring special playback effects, such as still field, pause, and slow motion playback, place special demands on the servo circuitry. The capstan servo, in response to commands from the system control circuitry, performs most of these special effects.

When playback pause is engaged, the capstan motor must be stopped at a position that provides the best playback signal from the recorded tape. Some machines stop the tape with the capstan motor and allow the system control circuits to sample the RF envelope coming from the video heads. If the system control circuitry determines that the RF envelope is insufficient to produce a good still frame picture, it commands the capstan motor to pulse forward until a good RF envelope is detected. The system control circuitry provides most of the commands of the full electromechanical operation of the VCR. Functions

and features of the system control are described in another chapter.

Slow motion forward and reverse speeds are also controlled by the system control unit. System control directs the capstan servo to jog the tape in the forward or reverse direction to provide good slow motion video playback. In many machines slow motion playback causes rapid starting and stopping of the capstan motor. The tape is stopped for an instant, a video frame is generated on the screen, and then the tape is jogged forward or in reverse until the next field can be displayed.

SYMPTOMS OF SERVO FAILURE

As you diagnose a suspected servo failure, be sure to study Table 8.1. A review of each servo section's function in the record and playback modes will assist you in determining what feedback and reference signals to check. Also, comparing the failure symptoms with the table's stated responsibilities for each of the various servo sections, will help direct you to the correct servo section for troubleshooting.

Table 8.2 lists some common servo related symptoms and associated causes, which should be investigated when troubleshooting servo failures. Be sure to use this table in conjunction with Table 8.1 as you investigate the cause of a failure.

When verifying the symptoms you observe with those listed in the chart, you must first decide whether the failure is occurring in the playback or record mode or both. A tape with a known good recording can be used to verify the condition of the playback servos. A second VCR should be used to play back recordings made on the defective unit to determine the status of the record servos. Once you have narrowed the symptom down to the record or playback mode, the chart in Table 8.2 will make isolation of the failure much easier.

Table 8.2. Servo Troubleshooting Chart

Symptom	Possible Circuit	Possible Cause (Playback)	Possible Cause (Record)
Fine horizontal line drifts vertically through picture during playback. (See Fig. 4.10.)	Drum phase control	1) Loss of drum PG 2) Loss of 30 Hz reference	1) Loss of drum PG 2) Loss of input separated vertical sync signal
Loss of horizontal sync. Picture may bend or sway from side to side (see Fig. 4.11) or have a series of black and colored diagonal lines. Looks like someone has turned the horizontal hold control on the television.	Drum speed control	Loss of drum FG	Loss of drum FG
One or two horizontal bands of snow float vertically through the playback picture. (See Figs. 8.27 and 8.28.)	Capstan phase	**VHS** 1) Loss of control track 2) Loss of crystal oscillator (30 Hz) reference 3) Defective pinch roller **8mm** 1) Loss of playback ATF signal 2) Loss of 30 Hz reference 3) Loss of reference pilot frequency 4) Tape path alignment error	**VHS** 1) Loss of record CTL signal 2) Loss of input vertical sync (30 Hz) reference 3) Tape path error around pinch roller/AC head area **8mm** 1) Loss of playback ATF signal 2) Loss of 30 Hz reference 3) Loss of reference pilot frequency 4) Tape path alignment error
TAPE SPEED PROBLEM. Possible symptoms include: 1) Tape runs too fast 2) Tape runs too slow 3) Normal audio is playing back at the wrong pitch indicating a tape speed problem (VHS ONLY) 4) Scan lines may be present in the picture—as if the machine is in the search mode	Capstan speed	Loss of capstan FG	Loss of capstan FG
INTERMITTENT TAPE SPEED PROBLEM. Symptoms may include: 1) VCR intermittently jumps between CTL speeds during playback 2) Will not select correct speed when recorded tape speed has changed	**VHS** 1) Auto tape select 2) Capstan speed **8mm** 1) ATF Lock 2) Servo microprocessor 3) Capstan speed	**VHS** 1) Intermittent loss of CTL 2) Weak CTL signal amplitude 3) Failure in the auto speed select circuit 4) Intermittent loss of capstan FG **8mm** 1) Loss of ATF Lock signal to microprocessor 2) Tape path alignment error 3) Loss of reference pilot frequency	**VHS** 1) Intermittent loss of record CTL 2) Weak record CTL signal amplitude 3) Intermittent loss of capstan FG during record **8mm** 1) Intermittent loss of capstan FG during record 2) Tape path alignment error 3) Loss of record reference pilot signal

Table 8.2. *Continued*

Symptom	Possible Circuit	Possible Cause (Playback)	Possible Cause (Record)
Motor won't start (may be either capstan or drum)	1) System control 2) Motor drive	1) Motor direction command voltage to motor drive IC 2) Missing or insufficient supply voltage to motor drive coils or motor drive IC 3) Defective motor drive IC 4) Open motor coil 5) Defective Hall IC in motor	Same as playback

Notes:

1. With the exception of the tape path errors indicated in Table 8.2, defective connections, poor soldering, and open wires should be suspected as a possible cause for missing signals.

2. Be sure to clean all video heads, audio heads, tape guides, and the lower drum before troubleshooting any servo failure. Dirty heads and tape path can cause incorrect servo operation.

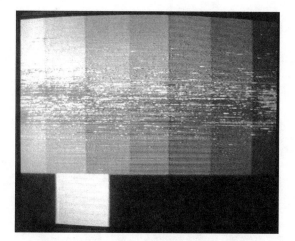

Fig. 8.27. *Drifting band of snow caused by loss of capstan servo phase lock. Note:* If the capstan servo loses phase lock, bands of snow, such as those shown in Fig. 8.27, drift vertically through the picture. This display looks initially like a tracking problem, but with tracking errors, the snow band does not drift nearly as much. Also, for a VHS machine, the tracking control has no effect in reducing the appearance of the drifting snow band when the capstan phase control is defective. If the snow band is stationary and the VHS tracking control can make the snow band go away, the capstan phase servo is functioning properly.

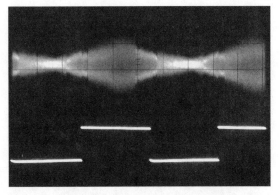

Fig. 8.28. *Display confirming a capstan servo problem. Note:* A capstan servo problem can be confirmed by monitoring the output RF waveform from the video head preamplifier and switching pulse. A display like that in Fig. 8.28 definitely confirms that the capstan servo requires alignment or repair.

CHAPTER REVIEW QUESTIONS

1. The VHS format units use a CTL pulse for playback tracking. What does the acronym CTL stand for?

2. What is the record source of the CTL signal?

3. What is the purpose of the ATF pilot frequencies in the 8mm format?

4. Does the 8mm format ever use a PWM signal for capstan phase control?

5. What is the purpose of the FG signal in the drum and capstan servo?

6. What is the function of the drum phase control in the record mode?

7. What is the function of the capstan phase control in the record mode?

8. What is the source of the capstan PG signal in record?

9. Describe the symptoms that you would observe if the FG signal were to disappear in the playback mode for:
   (a) The capstan motor.
   (b) The drum motor.

10. What are two functions of the ATF LOCK signal?

11. What is the source of the 30 Hz reference signal used by the capstan and drum phase servos?

12. A loss of the video head switching pulse would cause what symptoms?

VCR Audio Processing: Theory and Troubleshooting

Chapter 9 gives you an in-depth look at audio recording and playback. As you'll soon discover, audio and video signals are processed differently in a VCR. In the case of video, you're concerned about color, color-under, and luminance. In a VCR's audio section, the technology deals with conventional linear track audio used in some machines, and the frequency modulation and demodulation, companding, dropout and the signal processing capabilities of the Hi-Fi and pulse code modulation audio circuits used in other models.

AUDIO RECORDING TECHNIQUES

The first VHS VCRs used only one method for recording audio onto tape. This method of sound reproduction is identical to that used today in conventional reel-to-reel and cassette audio recorders. The incoming signal is recorded onto a longitudinal track on the tape using a stationary audio head. The audio signal must be amplified and equalized during record and playback to overcome undesirable frequency characteristics common to all magnetic read/write heads. This conventional audio record/playback method (known as *normal audio, longitudinal audio,* or *linear track audio*) is still a standard feature on current VHS machines.

In the early 1980s, Sony introduced a moving head audio recording technique that provides improved sound reproduction. This advanced design uses rotating video heads mounted on a head drum. It was first introduced in selected Sony Betamax VCRs—Betamax was a popular format several years ago. Sony called this improved sound method *Beta Hi-Fi.* The Hi-Fi recordings applied FM modulation to the recorded sound. After modulation, the audio FM (*AFM*) was recorded on the video track along with the FM video and color-under signals. For compatibility, these Beta Hi-Fi units maintained the normal audio recording capability in addition to the AFM.

Shortly after Beta Hi-Fi was introduced, VHS manufacturers introduced their own AFM-capable models. Some VHS manufacturers call their AFM feature *HD Audio* for high definition audio, while others refer to this option simply as *VHS Hi-Fi audio*. For compatibility with non-Hi-Fi (non-HD) units, all VHS format VCRs still provide conventional stationary audio recording and playback, even if they include the Hi-Fi audio feature.

Although the 8mm format includes space on the tape for linear track audio recording with a stationary head, current models do not use this feature. Stationary heads are not found in most 8mm models. These machines use AFM as their standard audio record/playback method. However, some 8mm models offer optional pulse code modulated (PCM) audio record and playback in addition to AFM. While AFM sound reproduction yields excellent audio, the quality of PCM audio rivals CD audio and is considered superior to AFM. PCM audio will be described later in this chapter.

In this chapter, we first explain linear track audio technology as used in VHS machines. Then we describe the different Hi-Fi and AFM methods associated with the 8mm and VHS formats. Finally, we explain PCM audio.

LINEAR TRACK AUDIO RECORDING

Linear track audio is written onto a tape by a stationary audio head. The stationary head writes the normal audio on a longitudinal track located on a section of the tape reserved for this signal. Until stereo track audio was developed, monaural audio was the only option available on VHS machines. Monaural has also been called *mono* or *single channel* audio. Stereo audio uses the same recording techniques as monaural recording and playback but has two separate channels for left and right (channel 1, channel 2) audio. The 1 mm wide track of video tape space designated for the audio information was divided in half enabling two audio tracks to occupy what had previously been the space of one track (Figs. 5.50A and B).

The following description applies to monaural linear track audio systems. VCRs with stereo linear audio are identical in principle to the monaural units except they have two identical audio record and playback circuits. Also, in stereo linear track audio machines, the stationary audio head has two sections in the same linear track space as the entire monaural audio head for recording both the left and right channels.

Figure 9.1 is a block diagram of linear audio recording. Incoming record audio enters the audio processing circuitry through a common line. Audio from the tuner, an external microphone, or the line level input jack is applied to a common connector at the input to the audio amplifying circuitry. Typical audio levels from the tuner section reach the audio circuit at standard line level voltages of around 0.7 V peak-to-peak (p-p), as does audio from an external source entering through the Audio In or Line In jack in the VCR. VCRs can also have an external microphone input. Typical microphone (mic) levels range from 1.0 to 10 mV p-p. Tuner audio and line input audio levels are attenuated (reduced) to mic level by resistor networks prior to the first audio amplifier. This normalizes all inputs to the first amplification stage to mic level regardless of the source.

Audio entering the audio processing circuitry is selected by a series of switches. The switch in the microphone jack gives priority to the mic input. When a microphone is plugged into the jack, the line and tuner audio sources are disconnected. The line and tuner switch shown in the figure routes the audio signals to the processing circuitry from the selected source. Many VCRs don't have an external switch for selecting line or tuner inputs. The switching automatically occurs inside the Line In jack when a plug is inserted.

The selected incoming audio to the first audio amplifier is amplified and sent in three directions. The first direction is into the audio AGC (automatic gain control) circuit where the level of the amplified audio is sampled and au-

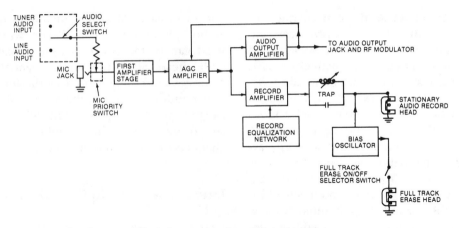

Fig. 9.1. *Conventional audio record block diagram used in VHS machines.*

tomatically adjusted for the optimum recording level. Audio AGC acts like an automatic volume control. It reduces the amplification if levels are too high and increases amplification if levels are too low. Typical audio AGC is not as sophisticated as the AGC and ACC (automatic color control) found in chroma and video processing circuitry. If incoming audio has long periods of very quiet sound, AGC will attempt to increase those levels. Programs with very loud audio will have the signal peaks limited by AGC.

The second direction that amplified audio takes is to the audio output circuitry. This line level audio out is passed to the Audio Output jack and the channel 3 and 4 RF modulator for E-to-E audio monitoring.

The third audio signal direction is into the record amplifier. The record amplifier increases the audio level and passes the signals to the audio record playback head. The record amplifier has equalization amplified to correct for the nonlinear playback response of the audio head, as described in Chapter 5. Equalization is performed using discrete resistors, capacitors, and inductors in the amplification process causing different amplifier response to various frequencies. In this case, low frequencies are amplified more than high frequencies.

The audio from the record amplifier passes through a bias trap that is placed in the circuit to stop high-frequency audio bias from being fed back into the record amplifier. The audio bias oscillator operates at a frequency of about 60 kHz. Enough of the bias oscillator is mixed with the audio during record to cause operation in the linear region of the hysteresis curve, as explained in Chapter 5.

Bias is also applied simultaneously to the full track erase head and to the audio erase head during normal audio recording. Using a high-frequency oscillation signal, the full track erase head eliminates all audio, video, and control track signals from the tape during record. Bias fed to the audio erase head eliminates previously recorded audio from the tape.

The switch for the full track erase head eliminates head bias during insert audio and insert video recording modes because it's not good to turn on the full track erase head during these functions.

VHS VCRs with the stereo normal audio feature may also have the capability to rerecord channel 2 audio (right channel) after the original audio and video program has been stored on the tape. This form of audio insert editing is provided to give the operator two separate audio channels for special applications. For example, in videotaping a special family event, the original audio program could be recorded on channel 1, while background music could be recorded on channel 2 without interfering with the channel 1 audio or the recorded video. During playback, the two channels of audio could be mixed together producing the original audio with dubbed-in background music. In these types of VCRs, channel 1 is used to record the main or primary information because of its placement away from the upper edge of the tape. Because the channel 2 audio track is placed closest to the top edge of the tape, it is subject to minor variations in playback level. Such variations are caused by minor tape edge damage from improper tape handling by the VCR tape guides and mechanism. Therefore, the channel 1 main audio is more protected than the channel 2 dubbed-in audio. Professional editors use this same channel allocation method when dubbing in audio to be mixed with the main part of a program.

Linear Track Playback Audio

In linear track playback, audio information is read off a tape using the stationary audio head. The low audio signal (about 1 mV p-p) is amplified and equalized as shown in Fig. 9.2. The equalization amplifier boosts the playback signal producing an opposite effect to the incoming frequency record equalization amplifier. The output from the equalization amplifier is passed to an audio output amplifier that boosts the audio signal to a level suitable for external use via the Audio Line Output jack and the channel 3, 4, RF modulator.

Monaural audio VCRs are capable of playing back both channels of a stereo linear track recording. The nonstereo VCR reproduces both tracks by combining the two channels into a single signal for playback amplification

and processing. Monaural mixing of the two audio channels is accomplished within the head gap of the single channel linear track audio head. Since the single channel audio head is wide enough to cover the full 1 mm linear audio head track, the monaural head is capable of detecting both the left and right stereo audio tracks and combining the two channels of sound into one signal.

Linear Track Audio Specification

Linear track audio record and playback modes perform well for most applications. Specifications are published by VCR manufacturers that describe the audio, record, and playback limitations of each machine. These specifications reflect three major areas of audio reproduction concern: (1) harmonic distortion, (2) frequency response, and (3) wow and flutter. Harmonic distortion is a measurement of signal deformity associated with normal recording, playback, and signal amplification. In distortion, the original signal is not perfectly reproduced during playback. For example, if a perfectly shaped sine wave is recorded on a tape, that same sine wave may appear normal during playback, but when compared on an oscilloscope with the original signal some minor variations may be visible. This is called *harmonic distortion*. It occurs because frequencies generated in record and playback produce small amplitude harmonics that combine to distort the original signal.

The measurement of *frequency response* determines the upper and lower frequency range for the machine. Writing speed limits the

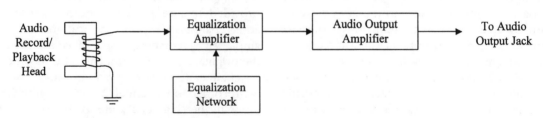

Fig. 9.2. VHS linear track playback audio.

audio frequency response. Linear track audio frequency response typically ranges between 50 Hz and 10 kHz. Slow speed audio recording reduces the upper frequency limit. The SLP frequencies rarely extend above 6 kHz. Lower frequency response specifications are not affected by recording speed.

Audio *wow* and *flutter* occur when the video tape is not pulled through the machine at a steady speed. This distortion affects the original recording signal and is noticeable as a change in the pitch of constant tones. If a single note is recorded on a tape and sustained for several seconds, wow and flutter are heard during playback as a slight change in pitch. Wow and flutter can be compared to the vibrato sound in a vocalist's sustained note.

Wow, a measurement of the rate of change of pitch, received its name from the sound produced by a continual, yet slow change in pitch of a tone. *Flutter* measures the fast rate of change in pitch of a constant tone. The rapidly changing audio pitch produces a fluttering sound. Both wow and flutter are results of video tape not being drawn through a machine at steady speed. Both can be present during playback.

Wow and flutter are not changes in volume. They are changes in pitch or frequency. Imagine a 1 kHz tone recorded on a tape without wow or flutter and then played back on a VCR with a mechanical misalignment causing slight variations in playback tape speed. The ear would hear the 1 kHz change in frequency. This change is directly related to the variation in playback speed. As the tape is drawn faster through the machine, the pitch frequency shifts upward. As the tape movement slows down, the playback pitch decreases. The rate at which this occurs differentiates between wow and flutter. Flutter occurs many, many times a second. Wow is a much slower shift in frequency.

Wow and flutter prevention is less effective as the tape speed is reduced while recording because it's increasingly difficult for the mechanical parts of the VCR to steadily pull the tape through the machine as tape speed is reduced. Unless specifically stated otherwise, published specifications for VCR wow and flut-

ter are defined for the SP mode only. Most VHS machines have wow and flutter that can be detected at the slowest recording speeds. The best audio recording occurs during faster recording speeds. Frequency response and wow and flutter are optimized when the audio is recorded in standard play (SP) speed.

8MM AUDIO FREQUENCY MODULATION (AFM)

The 8mm AFM audio system was developed around the original Beta Hi-Fi AFM design. An advantage of the 8mm AFM system is its design simplicity when compared to the Beta Hi-Fi circuitry. The first 8mm AFM design provided only monaural (single channel) sound. The Beta Hi-Fi design records in stereo. Beta Hi-Fi recordings implement a complex AFM frequency shifting technique to produce four AFM frequencies. Four frequencies are needed to reduce the adjacent channel crosstalk in the Beta format. Only a single 1.5 MHz AFM carrier is required in the 8mm format for suitable monaural AFM audio reproduction with minimal crosstalk. It is the combined use of monaural AFM sound and the increased 8mm gap azimuth angle difference between heads (Betamax only has a ±7 degree offset between heads) that allow a simplified 8mm AFM circuitry design to be used and still obtain reasonable crosstalk specifications.

Figure 9.3 shows the recorded RF frequency spectrum for the conventional 8mm format with the AFM and pilot carrier frequencies. Notice that Fig. 9.3 reveals two frequencies—one at 1.5 MHz and another at 1.7 MHz. The 1.7 MHz carrier frequency was added later for stereo AFM audio.

Some bandwidth is sacrificed in the lower sideband to make space in the spectrum for the AFM signal, but this has minimal effect on the reduction of picture detail. The advantages in the quality of the AFM sound reproduction far surpass the minimal loss of detail created by reducing the lower bandwidth passband. Some advantages of AFM audio include

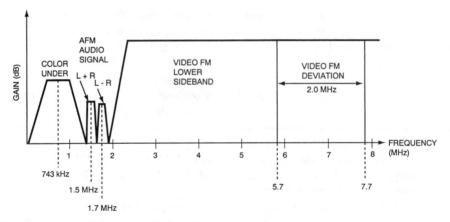

Fig. 9.3. *The 8mm RF frequency spectrum showing the AFM carriers.*

the absence of wow and flutter, a wide dynamic range, flat frequency response over the human hearing range and a high signal-to-noise ratio. The advantages of AFM sound are quite obvious when comparing these qualities to the linear track audio specifications.

Stereo 8mm AFM and L+R/L–R Multiplexing

As the design of the 8mm format evolved, stereo AFM was developed and is now available in selected models. The stereo AFM system uses a two channel sound multiplexing method of left (L) and right (R) channel audio that is similar to the commercial FM radio music you receive in your home stereo. The original FM radio broadcasts were all done in monaural sound. Later a stereo FM broadcast method using multiplexed L+R (left plus right) and L–R (left minus right) audio programming was developed for these broadcasts. A multiplexed stereo broadcast was required so that the original monaural FM receivers could reproduce a combination of both the left and right channel sound on a single speaker, while still providing individual left and right program material for the stereo radio receivers.

The 8mm format took the concept of FM radio stereo and monaural compatibility and

applied it to the stereo AFM machines. The concern was that the "monaural only" AFM 8mm units must be capable of reproducing both left and right recorded channel information on the single carrier AFM signal they are capable of detecting. This is accomplished by using L+R and L–R multiplexed signals in the 8mm stereo AFM design.

To understand how L+R/L–R multiplexing works, you will need to know how the entire AFM record/playback system operates. The multiplexing and demultiplexing process requires that the signal pass through the entire record and playback process.

8mm AFM Record Path

Take a look at the record AFM process shown in Fig. 9.4. The selected audio record signal (line, tuner, or mic) is applied to the left and right inputs of the record circuit. The input audio is applied to AGC control to prevent high input volume levels from overdriving the circuit.

The input source is then applied to a 15 kHz LPF (low-pass filter) to eliminate high frequencies above 15 kHz. Blocking these frequencies prevents interference with the noise reduction process occurring later in the signal path.

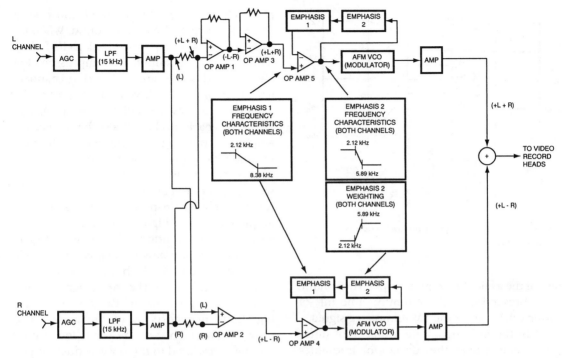

Fig. 9.4. 8mm AFM record.

L+R and L–R multiplexing

The audio signal is amplified after low-pass filtering and then applied to a resistor. Notice the L and R signal mixing that takes place for both channels in the vicinity of the resistor and the first operational amplifier (op amp) following it. This is the point where the L+R and L–R multiplexing begins. The L and R signal phase relationships are indicated at key points in the signal paths.

The op amps shown in Fig. 9.4 not only amplify the signal but also invert the phase of signals applied to the negative (–) input. For example, look at the inputs to op amp 2 in the drawing. Notice how the R signal is applied to the inverting (–) input and the L signal from the other channel is applied to the noninverting (+) input. The op amp function inverts the R signal 180 degrees and mixes that signal with the noninverted L signal. Thus, the output is defined as L–R. By looking at the various op amp inputs and the L and R phase indications

along the record AFM signal paths of both channels, you will see how the L+R and L–R signals are derived.

Record Emphasis, Dynamic Range Compression, and Noise Reduction (COMPANDING)

After the signals of both channels have been multiplexed to form L+R and L–R products, they are applied to two emphasis circuits for pre-emphasis and noise reduction. All AFM noise reduction methods apply signal volume compression during record and volume level expansion during playback. This is known as *companding*. Figure 9.5 shows the companding signal levels for both record and playback.

Companding the AFM audio provides two benefits. First, compressing the record signal's dynamic range means that the modulator frequency modulating the AFM signal will deviate less from the carrier frequency. Thus, the dynamically compressed signal will occupy

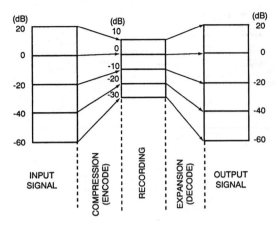

Fig. 9.5 AFM signal compression and expansion (companding).

less of the available frequency spectrum on the tape preserving the maximum amount of lower sideband for the FM luminance signal.

In the AFM signal, the number of times per second that the deviation frequency changes from the carrier frequency to the deviation value equates to the frequency of the reproduced audio signal. Also the amount of frequency deviation (frequency change from the carrier frequency) equates to the volume level of the reproduced sound. This means that an increase in the deviation frequency swing from the carrier frequency produces a louder volume than a narrower deviation swing. Therefore, if the dynamic range, which is an expression of volume level from the softest to the loudest recordable level, can be compressed to a smaller ratio, the deviation range required of the AFM modulator can also be reduced.

A second advantage to companding the AFM signal is that noise reduction is possible as part of the process. The most common form of noise is a low level hiss that comes from playback head-to-tape contact errors and the high gain playback preamplification stages. Figure 9.5 shows that the low level audio signals (those levels below –20 dB) are amplified as part of the record input audio signal dynamic range compression process. During

playback, these compressed levels must be restored to recreate the correct sound. When the playback signal is expanded back to its full dynamic range, the amplified (compressed) low level record tones are reduced (expanded) back to their original low levels. As part of this expansion, the low level hiss (noise) produced in the playback mode will also be lowered in amplitude pushing the volume of the unwanted hiss below our hearing range.

The emphasis 1 and emphasis 2 stages work together to compress the signal with the response characteristic shown in Fig. 9.5. The graph shows that high level input signals are reduced in amplitude while low level signals are amplified. Not only is the companding circuitry sensitive to levels, but it's also sensitive to the frequencies of the input signals. Since noise typically occurs in the low level and high-frequency portions of the audio spectrum, the frequency of the input levels must also be considered in the noise reduction portion of the companding process. The graphs near op amps 4 and 5 in Fig. 9.4 show the frequency weighting characteristics of the emphasis 1 and emphasis 2 feedback paths. For noise reduction to be effective, the emphasis 1 and emphasis 2 circuits must process high-frequency and low-frequency signals with the same low amplitude differently. This is accomplished by the frequency selection (weighting) inside the companding emphasis circuitry.

During playback, the AFM signal is expanded so the de-emphasis circuits must perform in the opposite (mirror image) manner to return the volume and frequency response back to the original input values.

8mm AFM modulation

After level compression and emphasis, the two channels are applied to their respective FM modulators. The modulation characteristics of the stereo AFM modulators are shown in Table 9.1.

The AFM modulators operate similar to the video FM VCO described in Chapter 7. If a 1 kHz audio tone is applied to the frequency

Table 9.1.
8mm AFM Frequency Modulation Characteristics

	L+R	L–R
Carrier frequency	1.50 ± 0.01 MHz	1.70 ± 0.01 MHz
Maximum frequency deviation	± 100 kHz	± 50 kHz
Typical frequency deviation	± 60 kHz	± 30 kHz

modulator, the carrier will swing back and forth (deviate) from the center frequency 1000 times per second. If there is no audio signal input to the modulator, the VCO signal will remain at the center (carrier) frequency. The volume of the AFM signal is represented by the amount of deviation. In other words, an AFM signal that deviates ±60 kHz from the center frequency at a rate of 1000 times per second will produce a loud 1000 Hz tone while an AFM signal deviating 15 kHz at 1000 times per second will generate a much quieter 1000 Hz tone.

One difference between AFM and video modulators is the deviation range. The audio deviation range is narrower than the video. Also, the AFM deviation range extends above and below the center frequency while the video FM deviation only extends above the carrier frequency. Table 9.1 shows that the L+R deviation range is ±100 kHz, indicating that the FM frequency can swing 50 kHz above and 50 kHz below the center frequency.

The table also shows that the L–R deviation range is narrower than the L+R signal. A narrower deviation was required for the 1.7 MHz FM modulator to prevent interference with the lower sideband of the Y FM signal. A narrower deviation range for this channel does reduce the overall dynamic range of the reproduced stereo signal slightly and also decreases the stereo channel separation. However, the AFM stereo reproduced by this format is more than adequate for the average listener. People who require higher quality stereo sound should use the PCM stereo audio available on some models.

Figure 9.3 shows that the recorded level of the 1.7 MHz carrier has a lower amplitude than the 1.5 MHz signal. By design, the 1.7 MHz carrier level is set to –2 dB, ±2 dB referenced to the 1.5 MHz carrier. The reduction of the recorded signal strength of this AFM carrier assists in reducing the appearance of carrier beat in the playback video produced by the proximity of the added 1.7 MHz AFM stereo signal and the lower sideband of the Y FM signal. Thus, the combination of the reduced 1.7 MHz deviation range and the reduced recording strength of that signal work together to minimize carrier beats produced by the added stereo AFM signal.

During monaural recordings, the 1.7 MHz carrier is cut off and only the 1.5 MHz signal is recorded.

After modulation, the two channels are amplified and sent to the video head record circuitry for mixture with the Y FM and color-under signals for recording onto the tape.

8mm AFM Playback

Figure 9.6 shows the 8mm AFM playback path. The playback Y FM, color-under, ATF, and AFM signals are read by the playback video heads and amplified and equalized. After amplification and equalization, the composite RF signal is applied to the band-pass filter (BPF) in the AFM circuit. The band-pass filter has a center frequency of 1.6 MHz and a passband of ±300 kHz. Therefore, this BPF will only pass frequencies ranging from 1.3 MHz to 1.9 MHz. Only the AFM frequencies are in this range and are passed on to the AFM circuits.

The separated (band-pass filtered) AFM signal is split into two paths that feed the signal to either a 1.5 MHz or 1.7 MHz band-pass filter for a second stage of filtering. These filters separate the playback AFM signal into the L+R and L–R paths for stereo processing.

With the exception of the 1.7 MHz frequency detection circuit and a demultiplexing circuit consisting of op amps 1, 2, and 3, the

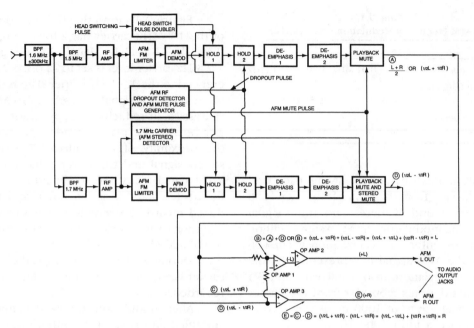

Fig. 9.6. *8mm AFM playback.*

stereo channel processing is identical for each channel.

The RF amp stage following the 1.5 MHz and 1.7 MHz BPF filters restores signal losses occurring in the band-pass filters.

AFM dropout detection

The AFM RF dropout detector monitors the playback AFM signal for losses. Short duration RF dropout generates a popping sound in the playback audio. The long-term absence of AFM RF signal creates audible static. Momentary AFM RF dropout is masked inside the Hold 2 block and will be described following the AFM demodulation. When the AFM RF signal disappears for long periods of time, the AFM RF dropout circuit mutes the audio output to prevent the static from being sent to the audio outputs. The RF dropout detector is a level threshold type of circuit. It activates once the playback AFM RF signal has dropped below a preset level and doesn't deactivate until the RF level returns.

AFM FM limiting

After RF dropout detection, the AFM signal is applied to an FM limiter where RF level fluctu-

ations are removed from the signal. If not removed, these unwanted level fluctuations will amplitude modulate the FM signal and create distortions in the demodulated sound. The source of the unwanted playback AFM RF level fluctuations is the same as the luminance playback RF level fluctuations described in Chapters 5 and 7.

AFM demodulation

After limiting, the AFM signal is applied to the demodulator where it is returned to its original form. The AFM demodulators in each path are phase comparison demodulators. In addition to the playback AFM signal input, phase comparison demodulators require a frequency reference. A phase-locked-loop (PLL) VCO oscillator provides the phase comparison reference in this circuit. The L+R demodulator demodulates the 1.5 MHz AFM signal so its PLL reference frequency runs at 1.5 MHzs. The L–R demodulation PLL frequency is 1.7 MHz.

The demodulator operates by comparing the frequency of the playback AFM signal with the reference frequency. Since phase error detectors are sensitive to both phase and frequency discrepancies between the reference and input signals, differences in frequency and/or phase between the AFM playback and reference signals produce an error voltage at the comparator's output. The comparator is part of the FM demodulator. This error voltage is the demodulated audio signal.

As the playback AFM deviation swings above the center frequency, the comparator's error voltage increases in amplitude. When the AFM deviation frequency begins to swing back to the carrier frequency, the error voltage begins to drop correspondingly. The error voltage will return to zero volts once the deviation frequency returns to the carrier frequency. However, if the deviation frequency continues to swing past the carrier frequency and deviates below the carrier frequency, the comparator's output error voltage will go negative. An increase in the negative deviation of the playback AFM signal will push the error voltage further negative.

Therefore, large signal deviations in the playback AFM signal generate increased amplitudes at the comparator's error voltage output. Also, the rate at which the signal deviates above and below the carrier determines the frequency at which the comparator's error voltage swings negative and positive. If the deviation frequency swing extends above and below the carrier frequency in equal quantities, and the deviation swing occurs linearly, the error voltage output from the phase comparator will be a sine wave with equal positive and negative peaks. Sine waves form the signal that produces the sound we hear. The amplitude of the sine wave corresponds to the volume of the sound that we hear. The frequency of the sine wave produces the pitch of the tone of the sound that we hear.

To summarize, the AFM demodulator is a phase error detector that compares the AFM playback signal with a stable reference. Differences between the reference and playback AFM signals produce an error voltage that is the demodulated audio signal. The rate at which the playback AFM deviation frequency sweeps back and forth causes the error voltage to oscillate at

that same sweep frequency. The oscillation of the detected error voltage produces the frequency (pitch) of the tones we hear. Increases in AFM frequency deviation (wider sweep ranges) produce louder volume levels. Smaller AFM frequency deviation (narrower sweep ranges) produces quieter sound levels.

Playback head switching noise elimination and dropout correction

The demodulated audio is applied to the Hold 1 and Hold 2 blocks in Fig. 9.6. The Hold 1 block eliminates the popping sound that occurs during each head switching point. For the video signal, head switching is performed at a point on the TV screen 6.5 lines before vertical sync. This is not seen by the observer because switching occurs below the normal viewing portion of the TV image. Video head switching causes interference in the TV picture, but it occurs in an unseen portion of the screen.

With the AFM head signal, however, there is no convenient time to switch between heads that will keep the head switching noise out of the demodulated sound. Audio is continually available at the speakers so the head switching interference in the AFM signal must be carefully masked to prevent the 60 Hz audible popping noises that would otherwise be produced by the head switching interval.

The source of Hi-Fi audio head switching noise comes from the signal distortions created in the continuous AFM carrier at the head switching point. Figure 9.7 shows how this distortion develops. When the Hi-Fi FM carrier is switched between playback heads, the carrier phase is distorted as the FM signal is changed from one head to the other. This change in phase at the head switching point results in an unwanted frequency deviation change that is detected by the demodulator. The demodulator outputs a pulse when it demodulates this distortion. We would hear this as a popping sound. The head switching distortion and resulting pop sound occur at each head switch point or 60 times each second (60 Hz rate).

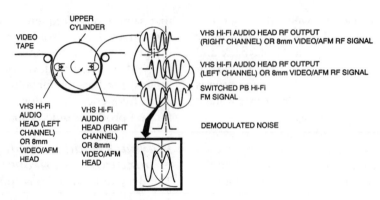

Fig. 9.7. 8mm AFM and VHS Hi-Fi head switching noise.

As shown in Fig. 9.6, head switching noise reduction is performed in the Hold 1 block after demodulation. Figure 9.8 shows how the head switching interference is masked in the Hold 1 circuit. The Hold 1 circuit receives the demodulated AFM signal and the doubled head switching pulse inputs. The head switch pulse doubler causes positive pulses to occur each time a head switching pulse transition happens. The doubled head switching pulse transitions mark the point where the distorted AFM carrier signal needs masking.

Head switch noise masking is accomplished by taking an average voltage of the signal at the moment the head switching pulse is present and inserting the average voltage into the demodulated audio. For this action to be successful, the period of the demodulated audio containing the head switching noise must be muted prior to insertion of the averaged playback signal. After the noise has been muted out and the average signal has been inserted, the Hold 1 circuit outputs a masked (popping noise free) audio signal. Some minor waveform distortion occurs at the masking position of the audio signal, but this distortion is much less objectionable than the popping noise that would otherwise be present if it were not masked out.

Notice that the Hold 1 circuit in Fig. 9.8 also receives a dropout pulse through an OR gate whenever dropout occurs. The dropout detector produces a hold signal when the AFM RF waveform drops below a preset threshold level. This hold pulse is sent to the Hold 1 and Hold 2 circuits and the RECORD/PLAYBACK/DROPOUT select switch shown in Fig. 9.8. The hold pulse cause the Hold 1 circuit to capture the demodulated audio level present just prior to the occurrence of dropout. The combination of the last moment of noise-free demodulated audio sampled by the dropout detector's hold pulse and the doubled head switching pulse produces the masked output from the Hold 1 circuit.

Dropout correction takes place inside the Hold 2 block. The output of the Hold 1 circuit is fed to a RECORD/PLAYBACK/DROPOUT switch and then to a low-pass filter. The LPF blocks frequencies above 15 kHz to prevent improper operation of the noise reduction processing that occurs later in the path. This LPF is an active circuit that also delays the demodulated sound slightly as it is filtered. The Hold 2 circuit takes advantage of this slight audio delay when dropout correction is necessary. Figure 9.9 shows the timing of the AFM dropout signals.

When dropout occurs, the hold signal from the dropout detector is applied to both the Hold 2 block and the RECORD/PLAYBACK/DROPOUT selection switch. Since the audio input to the Hold 2 circuit coming from the low-pass filter is delayed audio, this input contains the last good sound present before

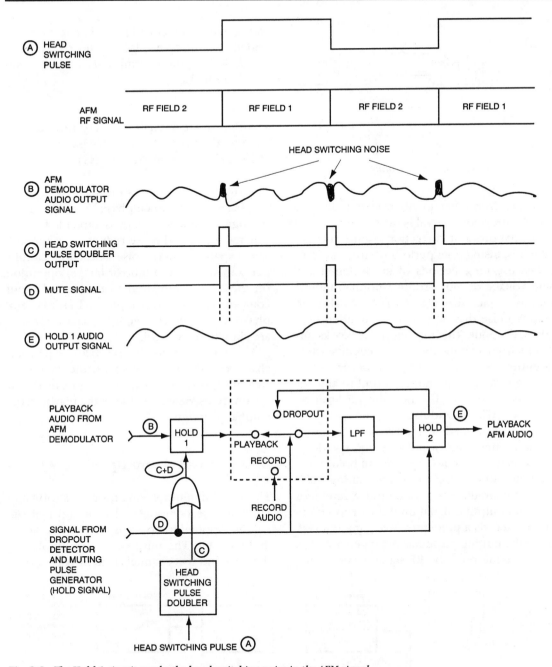

Fig. 9.8. The Hold 1 circuit masks the head switching noise in the AFM signal.

the dropout occurred. This delayed audio voltage is held by the Hold 2 circuit and fed back to the RECORD/PLAYBACK/DROPOUT selection switch at the DO pin. The selection switch is forced into the DO position whenever the

dropout hold pulse is present. As long as the switch is in contact with the DO pin, the selection switch will supply the Hold 2 circuit with the held voltage via the low-pass filter. This voltage is held until the dropout hold pulse

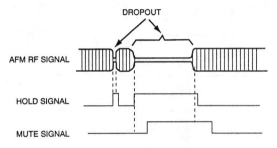

Fig. 9.9. AFM dropout correction timing.

from the drop detector disappears indicating that the dropout interval is over.

AFM dropout correction is only successful in disguising brief periods of dropout. This is because long periods of an averaged and held voltage do not produce normal audible sounds. Notice that Fig. 9.9 shows a delayed muting pulse that occurs after extended periods of dropout. This muting pulse blocks any sound from exiting the audio circuits when extended periods of dropout occur. When muting is active, the output sound totally disappears. Muting ends when the RF level returns and AFM demodulation resumes.

The dropout detector circuit is the source of the muting control pulse. When the dropout detector generates the dropout hold pulse, the hold pulse is applied to the muting pulse generator circuit. The muting pulse generator does not output a signal until the dropout duration exceeds a predetermined time interval. Once the muting pulse has been generated, it will continue until the RF signal returns indi-

cating that the extended period of dropout has ended. Muting takes place in the PB Mute circuit following the De-emphasis 1 and De-emphasis 2 blocks.

Playback De-Emphasis, Dynamic Range Expansion, and Noise Reduction (COMPANDING)

The Hold 2 block passes the signal to the De-emphasis 1 and De-emphasis 2 blocks. Here the signal's dynamic range is expanded out to original levels and noise reduction is applied. The expansion and noise reduction circuits perform an inverted (mirror image) operation on the signal compared to the record input compression. The De-emphasis 1 and De-emphasis 2 frequency weighting characteristics are shown in Fig. 9.10.

By comparing the playback expansion characteristics of Fig. 9.5 with the frequency weighting shown in Fig. 9.10 you can follow the entire expansion action of the playback de-emphasis circuits.

L+R and L–R Demultiplexing

The signal demultiplexing process is shown in Fig. 9.6. The demodulated L+R and L–R signals are applied to op amps, 1, 2, and 3 for demultiplexing. The purpose of demultiplexing the L+R and L–R channels is to provide a pure

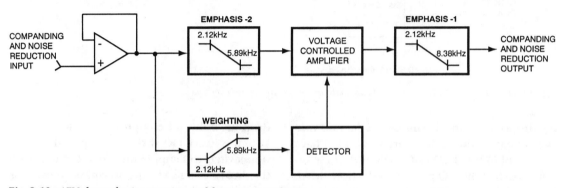

Fig. 9.10. AFM de-emphasis operation and frequency weighting.

L (left) channel audio signal on one output and pure R (right) channel on the other line.

The amplitude of the L+R signal is reduced by one half in the L+R demodulation process to become ½L + ½R (labeled as signal A in the path) and applied to the inverting input of op amp 1 at point B. Also at point B, the output of the L–R playback channel is mixed in the resistor network to form the complete signal feeding op amp 1. Note that the L–R channel signal (shown at point D) was also reduced to become ½L – ½R. The output of op amp 1 becomes the –L because:

$$\frac{point\ A\ signal\ +\ point\ D\ signal}{1} = -L$$

where A = (½L + ½R) and B = (½L – ½R) and like terms of L and R are added together. Op amp 2 inverts the op amp 1 output to provide a noninverted (demultiplexed) L signal as one of the stereo AFM outputs. This is the "left channel only" signal.

Point C at op amp 3 is the ½L + ½R demodulated input. Op amp 3 does not invert this signal at its output. The input at point D comes from the L–R demodulator and is inverted at the op amp output. This input is ½L – ½R. Op amp 3 mixes the inverted and noninverted input signals to produce +R at its output. This is because:

$$point\ C\ signal\ +\ point\ D\ signal\ =\ R$$

where C = (½L + ½R) and D = –(½L – ½R) and like terms of L and R are added together. Thus, op amp 3 demultiplexes its inputs to provide the R only signal at its output.

Notice that the L–R channel in Fig. 9.6 contains a stereo detection circuit. This circuit blocks the L–R channel from outputting a signal when a monaural program is being played back. A monaural AFM signal will be played back when the tape was recorded on a single channel 8mm unit. Monaural 8mm units record AFM only on a 1.5 MHz carrier. Thus a stereo 8mm VCR can detect a monaural signal when the 1.7 MHz AFM carrier is absent. A lack of 1.7 MHz carrier causes the stereo mute circuit to activate and block any audio output from leaving the L–R AFM demodulation channel. Signal blocking takes place at the L–R channel PB mute circuit in the block diagram.

When the L–R channel is muted by the stereo detection circuit, there is no signal at point D in Fig. 9.6. Therefore, the L+R audio at point A is passed out of op amps 1 and 2 without canceling the R signal. The R signal is not eliminated because there is no cancellation effect occurring at point B in Fig. 9.6.

When the L–R stereo channel output is defeated, the monaural audio signal is applied to both L and R audio outputs through a switching circuit so that both speakers receive the L+R monaural signal. However, in this case, both speakers will produce the same signal.

Adjacent RF track AFM crosstalk

Regardless of the format, all AFM systems use a single head to record program material for both left and right channels. The head recording the two channels of sound at any given moment in time is simply the head contacting the tape at that instant. A common misconception is that one head writes and reads one channel while another head writes and reads the other channel. This is not true.

The concern for adjacent channel crosstalk in an AFM recording is that the playback head may detect some of the AFM program recorded on an adjacent track. If AFM crosstalk is detected by the playback head, the crosstalk audio will be demodulated by the playback circuits creating a weak echoing effect in the output sound. Recall that the azimuth angle offset between heads is sufficient for crosstalk cancellation of the high-frequency Y FM signal but is less efficient at lower frequencies.

Chapter 6 described the electronic processing required to eliminate the relatively low frequency color-under crosstalk signal. Color-under crosstalk could be eliminated with a comb filter because of the particular phase relationships of the color-under signal on adjacent tracks. In AFM, no stable phase relationship is present on any track.

The original AFM system developed for the Beta Hi-Fi units overcomes the crosstalk dilemma by shifting the two AFM carriers up 150 kHz (as compared to the AFM frequencies for the other head) when one head is in contact with the tape. The use of four carriers and four separate band-pass filters provides good isolation between tracks in these units.

The 8mm format does not shift the AFM carrier between heads. It depends entirely on the azimuth cancellation of the head gaps for crosstalk isolation. The 8mm units have adequate crosstalk cancellation occurring within the total 20 degrees of azimuth angle offset between the two heads. Later in this chapter we will describe how the VHS Hi-Fi system reduces AFM crosstalk.

VHS HI-FI (HD) AUDIO

VHS Hi-Fi audio (also known as HD or high definition audio) incorporates a separate pair of heads specifically designed for AFM recording. The head gaps on these Hi-Fi audio heads are placed at an angle of ±30 degrees. This large offset angle provides adequate crosstalk isolation from adjacent RF tracks during playback of the relatively low Hi-Fi audio FM frequencies. These audio heads are mounted 180 degrees apart and 120 degrees ahead of the adjacent video head as shown in Fig. 9.11. Figure 9.12 shows a VHS video head drum with four video heads and two AFM heads.

The VHS Hi-Fi audio frequencies are 1.3 MHZ for the left channel and 1.7 MHZ for the right channel. Unlike the 8mm stereo AFM de-

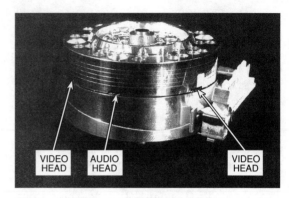

Fig. 9.12. Six head VHS drum assembly.

sign, VHS Hi-Fi does not multiplex the left and right channel audio signals. In VHS Hi-Fi, the left audio program sound is unique to the 1.3 MHz channel while the right program sound is found only on the 1.7 MHz channel.

Offsetting the Hi-Fi audio head gaps by 60 degrees provides adequate audio RF track crosstalk isolation for the comparatively low Hi-Fi FM audio frequencies. This also provides suitable isolation between the Y FM and Hi-Fi FM signals that occupy the same RF track. The difference in azimuth between the L video head and the AL audio head provides good isolation between the Y FM and Hi-Fi FM signals. This isolation allows the Y FM lower sidebands and the Hi-Fi FM signals to occupy the same space (overlap) in the frequency spectrum without causing picture interference (carrier beat).

Figure 9.13 shows the relationship of the video recording frequency spectrum and the position of the Hi-Fi FM signals.

The Hi-Fi FM signals occur in the lower sideband portion of the Y FM frequency range.

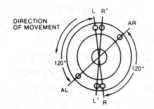

Fig. 9.11. The audio head precedes the video head by 120 degrees.

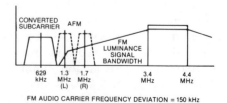

Fig. 9.13. Hi-Fi audio position within the audio-video frequency spectrum.

In the 8mm system, AFM and Y FM are intentionally isolated and do not occur in the same general frequency band.

Depth Multiplexing

The VHS Hi-Fi recording technique differs from that for the 8mm AFM system in that it is capable of placing the Hi-Fi FM audio signal within video FM lower sideband without causing any carrier beat interference or loss of high-frequency response in the picture. This is accomplished with special Hi-Fi audio heads and a recording technique called *depth multiplexing*. Figures 9.14 and 9.15 describe depth multiplexing. As the video head rotates, the AL (L Hi-Fi audio) head contacts the tape 120 degrees (5.53 milliseconds) before the L FM video head reaches the same spot. The AL head first magnetizes the tape with an FM audio signal. The record current amplified to the AL head is strong enough to create a magnetic field that fully penetrates the tape oxide coating. The L video head contacts the tape 5.33 milliseconds later, but now the record current in the L video head is about half that applied to the AL head. The magnetic field thus created is only strong enough to magnetize the top layer of the tape oxide, so the deeper portion of the oxide coating retains the audio magnetic information while the upper portion is remagnetized to the video FM signal.

In playback, the AL head detects just the Hi-Fi signal. Azimuth angle differences between the Hi-Fi audio and the video heads prevent the AL head from recognizing the L video head information on the top layer of the same track. These same azimuth angle differences prevent the AL audio head from detecting the L video track information. The R heads operate in a similar manner. Recall that VHS video heads have an azimuth angle offset of ±6 degrees. The VHS Hi-Fi audio heads have an azimuth angle offset of ±30 degrees. The greater angle offset is required for the Hi-Fi heads to provide good stereo channel isolation and isolation between the multiplexed audio and video signals.

Using depth multiplexing, the VHS Hi-Fi design has an advantage over other VCR formats by providing better isolation between the Y FM and Hi-Fi audio FM carrier frequencies. Another advantage of the VHS Hi-Fi system is that it is a true two-channel stereo system. VHS Hi-Fi does not multiplex the stereo signal into L+R and L–R components as is done in the 8mm format. Therefore, the VHS Hi-Fi system provides better stereo channel isolation than 8mm AFM stereo units. A third advantage of the VHS Hi-Fi design is that the adjacent RF track crosstalk is significantly lower than that for the 8mm units. This is because of the 60 degree head gap azimuth angle difference for the Hi-Fi heads compared to the 20 degree offset in 8mm heads. Notice that the 8mm and VHS units use audio FM frequencies that operate in the 1.3 MHz and 1.7 MHz range. Therefore, the increased azimuth angle offset between audio head gaps provides better adjacent track crosstalk isolation in VHS Hi-Fi units.

Notice that the head gap length for the Hi-Fi head in Fig. 9.15 is 1 micron. This is a substantially longer gap length than that found for the video head. One micrometer is necessary for the Hi-Fi audio head because of the longer recorded wavelength of the lower frequency Hi-Fi FM signal.

Another difference between the Hi-Fi audio heads and video heads is the width of the head core and corresponding recorded track width. The Hi-Fi head core and recorded track width may be narrower than the video RF track on some models. Narrower Hi-Fi head

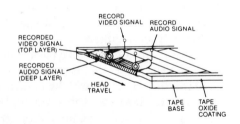

Fig. 9.14. Depth multiplexing (top view).

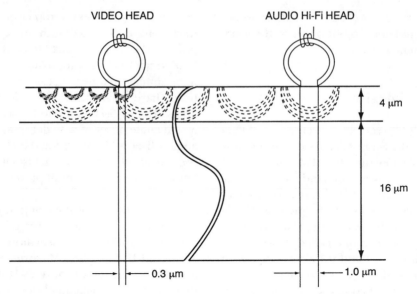

VIDEO HEAD   AUDIO Hi-Fi HEAD

4 μm

16 μm

0.3 μm   1.0 μm

Fig. 9.15. *Depth multiplexing (side view).*

core widths are required to accommodate the SLP (EP) record speeds, which require narrower tracks. Some consumer VHS Hi-Fi heads are only 19 microns wide.

One disadvantage of the VHS Hi-Fi design relates to the added cost of the video head assembly. The addition of the two Hi-Fi heads and the extra rotary transformers increases the production cost of both the upper and lower sections of the video head drum. Therefore, there is an increased repair cost to replace the video head cylinder.

When worn or defective heads are replaced, some VHS manufacturers specify replacement of the entire video head drum assembly, including the upper head and lower drum motor. This requirement makes head replacement for the VHS Hi-Fi units expensive. Manufacturers who specify this complete changeout indicate that replacing the entire drum is necessary to ensure tight mechanical tolerances between the upper and lower drum assemblies. This is because a possibility exists that if only the upper portion of the drum were replaced, the Hi-Fi and video FM signals may not properly record on top of one another and playback interchange would be lost.

For a VHS Hi-Fi unit, interchange not only means that the video RF signal on the tape should track correctly on all VHS machines, but the Hi-Fi audio should also track correctly. When tracking interchange is correct, both the video RF and Hi-Fi RF signals will produce peak amplitudes out of their respective playback heads when the tracking control is centered. If the video and Hi-Fi heads are not positioned precisely on the head drum, the RF for the video and Hi-Fi signals will peak at different tracking control settings. This happens because the video RF and Hi-Fi RF tracks are not laid perfectly on top of each other. When the depth multiplexed tracks are not perfectly on top of each other, the operator must compromise between optimum playback picture quality and best Hi-Fi audio reproduction by setting the tracking control so that both stereo Hi-Fi sound and an acceptable viewing picture can be obtained.

One symptom of lost playback interchange on a VHS Hi-Fi unit is static or distorted Hi-Fi audio when the tracking control is adjusted for best playback picture quality, or conversely, poor picture quality when the tracking control is adjusted for best Hi-Fi sound. Tape playback interchangeability becomes an important topic for owners of VHS Hi-Fi machines who rent video tapes and expect to hear Hi-Fi sound. When playback interchange becomes a problem, the tracking control must be adjusted to a position that compromises between the rental tape's video and audio playback quality. Precise placement of the video and Hi-Fi heads on the drum is critical to proper operation and tape playback interchangeability.

Another disadvantage to this design is in the useful life of the video head. Since the Hi-Fi signal is depth multiplexed, the Hi-Fi heads must operate efficiently to recover the Hi-Fi signal residing below the recorded video signal. As the heads wear with use, their record and playback efficiency drops off, making the recovery of the deeply imbedded Hi-Fi signal harder to detect. Head wear will often cause unacceptable Hi-Fi audio record and playback before the L and R video heads stop producing a viewable picture. Therefore, a VHS Hi-Fi video head cylinder may provide fewer hours of trouble-free operation than non-Hi-Fi VHS head cylinders.

Hi-Fi audio recordings, playback FM signal strength, and magnetic signal retention

One concern that many professional VHS users have is the reduction of retained magnetic field strength of the FM video and Hi-Fi audio signals when Hi-Fi audio is recorded. Reducing the video RF record level has the potential of slightly reducing the picture signal-to-noise ratio because of the decreased playback RF levels. As previously stated, depth multiplexing requires that the video record signal strength be reduced so it will not erase the deeper layer of previously recorded FM audio. A reduced video record signal strength means that the video signal's magnetic strength retained on the recorded tape is lower than that of a magnetic coating that has been fully penetrated by the record signal. A reduction in the depth of the recorded magnetic signal's energy reduces

the overall magnetic lines of force present within the playback video head's gap. This means that the video playback RF signal detected by the video head is weaker when playing back Hi-Fi recordings than it will be if it plays back a non-Hi-Fi tape whose magnetic coating has been completely magnetized by only a video signal.

In addition to the reduction of the video signal strength on the tape, the Hi-Fi audio signal strength is also reduced because the top layer of the Hi-Fi audio track (the section of the Hi-Fi audio's magnetic signal closest to the head) is erased by the recording video head. This has the same effect of slightly pulling the Hi-Fi playback head away from the recorded information. The increase in distance between the recorded Hi-Fi audio track and the playback head is one form of *tape thickness loss*. Tape thickness loss is a phenomenon that results from recording a magnetic signal in the direction of the tape's thickness. Tape thickness loss increases as the magnetic signal gets farther away from the playback head. It is one cause of lower output signal from the playback head. Tape thickness loss has a similar effect on the playback signal strength as spacing loss. Spacing loss was described in Chapter 5.

Thus, for VHS Hi-Fi recordings, the strength of playback FM audio carrier is decreased for two reasons:

1. The reduction in depth of the recorded Hi-Fi track's signal in the magnetic surface cuts back the total signal strength available on the tape for playback recovery.

2. An increase in relative distance between the playback Hi-Fi head and the remaining Hi-Fi signal on the tape is generated when the topmost layer of the tape is rerecorded with the FM video signal.

Remember that, for depth multiplexed recordings, the tape thickness loss experienced by the playback Hi-Fi head is increased because the recording video head erases the portion of the previously written Hi-Fi audio track that is closest to the playback Hi-Fi audio head. This partial erasure occurs when the video head writes the video FM signal onto the uppermost layer of the tape's surface. (See Figs. 9.14 and 9.15.)

We can calculate the attenuation factor of the Hi-Fi audio FM carrier's playback signal strength created by tape thickness loss and the decrease in depth of the recorded signal. First, determine the depth (expressed in wavelength) of the FM video and Hi-Fi audio FM signals as they are impressed into the magnetic layer during record.

For the VHS NTSC FM video signal, the signal recording depth is calculated as shown below.

$$l = \frac{V}{f} = \frac{5.8 \text{ m/sec}}{3.4 \text{ to } 4.4 \text{ MHz}} = 1.3 \text{ } \mu\text{m to } 1.7 \text{ } \mu\text{m}$$

where

l = the recorded wavelength for the FM video signal,
V = the VHS writing speed (see Chapter 5),
f = the frequency of the video FM signal.

Thus, wavelength, l, ranges between 1.3 μm and 1.7 μm depending on the frequency of the video FM signal at any given instant in time.

If you substitute the frequencies for f with the VHS Hi-Fi audio FM values, you will find that the recording depth of the Hi-Fi FM signal ranges between 3.4 μm and 4.5 μm (see calculation below).

$$l = \frac{V}{f} = \frac{5.8 \text{ m/sec}}{1.3 \text{ to } 1.7 \text{ MHz}} = 3.4 \text{ } \mu\text{m to } 4.5 \text{ } \mu\text{m}$$

where

l = the recorded wavelength for the VHS Hi Fi FM signal,
V = the VHS writing speed,
f = the frequency of the Hi-Fi audio FM signal.

These recording depths are not used in actual practice because most of the signal detected by the playback head (Hi-Fi audio or video) comes from the one-quarter portion of the recorded wavelength depth that is closest to the reading head. Tape thickness losses become significant and interfere with the reading head's performance after the recorded wavelength goes beyond the one-quarter wavelength depth. Thus, to optimize performance of the write and read operation of the head, recording currents are set so that the recorded signal does not impress into the tape's coating deeper than the one-quarter wavelength value. (The one-quarter wavelength record depth is true for all formats, not just VHS.)

Therefore, when calculating the depth of the recorded FM audio and FM video signals for VHS depth multiplex recording, the recorded wavelength depths are divided by four. The depth of the Hi-Fi FM and video FM signals are calculated as shown below.

$$d = \frac{l}{4}$$

and for the VHS FM video signal,

$$d = \frac{1.3 \text{ } \mu\text{m to } 1.7 \text{ } \mu\text{m}}{4} = 0.325 \text{ } \mu\text{m to } 0.425 \text{ } \mu\text{m}$$

while for the VHS FM audio signal,

$$d = \frac{3.4 \text{ } \mu\text{m to } 4.5 \text{ } \mu\text{m}}{4} = 0.85 \text{ } \mu\text{m to } 1.125 \text{ } \mu\text{m}$$

It may be easier to understand this depth recording process by reviewing Fig. 9.15 and inserting the appropriate wavelength values for the respective head.

The VHS Hi-Fi playback head signal strength degradation that occurs because of tape thickness loss can be calculated as shown by the formula below. The signal strength losses are expressed in decibels (dB). This formula requires the values from both the Hi-Fi audio FM and FM video calculations previously determined.

$$\frac{54.6 \times 1/4d}{l} = \text{signal spacing loss in dB}$$

where

d = the recorded depth (l) of the video FM signal,

l = the calculated recorded wavelength depth for the recording audio FM signal,

54.6 = a numerical constant used to calculate depth multiplex thickness loss derived from mathematical theory and scientific evaluation and testing.

Plugging the numbers into the formula:

$$\frac{54.6 \times 3.75 \ \mu\text{m}}{l} = 5.18 \text{ dB}$$

(*Note:* The value 3.75 μm was used because it represents one-quarter of the average value between 1.3 μm and 1.7 μm, which was previously determined to be the range of the record wavelength of the FM video signal.)

In actual measurements, the Hi-Fi playback signal loss is slightly more than 5.18 dB. Fortunately, tape thickness loss for the Hi-Fi signal is not enough to prevent good Hi-Fi reproduction. However, as the Hi-Fi heads wear, the effects of signal strength loss becomes more of a concern for good audio reproduction.

The long-term storage life of VHS Hi-Fi tape has also been a concern of professional users, especially where valuable recordings are involved. As the tape is used, the strength of the stored magnetic energy weakens slightly. Then as the tape ages, more of the retained magnetic strength is lost. As the playback RF level becomes weaker, the potential for snowy and unusable playback video increases. Obviously, if the video signal is recorded with only half of its potential strength to begin with, the possibility of weak playback RF becomes more of a concern, especially with tapes you wish to keep for long periods of time. One way to overcome this concern is to record tapes with the Hi-Fi audio turned off.

Machines providing this option will record the video RF at full record strength. These machines only cut the video RF record level in half when Hi-Fi recordings are made.

Fortunately, depth multiplexing designers considered some of these magnetic signal strength problems when they developed the Hi-Fi audio system. Figure 9.15 shows how these issues were addressed. Lower recording frequencies tend to penetrate (and then reside) deeper into the tape's magnetic coating than high frequencies. Also, most of the energy recovered by the playback head comes from the magnetic signal within a distance from the head surface equal to the first quarter of the recorded wavelength. Therefore, for the NTSC VHS system, most of the recorded video RF playback signal strength (which is the higher frequency component) comes from the top 0.3 micrometers of the tape's surface while the lower Hi-Fi frequencies are in the deeper regions of the magnetic coating. For the Hi-Fi audio signal, the first quarter of its recorded wavelength extends 1.0 micrometer into the magnetic coating of the tape surface. Since the video signal is recorded on top of the Hi-Fi audio, depth multiplex recording provides the best picture quality possible by maintaining a close distance between the video head and the top layer of the magnetic surface while retaining enough of the first quarter wavelength of the Hi-Fi audio signal to recover the Hi-Fi sound.

Since the top 0.3 micrometers of the recorded Hi-Fi signal is rerecorded with the video signal, some of the Hi-Fi audio's first quarter wavelength is eliminated. This reduction in the Hi-Fi track's magnetic signal is what causes the 5.18 dB signal strength loss described earlier. It is also this increased distance between the Hi-Fi head and the recorded information and the resulting spatial signal losses that cause the Hi-Fi heads to fail from wear sooner than the video heads.

Even with the appropriate selection of the depth multiplexed signal laid locations, the fact that the magnetic energy of both video and Hi-Fi signals is reduced as part of this record process does warrant valid concerns for the longevity of valuable Hi-Fi recordings. The best defense against Hi-Fi recording signal strength concerns is to select a high-quality tape that offers good coercivity and retentivity for your important recordings.

VHS Hi-Fi Signal Processing

The VHS Hi-Fi record/playback circuitry shown in Fig. 9.16 shares many similarities with the 8mm AFM circuitry. The major VHS Hi-Fi circuit differences include the record and playback electronics required to support the separate pair of heads (i.e., the additional record amplifiers and playback preamplifiers) and the lack of L+R/L–R multiplexing and de-multiplexing circuitry. Recall that L+R/L–R multiplexing is not required because the VHS system is a true two channel stereo recording system.

VHS Hi-Fi record

The left and right channel audio is applied to the Hi-Fi stereo circuit inputs. The source for these signals is the selected line input, tuner, or mic audio coming from the audio selection switch. The input sound is frequency-limited in the low-pass filter to prevent frequencies above the audible range from interfering with the noise reduction circuitry.

VHS record emphasis, dynamic range compression, and noise reduction (COMPANDING)

After low-pass filtering, the signal is sent to a companding circuit that compresses the dynamic range of the input volume and applies noise reduction. As with 8mm audio record and playback, all AFM and VHS Hi-Fi noise reduction methods apply companding signal volume compression during record and volume level expression. Refer back to Fig. 9.5 to see a typical companding signal level response graph for both record and playback.

Companding the Hi-Fi audio provides the benefit of compressing the record signal's dynamic range so the modulator frequency modulating the Hi-Fi signal will have to deviate less from the carrier frequency. Thus, the dynamically compressed signal will occupy less of the available frequency spectrum on the

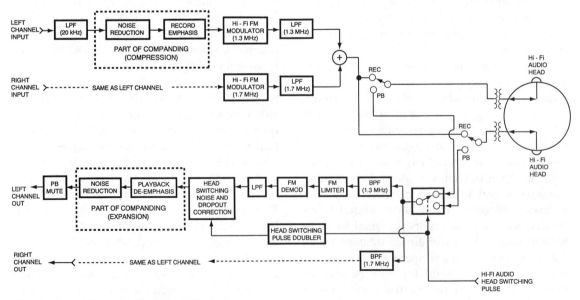

Fig. 9.16. VHS Hi-Fi record/playback block diagram.

tape. This action helps to reduce the appearance of carrier beat that occurs when the Y FM and Hi-Fi FM signals interact with each other to generate carrier beat in the video. Even though the azimuth angle differences between the audio and video heads eliminate most of this Y FM to Hi-Fi FM interaction, some of the resulting beat is still present. Reducing the Hi-Fi modulation deviation range helps to minimize this leakage from appearing on the screen during playback.

For the Hi-Fi signal, the number of times per second that the deviation frequency changes from the carrier frequency to the deviation value equals the frequency of the reproduced audio signal. The amount of frequency deviation (frequency change from the carrier frequency) equates to volume level of the reproduced sound. This means that an increase in the deviation swing from the carrier frequency produces a louder volume than a narrower deviation swing. Therefore, if the dynamic range (which is an expression of volume level from the softest to the loudest recordable level) can be compressed to a smaller ratio, the deviation range required of the Hi-Fi modulator can also be reduced.

A second advantage to companding the VHS Hi-Fi signal is that noise reduction is possible as part of this process. The most common form of noise is a low level hiss that comes from playback head-to-tape contact errors and the high gain playback preamplification stages. Figure 9.5 shows that low level audio signals (those levels below the −20 dB level) are amplified as part of the record input audio signal dynamic range compression stage. During playback these compressed levels must be restored to recreate the correct sound. When the playback signal is expanded back to its full dynamic range, the amplified (compressed) low level record tones are reduced (expanded) back to their original low levels. As part of this expansion, the low level hiss (noise) produced in the playback mode will also be lowered in amplitude pushing the volume of the unwanted hiss below our hearing range.

The emphasis and noise reduction circuits work together to compress the signal with the response characteristic shown in Fig. 9.5. The graph shows that high level input signals are reduced in amplitude while low level signals are amplified. Not only is the companding circuitry sensitive to levels, but the frequencies of the input signals are also considered. Since noise typically occurs in the low level and high frequency portions of the audio spectrum, the frequency of the input levels must also be considered in the noise reduction portion of the companding process. For noise reduction to be effective, the noise reduction circuits must process high-frequency and low-frequency signals of the same low amplitude differently. Just like the 8mm design (see graphs near op amps 4 and 5 in Fig. 9.4), the VHS Hi-Fi noise reduction circuits have frequency selection (weighting) inside the companding emphasis circuitry to handle this requirement.

During playback, the Hi-Fi signal is expanded so the de-emphasis circuits must perform in the opposite (mirror image) manner to return the volume and frequency response back to the original input values.

VHS Hi-Fi frequency modulation

After level compression and emphasis, the two channels are applied to their respective FM modulators. The modulation characteristics of the stereo Hi-Fi modulators are stated in Table 9.2. The Hi-Fi modulators operate much like the video FM VCO described in Chapter 7. If an audio tone of 1 kHz is applied to the frequency modulator, the carrier will

Table 9.2.
VHS Hi-Fi Frequency Modulation Characteristics

	Left	*Right*
Carrier frequency	1.30 ± 0.01 MHz	1.70 ± 0.01 MHz
Maximum frequency deviation	± 150 kHz	± 150 kHz
Typical frequency deviation	± 50 kHz	± 50 kHz

swing (deviate) back and forth 1000 times per second from the center frequency. If there is no audio signal input to the modulator, the carrier will stand still at the center (carrier) frequency. The volume of a Hi-Fi signal is represented by the amount of deviation. In other words, a Hi-Fi signal that deviates ±50 kHz from the center frequency at a rate of 1000 times per second will produce a loud 1000 Hz tone while a Hi-Fi signal deviating 15 kHz at 1000 times per second will generate a much quieter 1000 Hz tone.

One difference between the Hi-Fi and video modulators is the deviation range. The audio deviation range is narrower than the video. Also, the Hi-Fi deviation range extends above and below the center frequency, while the video FM deviation only extends above the carrier frequency. Table 9.2 shows that the left and right maximum deviation range is equal indicating that the FM frequency can swing as much as 150 kHz above and 150 kHz below the center frequency. You will recall that the 8mm format does not allow equal deviation ranges between channels. (See Table 9.1.)

After modulation, the two channels are amplified and passed to the Hi-Fi head record circuitry where the Hi-Fi signal is depth-multiplexed on the tape with the Y FM and color-under signals.

VHS Hi-Fi Playback

Playback preamplification, head switching, band-pass filtering, and dropout detection
Figure 9.16 shows the Hi-Fi playback signal flow. The azimuth cancellation effect causes the Hi-Fi heads to read only the Hi-Fi signals in the depth-multiplexed RF track. Each head reads both the 1.3 MHz and 1.7 MHz FM signals and transfers those carriers on to the Hi-Fi head preamp stage via the Hi-Fi head rotary transformer mounted in the video head drum.

After preamplification, the Hi-Fi signals from each head are passed to a head switching circuit that only allows the preamplified signal from the head in contact with the tape to pass on to the remaining circuitry. Head switching is

required to prevent the preamplified signal noise from the head not in contact with the tape from getting into the demodulation signal path and producing static in the playback audio.

After head switching, the playback Hi-Fi FM RF becomes a continuous RF signal that is applied to two band-pass filters. These band-pass filters are rated at 1.3 MHz and 1.7 MHz respectively. They separate the two FM channels into their respective stereo channels for demodulation. Each band-pass filter has a passband of ±150 kHz to allow the full deviation swing of each Hi-Fi signal to pass.

Playback Hi-Fi FM levels are monitored by the dropout detector circuit following each of the band-pass filters. Once the playback Hi-Fi RF levels drop below a predetermined threshold, the dropout detector generates a pulse that is passed to the dropout correction circuit for appropriate corrective action.

After band-pass filtering and dropout detection, the FM signal is input to an FM limiting circuit. The FM limiter removes any amplitude signal variations present in the playback FM signal. If not removed, these unwanted level fluctuations will amplitude modulate the FM signal and create distortions in the demodulated sound. The source of the unwanted playback AFM RF level fluctuations is the same as the luminance playback RF level fluctuations described in Chapters 5 and 7.

Hi-Fi audio demodulation
After limiting, the Hi-Fi signal is applied to the demodulator where it is returned to its original form. The Hi-Fi FM demodulators in each path are phase comparison circuits. In addition to the playback Hi-Fi FM signal input, phase comparison demodulators require a frequency reference. A phase-locked-loop (PLL) VCO oscillator provides the phase comparison reference in this circuit. The left channel demodulator demodulates the 1.3 MHz Hi-Fi signal so its PLL reference frequency runs at 1.3 MHz. The right channel demodulation PLL frequency is 1.7 MHz.

The demodulator compares the frequency of the playback Hi-Fi signal with the reference frequency. Since phase error detectors are sensitive to both phase and frequency discrepancies between the reference and input signals, differences in frequency and/or phase between the Hi-Fi playback and reference signals produce an error voltage at the comparator's output. The comparator is part of the FM demodulator. This error voltage is the demodulated audio signal.

As the playback Hi-Fi FM signal deviation swings above the center frequency the comparator's error voltage increases in amplitude. When the Hi-Fi FM deviation frequency begins to swing back down to the carrier frequency, the error voltage also reduces. The error voltage will return to zero volts once the deviation frequency returns to the carrier frequency. However, if the deviation frequency continues to swing past the carrier frequency and deviates below the carrier frequency, the comparator's output error voltage will go negative. An increase in the negative deviation of the playback Hi-Fi FM signal will push the error voltage further negative.

Therefore, large signal deviations in the playback Hi-Fi FM signal generate increased amplitudes at the comparator's error voltage output. The rate at which the signal deviates above and below the carrier determines the frequency at which the comparator's error voltage swings negative and positive. If the deviation frequency swing extends equally above and below the carrier frequency and the deviation swing occurs linearly, the error voltage output from the phase comparator will be a sine wave with equal positive and negative peaks. Sine waves comprise the signal that produces the sound we hear. The amplitude of the sine wave corresponds to the volume that we hear. The frequency of the sine wave produces the pitch of the tone that we hear.

To summarize then, the Hi-Fi demodulator is a phase error detector that compares the Hi-Fi FM playback signal with a stable reference. Differences between the reference and playback Hi-Fi FM signals produce an error voltage, which is the demodulated audio signal. The rate at which the playback Hi-Fi devi-

ation frequency sweeps back and forth causes the error voltage to oscillate at that same sweep frequency. The oscillation of the detected error voltage produces the frequency (pitch) of the tones that we hear. Increases in Hi-Fi frequency deviation (wider sweep ranges) produce louder volume levels. Decreases in Hi-Fi frequency deviation (narrower sweep ranges) produce quieter sound levels.

VHS Hi-Fi head switching noise reduction and dropout correction

Designing a pair of heads specifically for Hi-Fi audio signals requires an audio head switching pulse that is separate from the video head switching pulse generation. The Hi-Fi heads also require their own preamplifier. The audio head switching circuitry is similar to that for video head switching except that audio head switching is performed at a different time to compensate for the 120 degree advanced position of the Hi-Fi heads. Like 8mm units, VHS Hi-Fi head switching circuitry does produce a slightly audible head switching noise. This slight noise is most noticeable as a form of distortion that sounds like a muffled ticking or popping sound when a steady tone or a monaural sound is recorded and played back. Generally, the popping is not noticed when playing back normal stereo audio programs. Some higher priced recent VHS models have head switching noise reduction circuit designs that make this noise nearly impossible to detect, regardless of the reproduced audio signal.

The source of the Hi-Fi audio head switching noise comes from the signal distortions created in the continuous Hi-Fi FM carrier at the head switching point. Figure 9.7 shows how this distortion develops.

When the Hi-Fi FM carrier is switched between playback heads, the carrier phase is distorted as the FM signal is changed from one head to the other. This change in phase at the head switching point results in an unwanted frequency deviation change that is detected by the demodulator. The demodulator outputs a short duration pulse when it demodulates this distortion. We hear this pulse as a popping sound. The head switching distortion and resulting pop sound occur at each head switch point, which happens at a 60 Hz rate.

As shown in Fig. 9.16, head switching noise reduction is performed after demodulation in the Hold block. Figure 9.17 shows how the head switching interference is masked in one VHS Hold circuit.

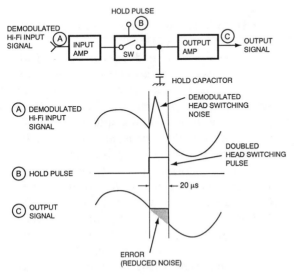

Fig. 9.17. *The Hold circuit masks the Hi-Fi head switching noise in the demodulated Hi-Fi signal.*

The VHS system in Fig. 9.17 is a pre-hold design and operates similarly to the 8mm method previously described. The term *pre-hold* comes from the action of holding the delayed previous good audio signal for insertion into the distorted portion of the demodulated sound. This is a form of waveform interpolation. Figure 9.18 shows how this design works.

The low-pass filter following the Hi-Fi demodulator blocks frequencies above 100 kHz to prevent improper operation of the noise reduction processing that occurs later in the path. This filter eliminates any residual Hi-Fi carrier leak that escapes from the demodulator. It is an active low-pass circuit that also delays the demodulated sound slightly as it is filtered. The Hold circuit takes advantage of this slight audio delay when either head switch masking or dropout correction is necessary.

The low-pass filter following the demodulator delays the demodulated audio. The Hold circuit switches the last good delayed sound voltage level back into the output sound path each time head switching distortion occurs. The Hold circuit receives the demodulated Hi-Fi signal and the doubled head switching (hold) pulse as inputs.

The head switch pulse has been doubled to produce a positive pulse at the point of each head switching transition, thus marking the distortion point of the FM carrier. The doubled head switching pulse becomes the hold pulse shown in Figs. 9.17 and 9.18. The duration of the VHS Hold pulse is about 20 microseconds, which is long enough to mask the head switching distortion noise.

The pre-hold circuit masks the head switching noise by averaging the voltage of the signal at the moment the head switching pulse occurs and by inserting the average voltage into the demodulated audio. This sampling and averaging action is called *interpolation*. For interpolation to be successful, the signal during the period when the demodulated audio contains the head switching noise must be muted prior to insertion of the averaged playback signal. After the noise has been muted out and the average signal has been inserted, the Hold circuit outputs a masked (reduced popping noise) audio signal. As shown in Fig. 9.17, some minor waveform distortion is still present after the masking has taken place (point C of Fig. 9.17), but this distortion is much less objectionable than the louder popping noise (point A) that would otherwise be present if not masked out.

Notice that the Hold circuit in Fig. 9.18 also receives a dropout pulse through an OR gate when dropout occurs. The pre-held signal from the low-pass filter is also used for VHS Hi-Fi dropout correction, which also takes place in this Hold circuit. The VHS dropout detector produces a dropout hold signal when the Hi-Fi RF waveform drops below a preset threshold level. The hold pulse causes the Hold circuit to capture the previous moment's demodulated audio level when dropout occurs. The combi-

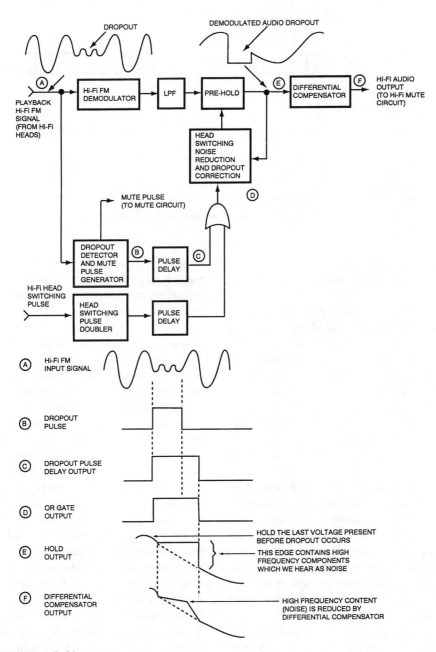

Fig. 9.18. *VHS Hi-Fi pre-hold circuit.*

nation of the last moment of noise-free demodulated audio sampled by the dropout detector's hold pulse and the doubled head switching pulse produces the masked head switching noise and dropped corrected audio output from the VHS Hold section.

Dropout correction takes place in this manner. The delayed output of the low-pass filter is supplied to the Hold circuit as explained before. When the dropout hold pulse is sensed by the Hold block, it simply outputs the last good signal voltage until the dropout disappears.

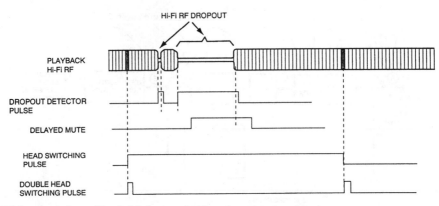

Fig. 9.19. *VHS head switch masking hold, dropout hold, and muting signal timing.*

Figure 9.19 shows the timing of the head switching masking hold, dropout hold, and muting signals.

Hi-Fi dropout correction is only successful in disguising brief periods of dropout. Long periods of an averaged and held voltage do not produce normal audible sounds. Notice that Fig. 9.19 shows a delayed muting pulse that occurs after extended periods of dropout. This muting pulse blocks any sound from exiting the Hi-Fi audio circuits when extended periods of dropout occur. When muting is applied, the Hi-Fi output sound totally disappears and the output audio is switched to the linear track sound. Hi-Fi muting ends and Hi-Fi sound reproduction resumes when the Hi-Fi RF level returns and Hi-Fi demodulation resumes.

Switching the Hi-Fi sound over to the linear track sound is one area where 8mm AFM muting and VHS Hi-Fi muting differ. Since 8mm has no linear track to resort to as a backup, the sound totally disappears when muting occurs in 8mm machines.

The source of VHS Hi-Fi muting is the dropout detector circuit. When the dropout detector generates a dropout hold pulse, the hold pulse is applied to the muting pulse generator circuit. The muting pulse generator does not output a signal until the dropout duration exceeds a predetermined time interval. Then, once the muting pulse has been generated, it will continue until the RF signal returns, indicating that the extended period of

dropout has ended. Hi-Fi muting and linear track audio backup switching occur in circuitry following the Hi-Fi de-emphasis and noise reduction blocks.

Closed Loop Interpolation and Head Switching Noise Reduction
Some recent VHS models incorporate improved head switching noise reduction designs. The improved circuitry uses a form of *closed loop interpolation* that works with the pre-hold circuit rather than relying solely on an average of the pre-held signal level. Closed loop interpolation is a fancy name for a technique that reduces the distortion caused by interpolating the head switching period of the Hi-Fi RF waveform.

Closed loop interpolation detects the slope of the demodulated Hi-Fi audio waveform immediately prior to the head switching period and causes the capacitor holding the pre-hold charge to discharge with a time constant that replicates the slope of the waveform until the head switching interval is over. This allows the noise masking circuit to more accurately reproduce the original waveform that occurred before the head switching distorted it.

Further advances in closed loop interpolation have made the circuit even more effective by changing (adapting) the hold capacitor's discharge rate to match the slew rates of either high or low demodulated audio frequencies present when head switching distortion appears. This improved interpolation circuit is called the *adaptive closed loop interpolation circuit.* The best masking result is obtained when the shortest possible interpolation time is used. When head switching noise is present in reproduced audio tones with a relatively low frequency, a longer interpolation time is required. When high frequencies are being played back, a much shorter interpolation time is needed. Thus by "adapting" the hold capacitor's discharge rate to the different slew rates of various frequencies, the interpolation time can be adjusted to effectively mask a broad range of playback frequencies.

Still another improvement on this advanced design goes a step further. This design uses *tangent closed loop interpolation.* This process examines the frequencies of the playback Hi-Fi audio and the head switching noise present in both of these channels. This design then averages both the head switching noise and the audio frequencies and then applies the averaged result to each channel independently. In this way, the best audio signal reproduction is separately selected for each channel based on the average of the audio in the two channels.

VHS playback de-emphasis, dynamic range expansion, and noise reduction (COMPANDING)

The Hold block passes the signal on to the de-emphasis and noise reduction blocks. Here the signal's dynamic range is expanded to original levels, and noise reduction is applied. The expansion and noise reduction circuits perform an inverted (mirror image) operation on the signal compared to the record input compression operation. The de-emphasis frequency weighting characteristics are shown in Fig. 9.10.

Comparing the playback expansion characteristics of Fig. 9.5 with the frequency weighting shown in Fig. 9.10 helps in comprehending the entire expansion action of the playback de-emphasis circuitry.

Since the VHS Hi-Fi signal was not multiplexed with L+R and L–R information, the signal is ready to be output on the audio output line.

Regardless of the format, VHS AFM and Hi-Fi sound provides operational performance exceeded only by that of PCM audio, DAT (digital audio tape) recorders, and compact disc players. Table 9.3 is a performance compari-

Table 9.3. VHS Audio Specification Comparison

Measurement	VCR Tape Speed	Typical Hi-Fi Specifications	High Quality Audio Reel-to-Reel Recorder Specifications	VHS Linear Track Audio Specifications
Frequency response (Hz)	SP	20–20,000	25–18,000	100–7,000
	SLP (EP)	20–20,000	—	100–5,000
Signal-to-noise ratio	SP	(+) ≥ 55 dB	60 dB	42 dB
	SLP (EP)	(+) ≥ 50 dB	—	40 dB
Total harmonic distortion	SP	(+) ≤ 0.5%	0.8%	1%
	SLP (EP)	(+) ≤ 0.5%	—	1.5%
Wow and flutter	SP	0.005%	0.04%	0.8%
	SLP	0.005%	—	(*) ≥ 1.3%

Notes:

1. (*) Normally not published (measurements well over 1% noted).

2. (+) These values are increased when the advanced closed loop interpolation and the tangent advanced closed loop head switching noise reduction circuits are employed in the design.

3. Frequency response is improved because writing speed is increased through the use of rotating heads. Wow and flutter are reduced because the slight tape speed variations that affect linear audio become insignificant. Minor errors in tape speed, which cause linear audio wow and flutter, have little effect on AFM and Hi-Fi performance.

son between VHS Hi-Fi and VHS linear track audio, and an expensive reel-to-reel audio tape deck. The Hi-Fi audio wow and flutter is practically nonexistent and Hi-Fi frequency response exceeds that of reel-to-reel and linear track audio.

8mm PULSE CODE MODULATION (PCM) AUDIO

Pulse code modulation (PCM) audio is an optional form of digital record and playback sound available on some 8mm units. It is not the purpose of this section to discuss all the details of PCM audio and related digital audio processing theory. Indeed, this topic alone can be a book of its own. Only items pertinent to 8mm troubleshooting and repair will be described.

For the recorded audio to be converted into PCM signal stream, the sound must be sampled and converted into a series of digital pulses. This action is accomplished by the A-D (analog-to-digital) converter shown in Fig. 9.20.

Analog-to-digital conversion is accomplished by taking a series of samples of the analog signal at specified intervals and changing each sampled voltage into a digital word that represents the sample. In this way each sample produces a discrete digital word. The sample clock in 8mm units runs at 31.5 kHz. Therefore, if a single sine wave of 1 Hz (one cycle per second) is sampled at this 31.5 kHz A-D converter clock rate, 31,500 samples of that analog oscillation are taken each second.

Each of the 31,500 samples will have an individual voltage representing the potential of the analog signal level at the sampling point. The samples are converted into a series of 10-bit digital words by the A-D converter. This form of digital conversion is called "quantizing."

Quantizing an analog signal creates some errors as shown in Fig. 9.21. These errors develop for two reasons. First, any signal information contained in the original analog signal between the sample points is lost in the A-D conversion. It can never be reproduced. Dur-

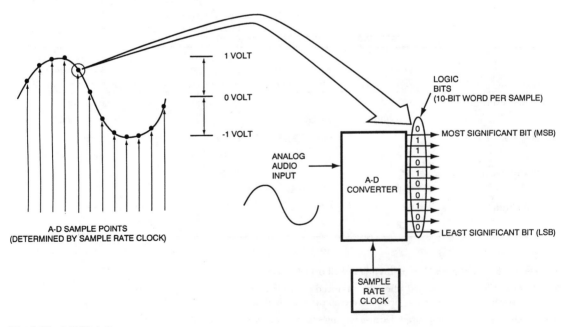

Fig. 9.20. A PCM A-D converter.

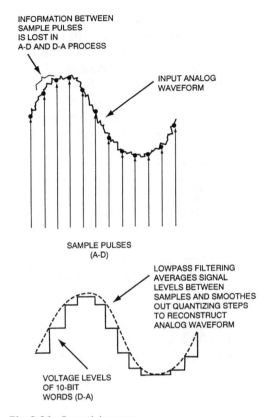

INFORMATION BETWEEN
SAMPLE PULSES
IS LOST IN
A-D AND D-A PROCESS

INPUT ANALOG
WAVEFORM

SAMPLE PULSES
(A-D)

LOWPASS FILTERING
AVERAGES SIGNAL
LEVELS BETWEEN
SAMPLES AND SMOOTHES
OUT QUANTIZING STEPS
TO RECONSTRUCT
ANALOG WAVEFORM

VOLTAGE LEVELS
OF 10-BIT
WORDS (D-A)

Fig. 9.21. Quantizing error.

ing playback the digital words are converted back into an analog waveform by the playback D-A (digital-to-analog) processing circuitry. The D-A conversion circuit interpolates the information between each of the sample points by averaging the signal levels between the two sample points and inserting that averaged level between the samples to reconstruct the analog sine wave. The loss of the original analog information between sample pulses and the playback D-A interpolation process is a form of *quantization error*. Quantization error is noise placed upon the signal during the A-D and D-A process. It's considered noise because, by definition, noise is a signal present in the reproduced audio that did not exist in the original sound.

The second source of quantization error (error noise) comes from sampling a wave-form comprised of an infinite number of possible voltages with a fixed quantity of words to represent those possibilities. This error occurs because each bit in a digital word represents a specific voltage level. If the sampled analog signal has a voltage that falls between two digital voltage possibilities, the A-D converter must choose which of the two digital words to use. This produces an error when the digital word is converted back into analog in the playback process. Thus, interpolation to digital is another source of quantization error noise.

Quantization error is reduced by 50% (6 dB) each time one more bit is added to the digital word length. For example, if a maximum analog input signal voltage of 1 volt peak-to-peak is converted into a 4-bit digital word, each bit represents 62.5 millivolts of that 1 volt signal. This is because there are 16 possible combinations of 1's and 0's in a 4-bit binary digital word, and one volt divided by 16 equals 0.0625 volts. But, by adding just one more bit to the word length, each bit in the word now represents 0.03125 volts. Further reductions in quantization error are made by increasing the sampling rate so that more samples are made within the same period of time.

For the 4-bit digital word, one form of quantization error happens when the sampling clock directs that a sample of the sine wave be made at a point where the analog voltage is not a precise multiple of 0.0625 volts. Since the A-D converter only has 16 possibilities of words to represent an infinite number of voltage possibilities, the A-D converter must decide which 4-bit word to assign to the sampled voltage whenever the voltage is something other than an exact multiple of 0.0625 volts. It does this by rounding off the sampled voltage and assigning the closest 4-bit word to represent that rounded number. For example, if the sampled signal has an amplitude of 0.71875 volts (halfway between 0.6875 volts and 0.75 volts, which are the 11th and 12th multiples of 0.0625 respectively), the A-D converter will round this value up to 0.75 volts since this is the closest 4-bit value to the sampled potential. This rounding up produces quantization error.

By adding a fifth bit to the digital word length, the quantization error generated by the 4-bit word is gone because 0.71875 volts is an exact multiple of the 5-bit word. Quantization error still exists in 5-bit words, but on a much smaller scale. Recall that quantization error is reduced by one-half (6 dB) with each added bit.

10-Bit to 8-Bit Nonlinear Quantization

The 8mm PCM system uses 8-bit words for recording PCM audio on the tape. The word length is limited to eight bits because of the 8mm width of the tape, the 30 degree PCM wrap on the tape, and the error correction bits which must also be added to the recorded audio words. If each bit reduces the quantization error by 6 dB, an 8-bit word will only provide 48 dB of dynamic range. Recall that dynamic range is a measurement of volume from the quietest sound above the noise threshold (including quantization error noise) to the loudest sound the system can reproduce.

To increase the overall dynamic range of PCM recordings, the 8mm system uses both a noise reduction and dynamic range compression scheme similar to the 8mm AFM companding process previously described, and a method of nonlinear 10-bit to 8-bit quantization in the A-D conversion process. The combination of companding and 10-bit to 8-bit conversion provides a dynamic range of 120 dB for 8mm PCM sound.

Nonlinear quantization records only 8-bit words on the tape, but provides the added benefit of greater dynamic range available from quantizing in input audio into a 10-bit word. Nonlinear quantization means that the analog audio input is A-D converted to create a 10-bit word, and that this 10-bit word is changed into an 8-bit word equivalent for recording onto the tape. This presents a "bit" of a problem because there are 1024 different word possibilities in a 10-bit word, while an 8-bit word only has 256 individual word possibilities. In actual practice, only half of the total number of possible bits is used to allow for additional head room in the record/playback process. Thus, an 8-bit word only uses 128 of the 256 possibilities and a 10-bit word only uses 512 combinations. Whether the full potential or only half the number of possible bit combinations is used, an 8-bit word still provides fewer possibilities than a 10-bit word. Therefore, in the bit-conversion process, some of the 10-bit words have to be given up. These sacrifices are made at the higher volume levels, where we are less likely to notice the errors.

The 10-bit to 8-bit quantization scheme works because quantization noise is much more obvious at lower volume levels than for loud sounds. At high volume levels, the errors are masked by the increased sound. At high volume levels, we are not as likely to notice the error noise produced by the 8-bit quantization, but at low audio levels, we are much more sensitive to the errors (noise). Therefore, by using a 10-bit digital word to represent low audio levels, quantization errors decrease, placing them beyond our perception. As the volume increases, the resolution of the PCM words is gradually shifted downward increasing the quantization error. As the volume increases so does the quantization error, but we don't recognize the increased noise because it is drowned out with the higher sound level.

Figure 9.22 shows how this nonlinear quantization works. The A-D converter actually outputs a 10-bit word for all volume levels, but the nonlinear 10-bit to 8-bit converter alters the bit output resolution quality as the input volume level increases. To accomplish this, the input audio volume range was divided into four groups ranging from low to high. The audio input level groups are indicated by the points on the graph.

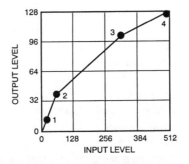

Graph Point	Encoding Law	Input Digital Level (Decimal)	Output Digital Level (Decimal)	10 Bit to 8 Bit Ratio	Added Quantization Bit Error	Signal to Error Ratio
0-1	10 bits ⇒ 10 bits	0 - 15	0 - 15	1:1	0	60dB
1-2	10 bits ⇒ 9 bits	16 - 63	16 - 39	2:1	+1	54dB
2-3	10 bits ⇒ 8 bits	64 - 319	40 - 103	4:1	+2	48dB
3-4	10 bits ⇒ 7 bits	320 - 511	104 - 127	8:1	+3	42dB

Fig. 9.22. 10-bit to 8-bit nonlinear quantization for 8mm PCM.

At low audio input levels (shown as points between 0 and 1 on the graph), the relationship of the 10-bit to 8-bit conversion is a 1:1 ratio. For every 10-bit word, there is a corresponding 8-bit word. In other words, at low sound levels, the digital words input range from 0–15 (in decimal) and the 8-bit output words also range from 0–15 (decimal). The relationship (stated in decimal) of the input 10-bit word value to the output 8-bit word value is shown in the chart in Fig. 9.22. According to the chart, there is a 60 dB signal-to-quantization error (quantization noise) ratio. This is based on the 6 dB-per-bit error ratio stated earlier. Since there is a 1:1 word ratio between the 10- and 8-bit words, there is no added quantization error at the low audio input levels. This is also shown in the chart.

As the input volume increases to something between points 1 and 2 on the graph, there are two 8-bit outputs for every 10-bit input. This means that there are two 10-bit words that will produce the same 8-bit word. This is a 2:1 ratio. The 2:1 ratio means that there will be an added error in the least significant bit of the 8-bit word for one of the 10-bit words that it represents, but the output will be an accurate 8-bit word when the other 10-bit word is represented. This added 1-bit error reduces the signal-to-error ratio to 54 dB.

When the sound level reaches point 3 on the graph, there will be four 8-bit outputs for every 10-bit input. This produces a 4:1 ratio and a 2-bit quantization error as it represents certain 10-bit words. The signal-to-error ratio is reduced to 48 dB at this volume level.

Finally, as the sound passes point 4 on the graph, there is an 8:1 correlation between the input and output bit relationship. There will be eight 8-bit words output for every 10-bit word. The last three bits of the 8-bit word will contain an error as they represent certain 10-bit words. This brings the signal-to-error ratio for this range in volume level to 42 dB.

During playback the 8-bit PCM words are restored to their original 10-bit from for D-A (digital-to-analog) conversion. The analog signal is then applied to the companding circuit for playback dynamic range expansion and noise reduction.

Bit Error Correction

When reproducing digital audio, missing bits create digital errors that distort the sound. When a large quantity of bits are missing, the distortion becomes severe. Bit errors produce popping sounds and a distinct form of static in the playback speakers. Some common causes of missing bits are tracking errors, tape drop-out, and head-to-tape contact problems as previously described. Worn heads and tape path misalignment are other common causes of PCM bit errors. Good PCM audio reproduction relies on maintaining a low bit error rate.

Regardless of how well a machine is maintained, bit errors will occur. The 8mm PCM format has three levels of error correction encoded into the PCM processing to keep the error rate low. These are parity error detection, cyclic redundancy check code (CRCC), and word interleaving.

Parity error detection

The lowest level of error correction is parity error detection. Figure 9.23 shows a simplified example of how a single-bit parity error detection scheme operates.

Assume that there are four 3-bit words ranging in decimal values from 1 through 3. This is shown as part A of the figure. For this example, a parity bit is added to each word according to the following rule: if the quantity of 1's in the word is zero or even, the parity bit added to the word is 0. If the number of 1's in the word is odd, the parity bit added is 1.

When digital errors develop during the record and playback process, some bits in the digital word are changed. Errors may also change the recorded parity bit in the PCM word. Parts B and C of Fig. 9.23 show some errors that occur and their effect on the parity check. (Errors are indicated by the shaded areas in the figure.)

Parity error correction takes place as part of the playback error detection and correction process. This is the lowest level of playback error correction performed on the 8-bit PCM word. A parity error check is made, and the parity check value is assigned to the word in the error correction scheme. When the parity error check senses no errors, a zero is placed in the parity check column signifying that no parity errors were detected. When the parity error correction detects an error, a 1 is placed in the parity check column indicating that a parity error exists in the playback word.

The first word in Fig. 9.23B has an error as indicated by the shaded boxes in the table. The parity error detector recognizes the error by sensing that there is an additional 1 in the word. The parity check circuit senses the wrong quantity of 1's by looking at the recorded parity bit (now being played back with the 8-bit PCM word) and comparing that bit to the number of 1's present in the word. Since the recorded parity bit in the first PCM word indicates that there are either no 1's or an even quantity of 1's in this word, the parity check outputs a 1 indicating an incorrect quantity of 1's in the word.

Notice that the decimal equivalent of the first word in part B is 2 rather than 0. The analog volume levels of decimal 0 and decimal 2 are different and will distort the sound if no correction is made. This is a result of the playback bit error.

The second word in section B has no errors (compare the second word between sections A and B in Fig. 9.23), so the parity check column displays a 0 indicating no detected parity errors.

Now look at word three of sections A and B. Notice the playback word in section B is correct, but the parity bit has an error. When the parity error detector examines this word and the associated playback parity bit, it determines that there is an error in the playback word because the parity bit and the quantity of 1's and 0's in the word conflict. Therefore, the parity check bit erroneously becomes 1. In this case, the error correction will apply an unwanted correction to a correct word.

RECORDED WORD DECIMAL EQUIVALENT	MOST SIGNIFICANT BIT		_	LEAST SIGNIFICANT BIT	RECORDED PARITY BIT
0	0	0		0	0
1	0	0		1	1
2	0	1		0	1
3	0	1		1	0

(a) _Original 8 bit word recorded with added parity bit._

PLAYBACK WORD DECIMAL EQUIVALENT (WITH BIT ERROR)	MOST SIGNIFICANT BIT		_	LEAST SIGNIFICANT BIT	PLAYBACK PARITY BIT	PARITY CHECK BIT
2	0	1		0	0	1
1	0	0		1	1	0
2	0	1		0	0	1
3	0	1		1	0	0

(b) _Playback of recorded words with bit errors and added playback parity check bit._

PLAYBACK WORD DECIMAL EQUIVALENT (WITH BIT ERROR)	MOST SIGNIFICANT BIT		_	LEAST SIGNIFICANT BIT	PLAYBACK PARITY BIT	PARITY CHECK BIT
5	1	0		1	0	0
1	0	0		1	1	0
2	0	1		0	1	0
0	0	0		0	0	0

(c) _Same four playback words with different bit errors._

Fig. 9.23. _Parity error detection._

The parity and digital bit structure for word four in section B is correct and is not affected by the error correction.

Now let's consider a different set of errors as shown in Fig. 9.23C. Notice that the parity check column shows all 0's. This is because the quantity of 1's and 0's in the word agrees with the recorded parity bit. However, comparing the original words in part A with the playback words in part C, you find that there are bit errors in some of the words. Errors are again indicated by the filled-in boxes in part C. In these cases, the parity check does

not recognize the bit errors because the parity check reveals no discrepancies between the recorded parity bit and the incorrect quantity of 1's and 0's in the word. When errors like these happen, parity error detection is not powerful enough to accomplish the error correction required.

A single-bit parity error detection scheme has other shortcomings. It cannot resolve multiple bit errors in the same word. A single-bit parity error correction circuit is not able to identify which bit in the 3-bit word of Fig. 9.23 requires correction. Secondly, by itself, parity error correction cannot correct for burst errors. Burst errors are data bit errors that last for longer than one 8-bit word time. Burst errors can run for several milliseconds, destroying the integrity of many words. Playback signal loss from dropout is an example of burst error. When burst errors occur, several digital words in a serial sequence may completely disappear from the playback stream making parity error correction impossible. To correct multiple bit errors in the same word, identify which bits in the word require correction, and resolve burst errors, much more powerful correction techniques must be designed into the circuit.

Cyclic redundancy check code (CCRC)

Instead of just a single parity bit, cyclic redundancy check code (CCRC) uses a parity word and a check sum to locate and correct bit errors. The parity word and CRCC check sum combine to form a second, more powerful, level of bit error detection and correction. Figure 9.24 shows how this form of error correction works. The 8mm system uses this form of error detection and correction.

Part A of Fig. 9.24 shows the digital words arranged in a horizontal row. This is the order in which they are received into the circuit that adds the parity words and CCRC code. Part B of the figure shows that the nine 8-bit words from part a are arranged into a block of three rows and three columns. Once again, the decimal equivalent of each 8-bit word is shown. Part B also indicates that several parity words (not just a single parity bit) and check sum CRCC totals are added to the rows and columns. The decimal equivalent of the parity words and the check sum CRCC totals is indicated in the figure. The rows and columns containing the nine 8-bit words, the parity words, and the CRCC sums combine to form a "data block" that is recorded onto the tape.

Part C of Fig. 9.24 lists the sequential order in which the data block is output from the parity word and CRCC circuit for recording onto the tape. (Later you will see that in actual practice, the data block is interleaved before it is recorded onto the tape.) The head writes the information in the block onto the tape in a sequential form. The individual segments in part C indicate the decimal equivalent of each of the 8-bit binary words and the added parity words. Notice that the CRCC information is written at the end of the block. The procedure of assigning the 8-bit words into blocks repeats throughout the entire record process. Thus, all PCM words are recorded onto the tape in blocks similar to these.

During playback, part D of Fig. 9.24 indicates that an error has occurred while attempting to recover the sixth 8-bit word in the data block. The decimal equivalent of that wrong number is shown in the shaded box in section D.

Now look at section E in the figure. If you add the 8-bit words and the parity words in each row and column you will see how the CRCC sum check is able to indicate the column, row, and the quantity of the incorrect word. It does this by doing a cross-check of the rows and columns and examining the sum totals for each. By identifying which CRCC sums are not equal to 24, the system can cross-coordinate to locate the incorrect word in the block and identify the amount of the error. Once again, the erroneous data is indicated by shaded boxes. After identifying the exact location of the faulty word and determining the value of the error, the error correction circuit is able to apply advanced mathematical calculations and restore the erroneous word to the correct quantity.

Although the CRCC sum of 24 was selected for the data block in the figure, this will not be true for all blocks. The CRCC sum is different for each data block in the recording.

The ability to determine the exact location of an error and to make accurate correction to a faulty digital word makes this system more powerful than the simple one-bit parity method previously described. The check performed in part E of the drawing is done in the playback error detection and error correction circuitry.

| 1 | 3 | 5 | 7 | 4 | 6 | 4 | 3 | 1 |

(a) Recorded 8-bit words (decimal equivalent).

1	3	5	+	15	=	24
7	4	6	+	7	=	24
4	3	1	+	16	=	24
+	+	+

| 12 | 14 | 12 | ← Parity | CRCC Sum (Check Code) |

| 24 | 24 | 24 | ← |

(b) Recorded PCM word block and added parity and CRCC data.

PCM Data Block														
Nine 8-bit Audio Words									Parity Block					
1	3	5	7	4	6	4	3	1	15	7	16	12	14	12	CRCC

——→ (Direction of Head Travel)

(c) PCM block recording sequence as written onto the video tape.

PCM Data Block														
Nine 8-bit Audio Words									Parity Block					
1	3	5	7	4	2	4	3	1	15	7	16	12	14	12	CRCC

(d) PCM block playback with data error in one of the words. (Word highlighted in gray indicates the word containing the error.)

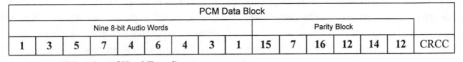

1	3	5	+	15	=	24
7	4	2	+	7	=	20
4	3	1	+	16	=	24
+	+	+

| 12 | 14 | 12 | ← Parity | CRCC Sum (Check Code) |

| 24 | 24 | 20 | ← |

(e) Playback 8-bit word block showing errors and the parity and CRCC sums. The CRCC check reveals the location of the word containing the error and the amount of correction required to restore the correct value. (Words and check sums with errors present are highlighted in gray.)

Fig. 9.24. *Cyclic redundancy check code and parity check.*

Word interleaving

Although CRCC parity error detection and correction is quite powerful, it still cannot correct for burst errors. Burst errors are digital word errors that occur for long periods of time. Sig-nal dropout is one cause of burst error. To overcome burst errors, the words (including the 8-bit audio words and parity words) are interleaved in the data blocks before they are written onto the tape. Figure 9.25 shows how this happens.

The digital words in Fig. 9.25A represent 20 words to be recorded. Before they are written onto the tape, however, they are interleaved (scrambled) in a predetermined order as shown in Fig. 9.25B.

Assume that the portion of tape containing the eight shaded words in Fig. 9.25B was recorded onto a defective portion of tape. This would create an 8-word burst error during playback. Error correction will be difficult, if not impossible, for this portion of tape because of the high quantity of missing data. By studying Fig. 9.24E, you can imagine how difficult error correction will be if eight of the nine audio words in the data block are missing.

However, since the words were interleaved prior to recording, the scrambled words in the burst error will come back as random errors after the interleaving is removed. Part C of Fig. 9.25 shows that after the playback recovery process has restored the word sequence back to its original form by removing the interleaving, the words present in the burst error have been scattered. Thus, the burst error now looks like a number of random errors rather than a long period of missing or incorrect data. The random errors are now able to be corrected by the CRCC parity correction method just described.

In summary, error correction is required to rectify bit errors in the record/playback process. If bit data errors are not corrected, the D-A process will reproduce improperly shaded analog sine waves which we will hear as noise and distortion in the playback PCM audio. Error detection and error correction are accomplished on different levels. First, word interleaving scatters burst errors so that they appear as random word errors. This makes it possible for the CRCC and parity correction system to correct burst errors. Second, CRCC uses a parity word and a CRCC check sum to identify bit and word errors occurring within a block of words.

Bi-Phase Bit Modulation

There are important requirements that must be met when recording digital data onto magnetic tape. The first of these is that the PCM system must be capable of reproducing the voltage of the logic representing the digital 1's and 0's in a word. In a digital word, logic one is represented by a high voltage level and logic zero is

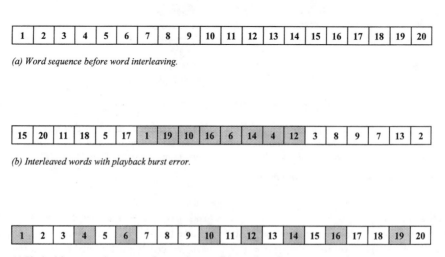

(a) Word sequence before word interleaving.

(b) Interleaved words with playback burst error.

(c) Playback burst error becomes random word errors when word interleaving is removed.

Fig. 9.25. *Word interleaving aids error correction in handling burst errors.*

represented by a low voltage. Therefore a word containing logic 1's and 0's will consist of a string of high and low voltages. In Chapter 5 you learned that recording a pure DC signal on tape will produce no output signal during playback. This is because the playback head only produces an output when it detects a change in voltage. Therefore, the recorded signal must be constantly changing for the head to output any information.

Applying this principle to PCM digital audio, imagine what would happen to the playback head output signal if the PCM audio word was all logic 1's. The recorded signal would be a sustained high voltage level—the same as recording a prolonged DC voltage. During playback of the sequential logic 1's, the head would produce no output. But the absence of head output would be interpreted as a logic low by the playback digital section. Therefore, the PCM system must not only be able to reproduce high and low logic levels, but it must be able to recognize individual bits within a word, even if the adjacent bits are of the same logic value (all 1's or all 0's). The recognition of individual bits regardless of their logic level is a second requirement of a PCM system. Bi-phase bit modulation enables a PCM audio system to meet both of these requirements.

Figure 9.26 shows how bi-phase bit modulation works. A logic 1 (one bit in the data word) is represented by a pulse containing both a high and low level. A logic 0 is represented by either a single high or low logic level. In other words a digital 1 has twice the frequency of a logic 0.

By comparing the bi-phase bit modulated signal to the DC component drawing in the middle of Fig. 9.26, you can see that for the bit stream shown, it is difficult to detect individual bits in a word containing logic levels of the same value when they appear in sequence. The digital signal waveform in the middle of Fig. 9.26 is called as a "nonreturn to zero" (NRZ) modulation code because the logic level does not return to zero until a logic low is generated by the digital system.

The bi-phase bit modulation scheme provides advantages over the NRZ digital signal that would otherwise be written onto the tape. One advantage is that it has a very low DC component. Second, it has a clock element. Finally, it provides a method of identifying each bit in the word, even if adjacent bits are of the same logic level.

The sync word indicated in Fig. 9.26 is written at the beginning of each data block to indicate its starting point. It is also used to synchronize the playback system clocks for bi-phase bit demodulation. Notice that the sync signal is high for $1\frac{1}{2}$ bits and low for $\frac{1}{2}$ bit. This waveform is unique in the data block making the sync signal easy to recognize.

During playback, the bi-phase bit modulated signal is demodulated back to a standard

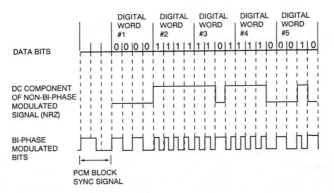

Fig. 9.26. Bi-phase bit modulation.

digital bit form for error correction and D-A conversion.

PCM Time Compression and Expansion

Chapter 5 described how the recorded PCM audio data is limited to a 28 degree section of the 221 degree tape wrap around the video head drum. The balance of the tape track is used for recording video and AFM information. This means that the PCM audio signal is only available for playback processing during a brief moment of each head scan interval. Since sound is continuous, this brief segment of playback PCM data must contain all the information necessary to create a continuous stereo audio program until the next head contacts the next PCM segment on the following RF track.

Reproduction of continuous sound will only occur if the input audio is compressed into a brief burst of data for recording onto the PCM portion of tape. This compression happens after the incoming sound has been converted into a digital signal.

The following example will explain how PCM time compression is accomplished. Imagine that an incoming audio tone of 1 kHz is sampled in an A-D converter at a clock rate of 10 kHz. There will be 10 samples taken of a single 1 kHz oscillation. Also assume that each sample generates a discrete 10-bit word representing the voltage at the sampled sine wave point. This means that there are ten 10-bit words to represent each sample of the 1 kHz waveform and a total of 100 bits in those 10 words. Finally, suppose that the 10 kHz clock directing the sampling rate is the same clock that causes the digital words to transfer through various stages of the digital circuits that follow for writing onto the tape.

Since the 10 samples were made at a 10 kHz rate, and the words are written sequentially (in serial form) onto the tape, a time period of 1/10,000 of a second will transpire from the start of one 10-bit word to the start of the next 10-bit word. It will take 1/100 of a sec-

ond to write the 10 bits representing a single cycle of the 1 kHz tone onto the tape. We can calculate the time it takes to write the 10 words onto the tape because we know the frequency of the sampling clock (also functioning as the data transfer clock in this example) and the frequency of the signal being sampled.

To compress the 1/100 of a second writing time of the single oscillation, it is necessary to condense the 1/100 of a second writing time into a smaller time period. In other words, the start of succeeding 10-bit words needs to appear for writing on the tape sooner than every 1/10,000 of a second. This is accomplished by increasing the clock that transfers the 10-bit words between circuit stages and eventually onto the recording heads. For example, if we increased the data transfer clock speed from 10 kHz to 100 kHz, each 10-bit word would be written onto the tape every 1/100,000 of a second. Thus the ten 10-bit words representing the cycle of 1 kHz tone would now be written onto the tape within 1/1,000 of a second instead of 1/100 of a second. By increasing the transfer clock by a factor of 10 we have compressed the writing time to 1/10 of the original period. This has the effect of increasing the analog audio frequency by a ratio of 10:1. The analog frequency ratio corresponds directly to the digital compression ratio.

Increasing the clock speed does not change the data or the number of samples. There are still 10 words and 100 bits. They are just compressed in time.

Note that the sampling clock for the A-D converter is not increased—just the data transfer clock. If the sampling clock is increased, more words will be generated, defeating the time compression ratio.

Although our example simplifies the compression process, the 8mm audio system works essentially the same way to compress the written audio signal onto a small section of tape. One might wonder how the audio recording time compression system can increase the data transfer rate while operating with a slower A-D clock speed. It may seem that the part of the system using the faster clock speeds would have to wait for the slower A-D clock to provide incoming data for the transfer and writing on the tape. In part, this is true. During record, the system does have to wait to write the data onto the tape. But, the waiting occurs during the time period that the video heads are away from the PCM tape zone.

During the time that the record heads are out of contact with the PCM track segment, the compression system is waiting to write the next series of PCM words. While waiting, the A-D converter is continually receiving analog audio and outputting 10-bit words. In the meantime, the PCM time compression circuit is receiving the data, compressing it, and storing it until it can be written onto the tape. Once the record head reaches the PCM section of tape, the time compressed words are dumped out of the circuit at the fast rate and written onto the tape.

The time compressed PCM data is the equivalent of approximately 1/30 of a second of stereo sound. It is written onto the 28 degree tape segment in less that 3 milliseconds. During playback, the compressed PCM data is expanded back to 1/30 of a second of sound data, which is enough audio information to provide continuous audio playback until the next head contacts the tape and reads another time compressed PCM audio segment.

During playback, the compressed data must be expanded back into the same time domain as it was received. This is done by slowing the data transfer clock to the same speed as the original record input A-D conversion clock. Time expansion is required to return the audio frequencies back to their correct pitch for output to the speakers and to provide continuous sound between head scans of the PCM audio track segments.

Frequency Modulating and Demodulating the Data

After error correction, bi-phase bit modulation and PCM time compression are applied to the record digital audio data, and the signal is sent to the PCM FM modulator where it is modu-

lated for recording onto the tape. PCM FM modulation should not be confused with the bi-phase bit modulation previously described. The PCM FM modulator is a VCO and operates just like the AFM and video FM modulators explained in earlier chapters. The PCM FM 2.9 MHz signal is output when a logic low (digital 0) is input. The FM frequency for a logic high (digital 1) is 5.8 MHz.

After FM modulating, the record PCM signal is passed to the video head record amplifiers for recording onto the tape.

During the playback demodulation, the FM frequencies are converted back into their original logic voltage levels and applied to the bi-phase bit demodulation, PCM time expansion, and error correction and D-A circuits for processing.

Multi-PCM Audio Record and Playback

Some 8mm models provide for multiple PCM audio segments of the tape. When this feature is selected, the unit becomes a "PCM audio only" machine. This is because in this mode, space normally used for video and the AFM RF signals is occupied by multiple PCM segments. Figure 9.27 shows the Multi-PCM segment footprint. The 8mm models with the Multi-PCM mode are capable of recording up to six separate segments of PCM audio. Using a 120 minute tape in the LP speed provides up to 24 hours of audio record time.

To achieve this extended audio record time, the section of tape normally used for video RF recordings is divided into five separate sections of 36 degrees of tape wrap around the head drum. The sixth segment comes from the PCM section at the bottom of the track already reserved for PCM audio.

Each 36 degree PCM segment is comprised of 26.32 degrees of audio and error correction data. The two degrees of tape wrap just prior to each segment is used for clock run in time, while the two degrees trailing the data is an after record safety margin. Each segment

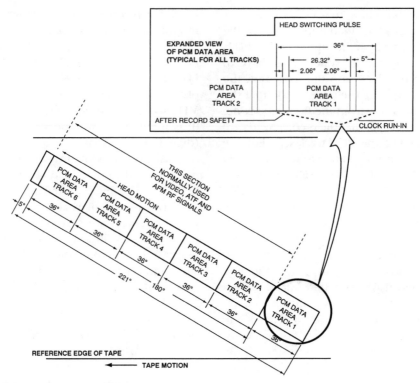

Fig. 9.27. *Multi-PCM segment tape footprint.*

is preceded and followed by a three degree guard band.

With the exceptions of the absence of the video and AFM RF signals and the different PCM segments, the Multi-PCM record mode is similar to the video record mode. To place the Multi-PCM signals at correct positions on the tape during record and properly retrieve them during playback, the video head switching pulse timing is altered according to the PCM track selected by the operator. Video head switching serves as the record and playback timing reference for each segment on the tape. The system control shifts the video head switching pulse by 36 degrees each time the Multi-PCM segment selection is shifted by one. The operator selects which segment on the track is to be recorded or played back and the system control changes the video head switching pulse automatically.

PCM Audio Block Diagram

Now that we've covered the operation and purpose of the various PCM audio record and playback sections, we'll address the PCM audio record and playback diagram in Fig. 9.28. Notice how the playback sections perform a reverse operation on the digital signal as compared with how the signal is processed during record.

Notice also that the record A-D converter receives both left and right analog audio channels but outputs only one digital pulse stream. The PCM record/playback system is able to keep the two channels separate by using signal channel identification pulses that are encoded into the data blocks. This data identification also marks monaural sound, when it is received, so the PCM playback system can recognize stereo and monaural PCM programming.

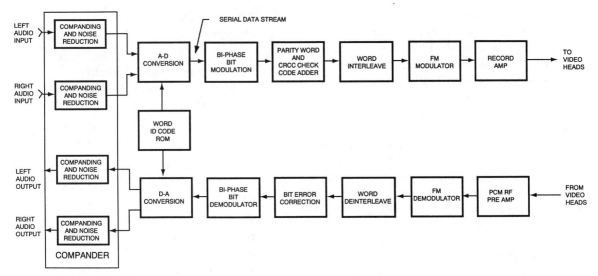

Fig. 9.28. *PCM audio record/playback block diagram.*

Marking of the two channels is performed in the A-D processing section.

During playback, the identification data is used by the D-A section to separate the data stream back into left and right stereo channels.

TROUBLESHOOTING VCR AUDIO PROBLEMS

When searching for the cause of audio problems, always examine the VCR tape path first. Tape path problems will affect FCM audio, linear track audio, AFM (Hi-Fi), and PCM audio interchange and playback. Your tape path inspections should include examination of the RF waveform at the video drum entrance and exit points and tape flow at the pinch roller/capstan area.

Minor curling of the tape at the pinch roller will affect the VHS linear track audio. Severe tape curling at the pinch roller will reflect back in the video head and affect both PCM audio and AFM (Hi-Fi) sound.

VHS machines will switch over to linear track audio when the Hi-Fi sound becomes defective. VHS Hi-Fi losses may be caused by weak Hi-Fi RF levels or tape path errors. Some causes of weak Hi-Fi RF levels were previously described.

If you have a VHS unit that switches from Hi-Fi to normal track audio (either intermittently or constantly) when you are trying to play back a Hi-Fi recording, and the tape path checks out fine, examine the Hi-Fi RF levels and Hi-Fi muting circuit for defects. A common source of this type of trouble is low RF caused by weak Hi-Fi record current during the record mode or caused by worn heads. You can eliminate weak record current as the source of the problem by performing a Hi-Fi record current alignment (if available on your VCR) or by playing back a known good recording with a good Hi-Fi signal. If the good Hi-Fi recording plays back poorly and the tape path alignment is correct, suspect worn heads.

When troubleshooting PCM audio problems, you need to understand the digital processing circuits in the VCR. Most of the digital processing occurs inside a few ICs. All that we are left with is the verification of analog input signals, digital clock signals, and digital word flow and digital command data flow between ICs.

Normally, when PCM problems appear, they are the result of tape path errors caused by a defective tape guide near the video head drum. Electronic PCM failures are less common than mechanical faults. Technicians frequently identify the source of electronic PCM audio problems by locating a missing pulse stream or a failed reference clock.

When troubleshooting any VCR audio problem, don't forget to check both record (using another VCR) and playback (using a known good tape with the type of audio recording you are troubleshooting). Also, if you have one functioning channel in a VCR with defective stereo output, use the functioning channel to help you find the defect by making a series of comparisons between channels.

CHAPTER REVIEW QUESTIONS

1. What is the standard means of recording audio in the 8mm format?

2. What are the left and right FM frequencies used in VHS Hi-Fi?

3. Why must the stereo AFM signal be multiplexed?

4. In addition to assisting linear track audio recording, the audio bias oscillation performs one other function. What is it?

5. Describe the cause of normal track audio playback wow and flutter.

6. What are the two purposes for companding the record and playback signal?

7. What signal modulation characteristic relates to the reproduced sound volume in Hi-Fi and AFM Audio?

8. What signal modulation characteristic relates to the reproduced audio frequency in Hi-Fi and AFM audio?

9. Define quantization error.

10. In the PCM record process, where does quantization error occur?

11. What are the FM frequencies for the bi-phase bit modulated PCM signal?

Miscellaneous VCR Circuits

VCR System Control, Electromechanical Interfacing, Tuners, Audio/Video Demodulators, RF Modulators, Audio/Video Hook-Up Configurations, and Power Supplies: Theory and Troubleshooting

Chapter 10 provides you with a look at several different electronic circuits in the VCR including the System Control and the Electromechanical Interface that sends the electronic commands from the System Control to the mechanical section of the machine. You will also learn about VCR tuners, RF modulators, and the demodulation circuits that convert the tuner's output signal into separate baseband video and audio signals. Here you will also discover different ways to connect a VCR between an antenna or cable system and the television receiver for optimum performance. VCR power supply operational theory and troubleshooting are also part of this chapter. Chapter 10 also supplies details on how to troubleshoot many of these circuits, including electromechanical and System Control related failures.

SYSTEM CONTROL

The System Control circuitry is responsible for coordinating all of the electronic and mechanical functions of the unit. Performing many of the functions of a simple computer, this circuitry monitors the status of the machine, acknowledges and responds to commands from the operator, and initiates functions for correct operation.

System Control turns on and monitors the operation of the power supply. In most VCRs, when the power button is pressed, power is not actually supplied until System Control has checked the status and condition of the electronic and mechanical sections. Once satisfied that all is well, System Control generates command signals to the power supply and causes the complete unit to be energized.

When a VCR is turned off, System Control is still active. Power off does not shut down all voltages within the machine. Instead, the VCR shifts to a power-down stand-by mode waiting for the next power-on instruction. Power consumption is extremely low when the VCR power switch is in the OFF position, but some voltages are still present inside.

System Control is comprised of several components that work together to monitor and operate the VCR. The heart of System Control is a microprocessor integrated circuit. The microprocessor constantly monitors the electronic and mechanical operation of the machine, receiving status signals and generating commands that are sent throughout the unit. As the brain of the VCR, the System Control microprocessor (there may be several) makes decisions and coordinates the operation of all parts of the machine. Acting much like a busy taxicab radio dispatcher, it receives incoming requests from the operating control buttons and switches on the outside of the VCR chassis, checks the current status and operating mode, makes appropriate decisions, and sends command instructions to the VCR electronic and mechanical subsystems. If it senses an incorrect or improper condition, the microprocessor generates a stop command telling the machine to cease operating or to shut down until the situation can be resolved.

The microprocessor controls the power on and off sequence. When you depress the power button, you do not have direct control over the on and off functions. Instead, your action instructs the microprocessor to begin the power up (or power down) sequence.

The microprocessor accepts operator inputs from the front panel such as Play, Rewind, Stop, Record, etcetera, and issues commands to the electronic and mechanical sections of the unit. When Record is selected, the microprocessor checks the status of the record safety tab switch located within the VCR. If the record tab has been removed from the video cassette, the record safety tab switch will activate causing the microprocessor to prevent recording on the tape.

The microprocessor also changes operating states. If the VCR is placed in the Rewind mode and the Play button is pressed during tape rewinding, typical microprocessors are programmed to know that a stop command must be issued before initiating the play command. The microprocessor responds as if you had first depressed the Stop button and then depressed Play.

Many VCRs offer Cue and Review (search forward/search reverse) features that are activated by depressing the Fast Forward or Rewind buttons in the Play mode. The microprocessor can be programmed to recognize the difference between Cue and Review and the actual rewind and fast forward modes.

It also controls many special effects functions. Chapter 8 described how the microprocessor monitors the RF envelope coming from the video heads and issues appropriate commands to the capstan circuitry to provide proper still frame pictures when Play Pause is selected.

The System Control microprocessor also determines the direction of the reel table motor rotation. The two reel tables insert slightly into the bottom of a video cassette where the cassette tape hubs are located. The reel tables in most current models are driven by a series of belts, pulleys, tires, and gears via the capstan motor. In the case of direct drive reel tables, each reel table has its own individual motor as shown in Fig. 10.1. There are no belts or pulleys to drive these reel tables. System Control determines the direction and speed of the reel table motors. In Special Effects playback, a command for slow motion in the forward or rewind direction changes the motion of the reel table.

In units with direct drive reel table motors, System Control monitors the tape tension caused by the reel table motors as the tape is drawn around the head drum assembly.

Fig. 10.1. Direct drive reel motor with one hub removed.

Fig. 10.2. Front load mechanism.

REEL TABLES

Fig. 10.3. The cassette is lowered onto reel tables.

The video tape must move around the head under constant tension. VCRs without direct drive reel table motors use other means to control tape tension.

The microprocessor also controls tape load and unload functions. It prevents the video tape cassette tray from ejecting the tape until the tape has been fully unloaded and re-tracted back into the cassette. A motor and gear arrangement similar to the one shown in Fig. 10.2 is used in some front load VCR models to lower the cassette onto the VCR reel tables such as those shown in Fig. 10.3. One popular front load mechanism design raises and lowers the tape tray unit with the capstan motor using a series of gears and levers. Regardless of the front load mechanism design, the System Control senses when a video tape cassette has been placed into the cassette basket. It causes the tape cassette to be drawn into the machine and lowers the cassette into the correct operating position. The microprocessor in System Control prevents other mechanical operations until the cassette has been correctly positioned over the reel tables.

The microprocessor functions closely with the tuner and clock timer circuitry. The automatic timer IC operates an internal clock that notifies the microprocessor when a pre-programmed time has arrived so auto record-ing can begin. It notifies the microprocessor that it should begin recording and also selects the preprogrammed channel.

Separate microprocessors can be used to control the timer and the tuner. The System Control and tuner microprocessors work to-gether to produce lighted displays on the front panel of the VCR. These digital displays can in-clude the time of day, a tape counter, and the mode of operation currently selected. On VCRs having a fully electronic tuner (no me-chanical moving parts) a microprocessor con-trols channel changing. The operator cannot change channels on an electronic tuner while recording. System Control acts to prevent changing a channel until the unit is placed in Record/Pause or Stop. This control was pro-grammed into the microprocessor to avoid ac-cidental channel changes if you unknowingly brush against the channel change button.

The System Control circuitry also accepts commands from a remote control transmitter. Circuitry within System Control receives and amplifies incoming commands from a hand-held remote control. This information enters the microprocessor where it is translated into appropriate internal commands for the VCR circuitry.

Whether the selected mode of operation enters System Control from a remote control transmitter or from the front panel function buttons, the microprocessor accepts the commands and takes appropriate action.

System Control is responsible for engaging the pinch roller against the capstan after the tape has been threaded in the machine. It controls the engaging and disengaging of the brakes which lock the reel tables in place in the stop mode. Many gears, pulleys, and levers must be moved into correct position to perform tape thread, unthread, normal play, fast forward, rewind, and the other functions. Through a series of mechanical switches or optical sensing devices, the microprocessor monitors the mechanical state of each major portion of the VCR. These switches and sensing devices send DC voltage-level changes, or (in some cases) digital pulses to the microprocessor as state identifiers.

System Control regulates automatic features that are often taken for granted during machine use. When a video tape is fast-forwarded or rewound, the microprocessor senses the end of the tape and immediately stops tape movement. All VHS and 8mm video tape has a special leader on it for identifying the start or end of the tape.

End-of-tape sensing is performed using a small incandescent light or an infrared light-emitting diode (LED). This light sticks up into a recess near the tape door of a cassette that has been inserted into a VCR. To the right and left of the cassette tape are two light-sensitive photo detectors. When either end of tape is reached, the light inserted inside the cassette housing shines through the clear leader onto one of the photo detect devices. The photo detector receiving the light generates a pulse to the microprocessor signifying that the end of the tape has been reached.

In Play, Record, or Fast Forward, the tape travels from the left-hand spool to the right-hand spool as viewed from the front of the VCR. When the end of the tape has been reached, the clear leader is on the left side of the cassette tape. Light shining from the LED is detected by the photo sensor on this side of the chassis. Some VCRs perform an automatic rewind function upon sensing the end of tape.

During rewind, whether initiated automatically or manually, the tape travels from the right-hand spool to the left-hand spool. When the end of the tape is reached, light shines through the clear leader at the beginning of the tape on the left spool and is detected by the right-hand photo sensor. The photo detector output is interpreted by the microprocessor, and a stop command is generated causing immediate cessation of tape movement. This happens so rapidly that the tape stops within two inches after an end-of-tape signal is produced.

Memory stop is another feature controlled by the microprocessor in most machines. The operator can identify a particular place on the video tape by zeroing the tape counter, and depressing the memory button. During rewind, the microprocessor initiates a stop command when all zeroes are detected on the tape counter. Some VCRs have sophisticated tape counters that not only display a number representing tape usage, but also display the amount of tape remaining. This is achieved by detecting the speed of the two rotating reel tables and comparing their different rotation rates. FG signals from each of the reel tables are sent to the microprocessor where they are compared. A short microprocessor program calculates the amount of tape used. Tape remaining indicators have demonstrated accuracy to within a few seconds.

A major function of System Control is to monitor the condition and operation of the machine. If abnormal operation is detected, an auto stop or emergency stop command is generated causing the VCR to enter the Stop mode and in some cases power down. In this role, System Control monitors such things as reel table position, capstan and drum rotation, moisture condensation (moisture will cause the tape to stick to the drum), the amount of time required to thread and unthread the tape, and the end-of-tape sensor LED. If any of these functions are sensed to be incorrect, an immediate stop is initiated.

The microprocessor also controls the pause limit function. If the machine is left in Pause for an extended period of time, a possibility exists that the video tape may be damaged because video tape motion is stopped, and the tape is pulled against the rotating video head drum. The high-speed rotating heads create friction causing thermal stress in the tape. This heat may cause the magnetic coating binder to break down allowing the magnetic coating to come off, severely damaging the tape. Some tapes have been found with diagonal scars so severe that no magnetic coating remained, leaving only the tape backing. The microprocessor monitors the amount of time that the machine is left in Pause during record and playback. If a specified time is exceeded (usually around 5 minutes) the microprocessor places the machine in Stop, or it releases Pause, and the VCR resumes the mode it was in when Pause was initiated.

System Control design and application methods vary between manufacturers. Each attempts to be unique in its application of the microprocessor. Custom microprocessor ICs are common and their operation often varies substantially between models from the same manufacturer.

System Control and Mechanism Defects

The ability to distinguish between mechanism malfunction or misalignment of gears, pulleys, levers, position sensor switches, and an actual failure of the microprocessor is confusing for many technicians. Correct failure diagnosis has become increasingly difficult as new designs come into the marketplace.

To produce less expensive machines, VCR manufacturers have added more gears and levers than were used in earlier VCRs. These additional mechanisms allow a single motor (typically the capstan motor) to control more than one operation. For example, some models use the capstan motor to control the cassette tray elevator, to control tape load/unload, to engage the pinch roller in Play and Record, and to drive the reel tables in Fast Forward/Rewind Play and Record. This single motor design makes the less expensive machine more prone to mechanical failure. Gears can easily misalign or slip because of wear.

The symptoms of these failures may initially appear to be microprocessor-caused because of the erratic operation of the mechanism. An odd sequence of mechanical events may occur, or the unit may refuse to perform some mechanical function, such as accepting a cassette because a gear has slipped a single cog.

Another failure that may occur involves the mechanism position sensing switch. Microprocessors in most models read one or more mode position sensing switches to identify the location of arms and levers as they change to various positions when the tape is loaded and unloaded. A bad contact in the mode sensing switch can confuse the microprocessor, causing the unit to perform any number of inappropriate events.

For example, suppose a mode sensor switch gets corroded and intermittent contact occurs. At power up, many designs initialize the chassis by causing a motor to move the load mechanism back and forth briefly. This causes the mechanism gears and linkages to shift back and forth slightly, which in turn rocks the mode sensing switch in and out of position. This establishes a predetermined positional status that the microprocessor expects at initial start-up. If the mode sensor switch doesn't make contact when it should, or if contact is minimal, incorrect status information could be passed to the microprocessor causing it to power down the system. The technician may assume that the circuitry in the microprocessor that controls power on and power off has failed, when a defective mode sense switch is actually the cause.

A failure in a mode sense switch can also cause the unit to stop the tape during record or play and eject the tape without warning. The malfunction acts like a microprocessor failure, but it isn't.

Each VCR model has its own unique mechanical alignment requirements, and service literature must be carefully read to understand the idiosyncrasies of the different models and brands being sold today. The important message is that modern VCRs are not as durable as they used to be. Cost saving steps in VCR designs cause gears and levers to misalign easily from wear or customer abuse.

Alignment must be performed very carefully and specifically as described in the service literature. A gear misalignment by a single tooth will prevent proper operation and may damage other gears. If a visual inspection reveals that a gear or lever looks worn, replace the part. Worn gears are common and are a typical source for intermittent mechanical operation. Another cause can be cracked gears or levers. The inexpensive plastic gears and levers in VCRs of today are common failure items. Cracks sometimes appear as a brown or dark discoloration line. Other cracks are almost invisible. This makes your job more difficult. Be observant. If you suspect a mechanical component failure, remove and inspect it for wear or cracking. Be particularly suspect of plastic parts.

As your repair experience grows, you will gain an intuitive feeling and deductive reasoning to distinguish between mechanical malfunctions that look like a microprocessor defect, and actual IC failure.

Understanding Microprocessors

Knowledge of logic and digital processing is helpful in understanding the operation of System Control. System Control methods vary between VCRs, and diagnosing the information communicated within the system control circuitry requires sophisticated truth tables, so this discussion will be limited to a general overview using block diagram explanations and general principles.

Digital systems are inherently more reliable than mechanical systems with moving parts. System Control failure generally occurs in mechanical switches, but secondary sources for problems include electronic sensing devices, poor solder joints, and plug contacts.

Figure 10.4 is an overall block diagram of the typical System Control circuitry. All mechanical and electronic functions are monitored and controlled by this circuit. As shown, sometimes several microprocessors are used to interpret data and generate commands.

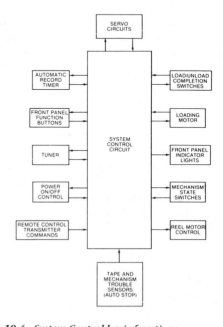

Fig. 10.4. System Control basic functions.

Figures 10.5 and 10.6 describe multiple microprocessor configurations. Figure 10.5 shows the communication paths for several microprocessors working in conjunction. The control communication between the system blocks is called *handshaking*, an appropriate word to describe the close relationship and mutual communications that occur in multiple microprocessor designs. Data lines are used to pass primary information between the microprocessors under control of Ready and Acknowledge handshaking signals between the microprocessor that is ready to receive data and the microprocessor transmitting data. The Ready/Acknowledgment line is also used to inform the transmitting microprocessor that the information has been received. This acknowledge signal is a shift in the output voltage level.

Some microprocessors use an input output expander (I/O expander) to enhance their capabilities. With the introduction of many additional VCR features, System Control responsibilities have increased. To handle the added functions, it was necessary to either increase the number of input and output pins on the main microprocessor chip, or to develop other means for receiving instructions and issuing commands, such as interleaved input/output where data is passed in serial streams (multiplexed) or is exchanged on bidirectional pins. The I/O expander is designed to accept additional information from many sources through pins on the IC and then to convert this information into data that can be transmitted to the main microprocessor on a few lines or circuit traces. The microprocessor can also send instructions to the I/O expander on a few lines while the commands are received and retransmitted out on many pins. Bidirectional data is passed to and from the main system control microprocessor via four data lines, as shown in Fig. 10.6.

A pair of microprocessors requires several things in common before they can communicate back and forth. They must be able to receive and transmit data which they both can understand. They must be connected by some common means, and they must be initialized or reset to a ready state when power is first applied so they can receive and transmit their first instructions. Reset is performed each time power is applied.

Digital information is transferred over data lines as binary waveforms or digital pulse trains. Microprocessors use a common high-frequency oscillation called a *clock* to move the data on the data lines and to open or close logic gates which interpret and process the data. The frequency of the serial clock is determined by the designer and the type of microprocessor used. Typically these frequencies are in the 1–5 MHz range. Figure 10.7 shows the relationship between the serial clock and the serial data pulses. The serial clock pulses are used to translate the serial data pulse trains.

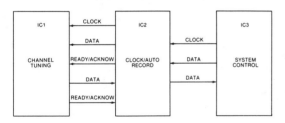

Fig. 10.5. *System Control handshaking.*

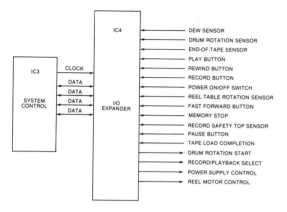

Fig. 10.6. *Input/output expander.*

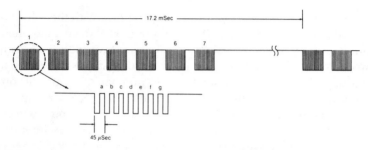

Fig. 10.7. Serial clock data.

Figure 10.8 is a block diagram of a typical microprocessor. As shown, the microprocessor is a single IC with many input and output pins. The number of pins can vary, but usually VCR microprocessors have between 40 and 128 pins. The microprocessor produces output pulses, called *scan pulses*, that are transmitted out through pins or ports to various switches and sensing circuits. When a switch closes, or makes contact, the scan pulses are returned to the microprocessor through an input port. The microprocessor is wired in the circuit so it knows which function is being requested or which sensor switch is activated. Using a code scheme, the microprocessor can recognize the output scan pulse that is being received by an input port. In this way, the microprocessor can accept different scan pulses in a single input port, and let four inputs and four outputs provide multiple control functions.

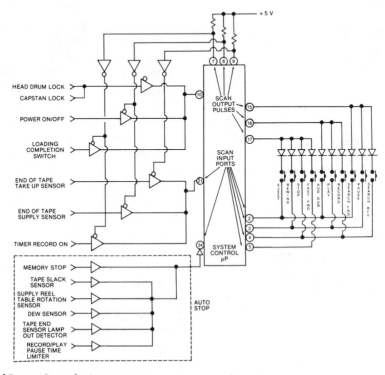

Fig. 10.8. Typical System Control microprocessor.

System Control seems complicated at first because it involves a lot of activities. But the process is simple if you remember that System Control must receive input (status) information and then produce outputs that control the individual functions in the unit. Put another way, System Control is a circuit design using ICs that must be prodded, pushed, and clocked into and through each task being performed.

Troubleshooting System Control Defects

When searching for the cause of System Control defects, you must first confirm whether the defect is the result of a mechanical failure, or a failure in a component peripheral to the microprocessor IC or the microprocessor IC itself. Peripheral components include sensors, switches, and miscellaneous electronic components that support the microprocessing action such as resistors, capacitors, transistors, oscillator clocks, motor drive ICs, and other ICs.

Many mechanical and electronic failures can cause the System Control microprocessor IC to appear defective. It is impossible to cover every possibility of failure in this book. The troubleshooting trees in Chapter 4 will help you differentiate between true System Control failures and defects caused by failed mechanical or peripheral electronic components. We have also developed a list of common failures to assist you in localizing System Control type of failures. These suggestions are listed below in the order of the most likely causes for microprocessor "appearing" type of failures. Your System Control troubleshooting should include items on this list and should be performed in the order suggested by this list.

1. *Mechanical misalignment or defect—* including gears, levers, cassette tray linkages and gears, and mechanical mode position switches.

2. *Defective VCR sensors—*including an internal condensation sensor called a

dew detector, reel rotation detector, mechanical position switches, cassette tray position and tape in/tape out sensing switches, and drum and capstan FG and PG sensors.

3. *Defective wiring or solder connections—* between the system control microprocessor and supporting sensors and components.

4. *Electronic components peripheral to the microprocessor—*including oscillator clocks, IC power supply, IC ground supply, motor drive ICs and transistors, switching ICs and transistors, defective control keys on the mode operation panel, and other electronic components such as capacitors, resistors, and transistors.

5. *The microprocessor IC—*Verification of microprocessor IC operation is done by checking the voltages and waveforms at IC pins listed below.

   a. All supply voltages and ground potentials.

   b. Microprocessor clock oscillation amplitudes and frequencies.

   c. Input and output data and scan pulse lines.

   d. Handshaking signals.

If you suspect that the mode operation button panel is defective, you may be able to verify that the microprocessor IC is functioning by using the remote control. It is safe to assume that the microprocessor is functioning if the VCR operates normally with the remote control when one or more of the front panel controls do not seem to function.

TUNERS

Most home VCRs have built-in tuners that allow them to operate completely independent of the TV. The incoming broadcast signal is connected to the VCR through a series of antenna connectors on the back. Using a standard rooftop antenna system, all local VHF (very high frequency) and UHF (ultrahigh frequency) channels can be received by attaching the incoming antenna signal to the appropriate VCR connector. The introduction of cable TV systems has caused confusion about proper hook-up of the VCR. It's useful to understand how the television broadcast and community cable company channels are arranged in the broadcast frequency spectrum.

VHF, UHF, Midband, and Superband Channels

Broadcast TV audio and video signals are transmitted as radio frequency (RF) information. These frequencies vary depending upon the selected channel.

Figure 10.9 shows the frequency spectrum for broadcast TV signals. Normal TV broadcast signals are divided into three blocks of frequencies known as VHS low band, VHF high band, and UHF. VHF low band covers Channels 2–6. VHF high band covers Channels 7–13; and UHF covers Channels 14–83.

Notice that a large frequency gap exists between Channels 6 and 7. This space is used for FM radio broadcast, police, fire, aircraft, and amateur radio signals. Cable TV companies also use frequencies in this gap to trans-

mit midband channels. Midband channels are between the VHF low and VHF high band channel frequencies.

Another large frequency gap exists between Channel 13 and the first UHF station (Channel 14). These frequencies are also used for radio communication. Cable TV companies use frequencies in this range just above Channel 13 as superband channels.

Confusion often exists between TV Channels 14 through 69, which are broadcast in the UHF range and cable TV channels which may also be numbered from 14 up. Cable Channel 14 is usually not the same as UHF Channel 14.

Cable TV companies rarely use the UHF frequencies to send signals into the home. The signal strength of RF signals from the cable company decreases at a faster rate for higher channel frequencies than for lower channel frequencies because capacitance in the connecting cables causes an increase in resistive reactance as frequency increases. Signal strength losses are much more severe for every foot of cable in the UHF band than they are for the lower VHF bands. The higher the frequency of the broadcast channel, the greater the amount of signal strength lost per foot of cable. This is why cable companies often convert UHF signals to VHF or midband channels before sending them into the home.

Since cable TV systems don't use the UHF band TV channel numbers, they designate midband and superband frequencies using these

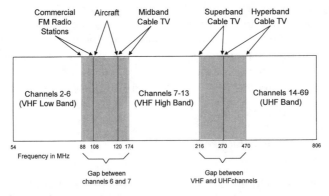

Fig. 10.9. Broadcast television frequency spectrum.

same numbers. The cable company Channel 14 may actually be a frequency found in the midband range; thus, it has no relationship to the original UHF broadcast.

Cable-Ready Tuners

Some older TV receivers and VCRs have tuners that can receive only VHF low, VHF high, and UHF channels. They cannot tune to the midband and superband frequencies. The popularity of cable TV systems was a catalyst for the introduction of special VCR tuner options called *cable-ready*. Unfortunately, until recently, the cable-ready designation was not standardized in the video industry so a number of forms exist. Generally, a cable-ready tuner is designed to receive the midband and superband frequencies. Nearly every VCR manufactured today comes equipped with a cable ready tuner.

Cable companies frequently offer a cable conversion box that can receive the cable company midband and superband channels as well as the VHF low and VHF high band channels. This box allows incompatible TVs to receive the midband and superband frequencies. The cable box receives the midband and superband frequency stations and the other VHF channels, and then converts these to a VHF low band frequency (typically Channel 3 or 4) for use by the TV or VCR. The TV or VCR tuner

receives these midband or superband stations by tuning to Channel 3 and selecting the desired station on the cable box.

Cable boxes also provide another useful function. Many of these boxes decode incoming special pay TV programs that are specially encoded at the cable company and are accessible for viewing by paying subscribers.

VCR cable-ready tuners will not decode these special pay channels, but they do offer the convenience of viewing all other non-scrambled cable TV channels without the use of the box. For this reason, a cable-ready tuner may be an option worth considering.

Later in this chapter we'll share various VCR hook-up configurations and explain the advantages and disadvantages of each.

Tuners and Audio-Video Demodulators

The tuner within a VCR is a specialized heterodyne frequency converter. All TV broadcast channel frequencies are simultaneously available at the input of the tuner. Figure 10.10 is a block diagram of a typical VHF tuner. The channel frequencies are applied to an RF amplifier within the tuner. The RF amplifier is a frequency selective circuit that amplifies only the frequencies selected by the tuner. Each time a new channel is selected, the response

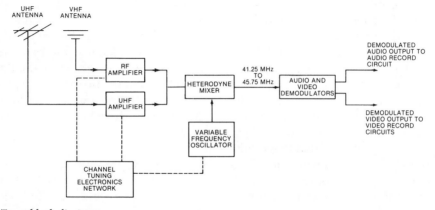

Fig. 10.10. Tuner block diagram.

characteristics of the amplifier are adjusted so the RF amplifier increases only the gain of the frequencies related to the selected channel. The amplified frequency is applied as one input to a heterodyne mixer circuit. The other input to the heterodyne mixer comes from a variable frequency oscillator that is adjusted to 45.75 MHz above the incoming selected channel frequency for the video carrier and 41.25 MHz above the audio carrier frequency.

Both sound and picture information are broadcast simultaneously using two separate carrier frequencies. The TV sound carrier frequency is always 4.5 MHz above the picture modulated frequency. The oscillator within the tuner is varied to the next correct frequency each time you select a different channel. Fine tuning adjustments make minor corrections to the oscillator in the tuner.

Picture and sound signals from the tuner are the sum and difference frequencies of the selected channel and the oscillator within the tuner. The tuner output is bandpass-filtered to pass the 41.25 MHZ audio and 45.75 MHZ video signals. All other frequencies are rejected. The signals of all broadcast stations tuned to by the tuner (whether UHF, VHF or midband frequencies) are converted to these new 41.25 MHZ and 45.75 MHZ frequencies.

The output from the tuner is processed, amplified, and applied to two demodulators— one for audio and one for video. Audio and video signals are separated by frequency (remember they are broadcast 4.5 MHZ apart) and applied to their respective demodulator circuits. The outputs of the audio and video demodulators are line level audio and 1.0 V peak-to-peak standard video. The audio and video signals are applied to their respective processing circuitry for recording onto the tape. The audio and video signals are simultaneously applied to the audio and video output jacks on the VCR as well as to the RF modulator, where the signals are converted to Channel 3 or 4 and available for output to the TV receiver. Many VCRs offer UHF and VHF tuners as well as the demodulator circuitry in a single

assembly, as shown in Fig. 10.11. Servicing the tuner demodulator sections is possible, but replacement of the entire assembly is the most common repair action conducted by service shops.

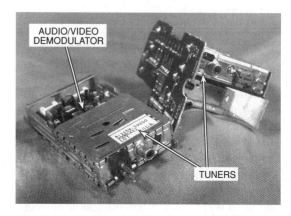

Fig. 10.11. *Tuner and audio-video demodulation assemblies used in some VCRs.*

RF MODULATORS

An RF modulator is a tiny TV transmitter built into the VCR. Audio and video signals are applied to the RF modulator where they are frequency converted to a VHF low band channel. Two channels are usually provided on the modulator. A modulator output channel is selected by a switch accessible on the outside chassis of the VCR. Two channels are provided so you can select a channel not in use in the local area. The RF modulator normally outputs on Channels 3 and 4. If a channel is selected which coincides with a broadcast station in the local area, interference may be noticed on the TV screen. This is caused by the transmission frequencies of the two stations beating against each other. If this occurs, the RF modulator should be changed to the alternate output channel.

RF modulators are carefully controlled by the FCC to prevent accidental interference to other nearby TV receivers. These low-power transmitters have been carefully designed to transmit signals within allowable frequencies.

RF modulators are not considered serviceable items and are replaced as an entire assembly.

HOOK-UP CONFIGURATIONS

The preferred way to connect any input or output to a VCR is through the baseband video and audio signal connectors. The baseband signals are passed over the non-RF-modulated audio and video cables. Whenever possible, the S-Video connection should be used for video input and output. This connection provides the best video possible for recording and playback. If S-Video connections are not possible on your VCR, the line video jacks are the next best choice.

For audio, the audio output signal should be connected directly into the television or audio amplifier's input jacks whenever possible.

Using the preferred connection configuration for audio and video will allow the signal to bypass much of the signal processing necessary when the Antenna In and RF Output connections are used. Bypassing the extra signal processing provides the best picture and sound reproduction possible.

When attaching your VCR to an antenna, or to a television receiver that does not provide the baseband signal connections, the antenna input and RF Channel 3/4 modulator output must be used.

VCRs connected to a normal rooftop antenna provide the opportunity to view one program while videotaping another by hooking up the VCR as shown in Fig. 10.12.

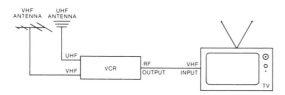

Fig. 10.12. Basic VCR hook-up using conventional antenna.

Incoming VHF signals are applied to the VHF antenna input terminal on the back of the VCR. UHF frequencies must be received by a UHF antenna and passed through associated cable to the UHF input on the VCR. Most current VCRs accept both the VHF and UHF antenna inputs at a single (common) connector on the back of the VCR. If your VCR has only one antenna input and you wish to receive both the VHF and UHF antenna signals, you may need to combine the two antenna signals externally to the VCR before attaching the wire to the VCR's antenna input.

The station that you want to record is selected on the tuner within the VCR. All channels arrive at the antenna input on the back of the VCR. Suppose that you want to record Channel 10 while watching Channel 8. By pressing the TV/VCR button on the VCR and placing it in the TV position, Channel 10 is received by the tuner within the VCR and recorded onto the tape. Another cable is hooked up from the VHF or RF output on the back of the VCR (this connector is labeled "Out to TV" on many VCRs) to the TV's antenna input terminals. This cable should be hooked up to the VHF input on the back of the TV.

With the TV/VCR button on the VCR placed in the TV position, all channels available from the antenna at the VHF input to the VCR are also available at the output from the VCR and sent to the TV. Remember that Channel 10 was selected for recording on the VCR. The TV tuner may be tuned to any station. By selecting Channel 8 on the TV, you can watch this station. By pressing the record button on the VCR you can record one station (Channel 10) while watching another (Channel 8).

If the TV/VCR button on the recorder is placed in the VCR position, only the station signal selected on the VCR tuner will be passed to the TV. In this case, the VCR is operating in the E-to-E mode described in Chapter 7. The signal source selected on the VCR tuner is converted to Channel 3 or 4 (same channel as the RF modulator) and is passed to the TV through the RF or VHF output connector on the back of the

VCR. In this configuration, the only channel the TV can receive is the one coming from the VCR RF modulator (Channel 3 or 4). In other words, when the TV/VCR button on the recorder is in the VCR position, only the program being received by the recorder can be viewed on the TV. When the button is in the TV position, all available channels are able to be selected using the TV tuner. In this mode, the VCR is independent of the TV.

With some limitations, this same feature may be used on cable TV systems. If the VCR doesn't have a cable-ready tuner, only Channels 2 through 13 can be recorded. These tuners cannot receive the midband channels. If the TV/VCR button is in the TV position, the VCR can record any station that the tuner can select while still allowing the TV to tune in on another station. The cable channels that the TV receives are also limited by the TV tuner. A TV without a cable-ready tuner can only tune in to the VHF low and VHF high bands from the cable system.

Figure 10.13 shows a hook-up configuration using a cable company conversion box with a VCR and TV that do not have cable-ready tuners.

All channels coming from the cable company are received by the cable box which converts each selected channel to a VHF low band frequency. In this example, Channel 3 is the output of the cable box. The following steps show how you can record Channel 10.

The cable box is tuned to Channel 10. It converts the channel to a frequency corresponding to Channel 3. Channel 10 audio and video is output as Channel 3 from the cable box and fed into the VCR. A cable is connected between the cable box and the VHF input on the back of the VCR. To record the Channel 10 program, the VCR must be turned to Channel 3 because the cable box has converted Channel 10 to Channel 3. In this configuration, only the VCR can receive Channel 3 from the cable box. You can tune the cable box to Channel 10, Channel 8, or any other station but the cable box always outputs as Channel 3.

In this configuration, you can only watch the station that is being recorded. Only one station is available at the input to the VCR and the input of the TV; this is the channel selected on the cable box.

With the system configured this way, you no longer have the option to record one station while viewing another. If the TV and VCR don't have cable-ready tuners, this configuration may be desirable, because the VCR and TV can receive and record the cable midband and superband frequencies. If a pay cable channel is coming into your home, the cable box will descramble these special channels and make them available for recording on the VCR.

Figure 10.14 shows another option for interconnecting a cable TV system. In this configuration, the incoming cable signal is first connected to the VHF antenna input on the back of the VCR. The output from the VCR is applied to the input connector on the cable box. The output of the cable box connects directly to the VHF inputs on the TV. This will allow the recording of any cable channels which the VCR tuner can receive. Again, if the VCR doesn't have a cable-ready tuner, the channels are limited to the VHF low and high band stations. All stations including those in the midband and superband range are avail-

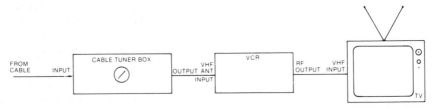

Fig. 10.13. Standard cable box-to-VCR hook-up.

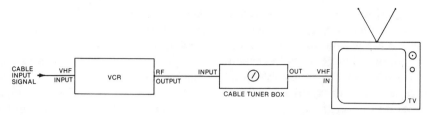

Fig. 10.14. *Standard VCR-to-cable box hook-up.*

able at the input to the VCR, but the recorder cannot be tuned to receive the special channels. If the TV/VCR button is placed in the TV position, all cable stations are simultaneously available at the input to the cable box.

This hook-up is desirable if you wish to record a channel on a normal VHF station while watching a scrambled pay channel program on TV. This is accomplished by tuning the VCR to the VHF station that you want to record, placing the TV/VCR select button in the TV position, tuning the cable box to the station you want to watch on TV, and setting the TV to Channel 3. Remember, the output from the cable box is always at the converted frequency (Channel 3). If you have a cable-ready VCR, the hook-up configuration in Fig. 10.14 is enhanced, and the VCR can easily record all cable channels except scrambled pay programs.

Figure 10.15 describes a cable system configuration with increased options using two accessories. The first of these components is called a *splitter*; the other is called an *RF switch* or *A/B switch*. In this system arrangement, you can record all cable stations (including pay stations), and watch those stations or a different station while you are recording a pay program.

The incoming signal from the cable is applied to a splitter that divides the signal equally and sends it in two directions. The first direction leads into the RF switch. If the RF switch is placed in the A position, the TV can select any station the tuner can receive. If a non-cable-ready tuner is in the TV, it will be limited to the VHF low and high channels.

All cable channels are available at the input to the cable box. The cable box is selected to the station to be recorded, the selected channel is converted to Channel 3, and passed to the VHF antenna input on the back of the VCR. In this configuration the VCR remains tuned to Channel 3 regardless of the station to be recorded because the output from the cable box will always be broadcast on Channel 3. The VCR can record any station selected on the cable box. The VHF or RF output jack at the rear of the VCR is connected to the B input on the RF switch. By placing the TV/VCR switch in the VCR position and the RF switch in position B, you can watch the program being recorded by the VCR.

There are some important things to remember when using this configuration. When the RF switch is placed in the B position, the TV must be tuned to Channel 3 (or Channel 4 depending on the VCR RF modulator) to watch the program coming from the VCR. If

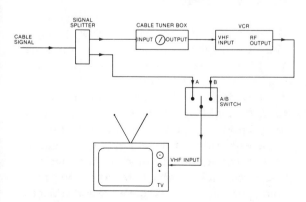

the VCR is to perform an automatic timer record function, it must be commanded to tune in Channel 3 at the time that the recording is to begin and the channel to be recorded must be previously selected on the cable box. If this is not done, the unit could record the wrong station or no station at all.

One disadvantage to this type of hookup is that multiple automatic record sequences using different channels can't be preprogrammed because the channel selector on the cable box must be tuned to the new channel. The VCR will automatically go on and off at its designated times but only the channel selected on the cable box will be recorded.

It is possible to use this configuration to view pay channels when the VCR is turned off. At that time, all signals available at its antenna input terminals are passed directly to the output VHF terminals and applied to the B input of the RF switch, so you can view any channel with the VCR de-energized using the cable converter box.

Many other configurations are possible by adding additional splitters and A/B switches, but picture quality deteriorates as cable splitters are added. Cable splitters divide signal strength in half for each dual output splitter. The number of splitters possible in a cable hook-up configuration varies according to the strength of the signal coming from the cable system. VCR accessories are available with a series of built-in A/B RF switches and signal splitters that provide flexibility and convenience.

Hard-of-hearing consumers can record closed-caption TV programs using a caption decoder. Two types of screen caption formats are used in the United States—Line 21 and Teletext.

Line 21 broadcasts can be recorded directly from the antenna and then replayed through the caption decoder to generate the captions on the screen. You should also connect your antenna directly to the decoder and the decoder output to the VCR and then the TV set. This will cause the decoded caption im-

ages to be stored on the cassette tapes so they will appear on any VCR-TV system without further need for the decoder.

Teletext broadcasts can be recorded by placing the decoder between the antenna jack and the TV set. The VCR is configured to receive the signal from the TV and record the captions on tape.

TROUBLESHOOTING TUNERS, DEMODULATORS AND RF MODULATORS

The easiest way to confirm a defective VCR tuner, demodulator, or RF modulator stage is to bypass the suspected circuit. The line video and audio inputs bypass the tuner and audio-video demodulation stages, and the line video and audio outputs bypass the RF modulator circuitry. Be sure to use the troubleshooting trees in Chapter 4 to assist you in localizing defects in this area.

Once you have localized the failure to a particular processing stage (either the tuner or demodulator for input problems or the RF modulator for RF output troubles) troubleshoot the circuitry in that stage. Use the troubleshooting trees in Chapter 4 to help you eliminate such things as incorrect hook-up configurations and bad cables.

POWER SUPPLIES

VCRs operate almost entirely on DC voltage. The DC voltage is obtained by converting the AC voltage present at a wall socket. AC power from the wall electrical outlet is applied to the rectifier circuit within the power supply, as shown in Fig. 10.16. Diodes are used to rectify the AC power by permitting only the positive-going excursions of the alternating current to pass. By configuring the diodes as a bridge rectifier circuit, it's possible to invert the negative-going portion of the AC sine wave and add it to the positive-going portion of the incoming sig-

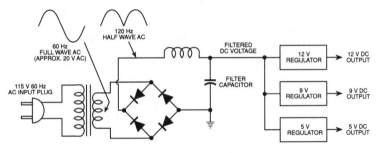

Fig. 10.16. *Typical VCR power supply.*

nal. This effectively converts the 60 Hz full wave AC signal into a 120 Hz half wave AC signal as shown in the diagram.

The incoming AC power is applied to a transformer that lowers the effective AC voltage from 120 volts to a much lower AC voltage (normally less than 20 volts AC). This lower AC voltage is then converted to DC by the diode bridge rectifier.

The rectified AC 120 Hz signal is applied to a filter capacitor that charges to approximately the peak voltage at each of the rectified AC peaks. This capacitor tries to maintain this peak voltage causing a filtering action which removes all of the AC component from the input power. The AC signal is thus converted to an unregulated DC voltage level.

The raw DC voltage is applied to several voltage regulators that reduce the filtered (approximately 20 V) DC source to specific 12 V, 9 V, and 5 V levels. Some designs use the unregulated 20 volts to drive motors and relays which require higher operating current. The 5 V supplies are typically used to drive the System Control microprocessor and other integrated circuits. The 12 V and 9 V supplies are used throughout the machine to operate various analog components.

Some power supply designs incorporate current limiting features. If a short occurs in the VCR, the power supply is designed to limit the amount of current that can flow through the short circuit. In effect, these power supplies shut down until the short circuit has been corrected. This feature reduces the dam-

age done to the power supply and other circuits by excessive current.

The switching power supply has replaced the more traditional voltage source in most models. Switching power supplies get their name from the high-frequency oscillation (normally between 40 kHz and 120 kHz) they create from the input AC source. They convert the generated oscillations to various low voltages used in the machine. Figure 10.17 is a block diagram of one switching regulator design. A high-frequency oscillation is produced at the primary input to the switching transformer causing the drive transistor to switch on and off.

The on and off action of the switching drive transistor produces a high-frequency, high-current pulse in the transformer's secondary windings. Diodes connected to the secondary windings at selected points receive pulses of selected peak-to-peak voltage levels. These pulses are rectified by the diodes and filtered by the capacitors to produce the various DC voltages. Each DC level depends on the peak-to-peak switching pulse voltage applied to its respective diode.

Switching power supplies are more efficient than conventional full-wave power supplies. Conventional supplies use larger transformers and filter capacitors. Using high frequencies for source inputs to the rectifying circuits in switching supplies can be accomplished with transformers a tenth the size (and cost) of those used in full-wave conventional supplies. In addition, these high frequencies equate to smaller capacitors. The size and

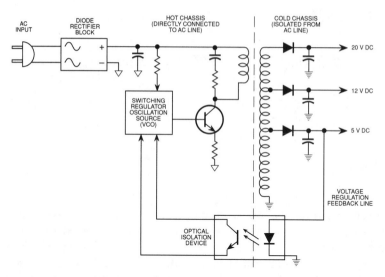

Fig. 10.17. Block diagram of a switching power supply.

weight reduction possible by using switching power supplies translates into lower production costs, hence lower purchase price.

Normally, circuitry in the switching supply monitors one leg of the DC output for proper voltage regulation. This single output monitoring concept is based on the theory that if one DC power output varies, the others will track the same error because they all derive from a common source. Any error detected by the regulation sensing circuitry on the single output is passed to the VCO as an error correction signal. If the voltage drops below acceptable limits, the VCO oscillates at a higher frequency causing more pulses in the secondary winding of the transformer. This drives up the average DC output. If the sensed output is too high, the VCO frequency is reduced causing a corresponding reduction in the output voltage.

Although much more efficient, switching power supplies do present additional danger to the servicer. The primary side of the transformer and its associated circuitry is typically not isolated from the input AC power. This pre-

sents a high-voltage danger to the technician. Test leads should not be placed in the primary circuitry, nor should you ever touch anything associated with the power supply without first connecting the AC power cord to an isolation transformer. Failure to heed the high-voltage safety precautions here can result in death, or serious damage to your test equipment.

CAUTION

Do not attempt to service a switching power supply without attaching the AC cord to an isolation transformer. Use only safety approved troubleshooting methods when working on any power supply. Lethal voltages may be present!!!

Power supplies are protected by fuses. The values of these fuses should **NEVER** be changed because a higher amperage fuse will affect the safety rating of the circuit and no longer protect the circuitry. If a fuse continues to blow, even intermittently, there is a definite problem.

CHAPTER REVIEW QUESTIONS

1. When troubleshooting a System Control problem, what are some items that should be checked before replacing the microprocessor IC?

2. Assuming that the power supply is providing correct stand-by voltages in the VCR OFF mode, why should the System Control be checked when a VCR refuses to turn on after pushing the POWER-ON button?

3. Explain handshaking.

4. Explain the function of the VCR/TV button on a VCR.

5. What is the purpose of the VCR RF modulator?

6. List the preferred order of VCR audio and video output signal hook-up to the television?

7. What is the function of the Mode Position Sensing switch?

8. Why do most VCRs cause the loading mechanism to shift position slightly at power on?

Chapter 11

Advanced Troubleshooting

In earlier chapters you learned basic techniques for troubleshooting most VCR failures. You discovered that there are several steps to successful fault identification and correction. You learned how to recognize the various components of your VCR, and you discovered that a VCR fails in two ways, electronically and mechanically. You also learned that for technical or electronic repairs, you use special tools to troubleshoot a circuit. This requires not only test equipment but some knowledge of electronics and VCR operating theory. Here in Chapter 11 you will learn advanced troubleshooting techniques. Concurrently, you'll be introduced to the repair technician's "tools of the trade."

Troubleshooting involves isolating a problem to a mechanical or electronic failure. If you conclude that the problem is not mechanical, you can use the techniques described in this chapter to test a VCR's electronic components in the area of a failure on a circuit board.

TOOLS OF THE TRADE

When a problem can't be solved using flowcharts and pictures, repair technicians reach for their tools. These include not only the tiny screwdrivers (tweakers), the diagonal cutters (dykes), and the soldering iron, but also various measurement meters (ACVM, VOM, DVM, DMM), NTSC video generator, monitor/receiver, pulse cross monitor, frequency generator, variable power supply, frequency counter, oscilloscope, and assorted gauges, jigs, and alignment tapes. Let's take a closer look at some of these tools.

Meters

Electronic measurement equipment has improved significantly over the years, greatly enhancing your ability to test and locate circuit troubles. Twenty-five years ago, a meter called a

VOM (volt-ohm-milliammeter) was used to measure the three parameters of an electric circuit—voltage, resistance, and current (Fig. 11.1).

Fig. 11.1. *A volt-ohm-milliammeter. (Courtesy Simpson Electric Co.)*

Then came the VTVM (vacuum-tube-voltmeter). It wasn't long before electric circuits made room for electronic circuits, digital replaced analog in gauges, and new meters were introduced for troubleshooting even more complex circuits. The DVM (digital voltmeter) and DMM (digital multimeter) quickly became the preferred measurement devices of technicians because they were better suited for electronic circuit testing, including increased accuracy. These meters have characteristically high input impedance (resistance) so they don't load down or draw down a circuit where the voltages and currents are low.

Two changes affected the types of tools used in troubleshooting and repair. First, vacuum tubes were replaced by solid-state devices such as transistors and integrated circuits (ICs), or chips. Second, circuits themselves became smaller with more components packed compactly into less board area. You need only

compare the early radios and televisions (standing four feet tall and weighing 40 pounds) with today's wrist radios and wrist televisions to recognize that electronic circuits are smaller, more complex, and more difficult to access by test probe.

AC Voltmeter

The AC voltmeter (ACVM) is a special purpose voltmeter. It can measure AC voltages ranging from only a few hertz to several hundred kilohertz. More expensive AC voltmeters will measure signal levels into the low MHz range. Figure 11.2 is a photograph of typical DVM and ACVM test instruments.

Fig. 11.2. *The DVM and the ACVM test instruments.*

An ACVM is used to measure audio voltages. It's required for the alignment of audio circuitry. AC bias at the linear audio head is also aligned with this meter.

To adjust audio AC bias, you need a meter with an AC voltage sensitivity of one millivolt, or less, so the ACVM must be able to measure very small voltages at high frequencies.

NTSC Generator

NTSC stands for *National Television System Committee*—the group who defined the specification for television signals. An NTSC video generator is a special purpose signal generator that produces precise video signals. Figure 11.3 is a photograph of an NTSC generator.

Fig. 11.3. NTSC video signal generator.

The alignment of video circuitry requires a generator that produces extremely stable video signals, horizontal and vertical sync, color burst, and peak white levels, and accurate color phase and saturation levels. Proper alignment of the video circuitry is not possible using broadcast TV signals because they vary from one channel to another in strength, color saturation, and peak-to-peak levels. If the VCR video circuitry were aligned using the incoming broadcast TV signal, the accuracy of the alignment would depend completely upon the quality of the incoming signal and on the calibration of the VCR tuner and video demodulator circuit.

A second reason that broadcast signals can't be used to align video circuitry is that TV broadcast signals rarely contain 100 percent peak white video levels, and when they do, it's only for an instant. To align video circuits, you need a one volt peak-to-peak video signal that can remain stable long enough for alignment adjustments to be made.

Monitor/Receiver

A monitor/receiver is a special purpose TV that can receive video signals through the standard cable or TV input connection and through special video input jacks connected by cable to either a VCR or video generator. The monitor/receiver demodulates the station selected on the tuner and makes that modulated signal available at the video output jacks. These handy receivers are convenient because the demodulated TV signal can be applied directly to the video line input jack in the VCR, bypassing the VCR tuner and demodulator circuitry.

If a failure occurs in the VCR tuner or demodulator circuit, the ability of the unit to record and play back can be checked by injecting demodulated TV signals from the monitor/receiver directly into the VCR at the line input connector. Some monitor/receivers provide S-Video jacks along with the line video connections. S-Video jacks allow you to bypass even more of the VCR's internal processing stages. Monitor/receivers with both types of video connections greatly simplify troubleshooting.

The monitor/receiver can also receive and display a video signal directly from the VCR. Either the composite video line output or the S-Video signal from the VCR can be connected directly to the corresponding monitor/receiver video input. The video signal can then be displayed on the screen bypassing the RF modulator of the VCR and TV antenna switching electronics. If a failure is isolated to the VCR RF modulator, it can be located rapidly using the monitor/receiver.

A monitor/receiver also accepts pure audio from the VCR audio line output. Demodulated audio from the station selected on the monitor/receiver tuner is also available at the audio output jacks and can be injected into the VCR audio line input. This is useful for testing VCR audio circuitry by enabling you to bypass the audio demodulator in the VCR tuner circuitry.

The monitor/receiver can also verify and isolate audio record failures in the VCR tuner and audio demodulation circuits, or verify that the main portion of the audio record playback circuitry is performing properly.

Pulse Cross Monitor

A *pulse cross* monitor is a special type of video display used in analyzing VCR performance. This monitor is named for the effect produced on its screen. Horizontal and vertical synchronization pulses are displayed in such a way that they produce a figure resembling a cross, as shown earlier in Fig. 5.29.

The pulse cross monitor provides information about the condition of the VCR by displaying the portions of a TV screen not normally seen. Displaying the horizontal and vertical synchronization pulses enables a technician to observe areas of the broadcast TV signal that cannot be seen on a standard television receiver. The display makes horizontal blanking, vertical blanking, and synchronization pulses, which are normally out of view on a TV screen, visible to the technician. The area of a TV picture in which video head switching occurs is also displayed on the pulse cross monitor.

Horizontal and vertical synchronization pulses and the time position of video head switching indicate the performance of the VCR during record and playback.

A pulse cross monitor aids in troubleshooting playback tape path error and in differentiating between playback disturbances caused by improper servo circuit operation and record input signal interference. All three factors can disrupt playback picture stability.

A common symptom of VCR malfunction is vertical instability (jumping) of the playback picture. This may be diagnosed as trouble in the servo circuitry, or tape path misalignment, but may actually be caused by worn video heads or the presence of RFI in the record input signal.

If RFI in the signal being recorded is strong enough, the machine may occasionally change speed or produce a vertically unstable picture during playback of that recording.

The trouble of video jumping can occur intermittently or only on certain channels. By viewing the E-to-E output signal from the VCR on a pulse cross monitor during VCR record, the vertical and horizontal synchronization pulses can be analyzed for interference. RFI mixed in with the synchronization pulses (especially the vertical synchronization pulses)

can cause servo instabilities. The servo circuit can become confused when RFI is present in the vertical synchronization pulse train. This can cause unstable picture recording.

The VCR uses the vertical sync pulse to reference the VCR head switching pulse and control track signal. The interaction between servo circuits and the vertical synchronization pulse was explained in Chapter 8. The pulse cross monitor is a valuable tool for quickly identifying servo difficulties.

As mentioned earlier, the pulse cross monitor can aid in troubleshooting tape path error and confirming the presence of worn heads. A fairly accurate preliminary diagnosis can be made by monitoring both the top and bottom portions of the picture (the scan lines just above and below the vertical blanking interval). Misaligned entrance and exit guides will cause distortions and horizontal noise bands in the pulse cross display near the vertical blanking period. Tape path misalignment is another source of vertical jumping in the playback picture.

Worn video heads contact the tape later at the entrance point and leave the tape earlier as the heads exit the tape. At some point head wear could become significant enough to reduce contact at the tape entrance and exit points producing noise in the vertical blanking portion of the playback picture. Worn video heads are another cause of vertical instability.

In each of the previous cases, the pulse cross monitor could be used to make a quick evaluation without ever removing the covers of the machines. Observations that suggest failure should be confirmed using an oscilloscope to examine the playback RF envelope and the signals at other key test points.

Frequency Generators

Frequency generators are used to align the audio circuits, the video record and playback circuits, and the video head preamplification circuit. The frequency generator used should have an oscillation range between 1 Hz and 10 MHz. Audio circuits need a steady tone during alignment of the record and playback amplification components. Audio frequencies range from 20 Hz to 20 kHz.

During alignment, a tone is recorded and then the amplifier electronics are adjusted to specifications listed in the VCR service manual. A second frequency is recorded, and more adjustments are made. Sometimes three or four frequencies must be recorded during alignment. The audio alignment instructions of some VCR manufacturers specify that these recorded signals be played back while adjusting the playback amplifier for equal output levels over all the recorded frequencies. Other manufacturers instruct you to first align the playback amplifiers using tones on an alignment tape and then to use a frequency generator to adjust the record circuits.

Audio sweep generators are popular because they can make audio circuit alignment fast and easy. With a sweep generator, all of the required frequencies are recorded simultaneously on the tape, so the electronics can be aligned rapidly. A sweep generator produces frequencies within a selected range. It can be adjusted to sweep the audio range (between 20 Hz and 20 kHz) several times per second. Once a sweep range has been selected, the sweeping is done automatically by the generator. The advantage of using a sweep generator for this procedure is that it eliminates the requirement for recording several frequencies in multiple operations.

Video record amplifiers also require equalization alignment. The need for equalization was covered in Chapters 5 through 7. During alignment of the high-frequency video record and playback amplification circuits, the top end frequency limits of the frequency generator become important. Typical video circuitry alignment procedures involve frequencies ranging from 3 MHz to 10 MHz.

It's not necessary to purchase a separate frequency generator for both audio and video. One frequency generator will do if the range

(bandwidth) is wide enough to cover the audio and video frequency requirements.

Some frequency generators include sweep marker identification, which is an electronic marker superimposed on the output of the frequency generator to identify specific frequencies. Sweep marker generators can sweep a selected range and identify the position of certain frequencies within that range using markers. For example, setting the frequency generator to sweep between 1 MHz and 5 MHz sixty times per second, you could also cause the generator to identify the 2 MHz, 3 MHz, and 4 MHz positions by placing an electronic marker on the output signal when each of these frequencies is reached. By using an oscilloscope to view the sweeping output of the frequency generator, you can see tiny notches in the oscilloscope presentation at the 2, 3, and 4 MHz points. This is helpful for accurately identifying events at certain frequencies in the operational spectrum.

Alignment procedures often require the adjustment of amplifiers for equal signal output levels at various frequencies. A frequency generator with sweep marker capability makes this alignment easy. The frequency generator is placed in the sweep mode and set to sweep the frequency range required in the alignment procedure. Markers are placed at the frequencies you are aligning. To perform the alignment, you adjust the controls as instructed for equal signal output at each of the marked frequencies.

Isolated Variable AC Supply

An isolated variable AC supply called a *Variac* is required for servicing some VCR problems. A Variac is an isolation transformer with one side connected through a rheostat. The isolation part of the transformer separates the VCR power supply from the incoming AC power supply.

Most VCRs have a hot chassis power supply that has one side of the AC line connected directly to the chassis. Test instruments (such as oscilloscopes) with ground probes that connect to the AC power line cannot be connected to machines with hot chassis power supplies without an isolation transformer because you could destroy both units. If you blindly connect test equipment to machines that have these types of power supplies, there is great potential for damage to the VCR, the test equipment, and you.

The variable AC part of the Variac is useful in troubleshooting VCR shorts. The AC voltage applied to the VCR can be varied by the rheostat control on the Variac. When the Variac is placed in series between the wall plug and the VCR, the power applied to the VCR can be increased slowly from zero to the nominal 115 V. Use extreme caution and an isolated Variac if you're going to troubleshoot the power supply section of your VCR.

Frequency Counter

A *frequency counter* counts individual pulses or sine wave cycles and displays the number of pulses or cycles that occur each second. Frequency is measured in hertz (cycles per second). Precise frequency is required for accurate adjustment of VCR color and servo circuits. An oscilloscope can also measure frequency, but not with the accuracy of the frequency counter.

Oscilloscope

The *oscilloscope* (frequently called a "scope") has been with us for years, although recent advances have added many capabilities to the instrument.

Simply put, an oscilloscope (Fig. 11.4) is an electronic display device that graphs signal voltage amplitude versus time or frequency on a CRT screen. A scope analyzes the quality and characteristics of an electronic signal by using a probe that touches a test point in a circuit. It also determines the voltage level of certain signals.

Fig. 11.4. *Oscilloscope with delayed sweep. (Courtesy Hewlett-Packard)*

Scopes come in all sizes, shapes, and capabilities. Prices vary between $500 and $20,000. Some scopes use a single test probe for displaying and analyzing a single trace signal. Others have two probes and display two different signals at the same time (dual trace). On some oscilloscopes, as many as eight traces can be analyzed simultaneously. Recently, color digital oscilloscopes have been introduced with LCD (liquid crystal display) displays. Colors make it possible to rapidly compare signals at different locations in the circuitry. Monochrome LCD scopes are also available. These LCD display scopes are small and use low power making hand-held battery powered oscilloscopes possible. Some scopes even have built-in digital memories to let the scope store a signal of interest for future evaluation. This type of scope, called a *digital storage oscilloscope* has become quite popular over the last couple of years.

Besides sensitivity and trace display, a major distinguishing characteristic of oscilloscopes is that a wide range of frequencies can be observed on the CRT screen as stationary images. We call this range of frequencies *bandwidth*. Bandwidths vary between 5 MHz and 500 MHz. The price of the oscilloscope is proportional to its frequency bandwidth.

Oscilloscopes display a signal as an analog or varying signal and as a static waveform on a CRT screen covered by a measurement grid. While it is time consuming to learn to use an oscilloscope, the analytical rewards are substantial. You can not only measure voltage amplitudes and frequencies of test signals, but also delay times, and signal rise and fall times. You can even locate intermittent glitches in a circuit during operation.

Some oscilloscopes also have a variable delayed sweep function that expands the frequency being measured. It permits a low-frequency signal to be displayed while expanding a small portion of that waveform for simultaneous viewing. Figure 11.5 shows a delayed sweep scope display. The upper portion of the figure is a vertical field containing 262.5 horizontal lines. The bright portion of the vertical field is the section identified by the delay sweep function. The lower portion of the figure shows the expanded individual horizontal lines identified in the vertical field.

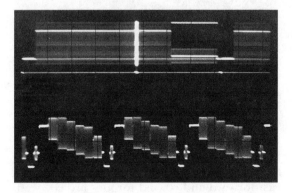

Fig. 11.5. *A delayed sweep scope display.*

For most VCR repairs you can usually get along fine with a dual trace, 25 to 30 MHz scope. Investment in an oscilloscope is a must if you intend to analyze and troubleshoot VCR component failures, or perform tape path alignments.

The nice thing about a dual-trace or quad-trace oscilloscope is that it allows you to look at different signal paths or different signals simultaneously. For example, you could look at the input and output of a gate and actually see and be able to measure the delay time for the signal passing from input to output of the chip.

Torque Gauges and Tension Meters

Proper tape handling in the VCR prolongs the life of the video cassette tape, the video heads, and the tape guides. Torques applied by the reel tables to the tape during play, fast forward, and rewind must be controlled within specified limits to prevent tape damage. And the video tape must be wrapped around the rotating video head drum assembly with a specified amount of play or forward holdback tension.

If the torques are too low, tape may spill into the machine during operation. If the torques are too high, the tape can be stretched and worn by excessive friction.

Tension and torques are measured in different ways depending upon the machine. Torque measurement will be covered first.

Torques are normally not adjustable. Correction of an unacceptable torque measurement typically requires the replacement of parts. Low torque readings may necessitate replacement of a clutch assembly, a motor, or rubber tires and belts. Excessive torques indicate the need to replace the torque limiting clutch. Brake torques that are out of specification normally require cleaning or replacement of the rubber brake shoes. Cleaning of the reel table assemblies is important for maintaining proper torques.

For some machines, torques are measured with a hand-held torque gauge that fits over the reel tables as shown in Fig. 3.19 back in Chapter 3. The machine is set to play with the torque gauge over the take-up reel table. As the reel table tries to rotate against the gauge spring, the meter needle deflects indicating the value of forward take-up torque.

One method for measuring rewind and fast forward torques is to place a torque gauge on the appropriate reel table and then engage the machine into the mode to be checked. When you check rewind torque, place the gauge on the supply reel table (left-hand reel table as you face the front of the machine). The right-hand or take-up reel table is used to measure fast forward torque.

Brake torques are measured by setting the machine to STOP, putting the same hand-held torque gauge used for the rewind and fast forward measurements on each of the reel tables, and gently twisting the torque gauge counterclockwise and then clockwise. The indications on the torque meter are noted for each of the reel tables and compared to specifications found in the service literature.

Tension is a factor that can be adjusted within your VCR. Tension is important for producing quality recordings and reproductions on a display screen.

Forward holdback tension is critical for correct playback. If the forward holdback tension is not within allowable limits, the TV will display a flagging or tearing picture at the top of the screen. Vertical objects in the picture will bend (in some cases severely) at the top of the screen as you saw earlier in Fig. 4.11.

Flagging results when the horizontal automatic frequency control within the TV reacts to time-base instabilities in the horizontal synchronization pulses from the video tape.

Holdback tension plays a significant role in reducing horizontal time-base errors. These time-base errors are discussed at length in Chapter 6. Video heads and tape guides will also wear faster if the forward holdback tension is excessive.

Using the tension gauge shown in Chapter 3, Fig. 3.18, set the VCR to PLAY (use a T120 tape for VHS), and measure the tape at the point described in the VCR service manual. On most VHS machines, the forward holdback tension is measured near the point where the full erase track head contacts the tape.

The tension meter has three legs. The tape is placed between the three legs and the meter indicates the amount of holdback ten-

sion. Adjustment of holdback tension is usually made by setting the tension of a spring, which holds a felt-padded brake band wrapped around the supply reel table, as shown in Fig. 3.15 (Chapter 3).

Adjustment procedures vary according to the manufacturer and model of VCR that you own. Specific instructions are provided with the service literature.

Some manufacturers have introduced cassette tapes like those shown in Fig. 11.6 for measuring tension and torque. Forward holdback tension and play torque are measured on one cassette; fast forward and rewind torques are measured on another. Separate gauges are attached to the hubs of the tape within these special cassettes. All 8mm tensions and torques are measured using these cassette type gauges.

Fig. 11.6. *Tapes are used to measure forward holdback tension, play torque, fast forward torque, and rewind torque.*

Holdback tension and forward play torque are measured during normal play operation. Fast forward and rewind torques are checked by placing the machine in the mode to be tested. These cassette gauges are quite expensive but convenient, especially for those machines without physical space for insertion of a hand-held gauge in the tape path. These units require the cassette tape gauges.

Tension and torque measurements on 8mm machines are made using tension and torque cassette tapes. Most VHS machines can be checked by using either hand-held gauges or the cassette tapes. For VHS machines, brake torque measurement requires the use of a hand-held torque gauge.

Tension and torques on machines with direct drive reel table motors are measured in the same way, but adjustments are made electronically.

Alignment Gauges and Jigs

When reel tables, tape guides, cassette tray assemblies, and (for some machines) video head drum assemblies are removed, they must be aligned with accurate gauges and jigs after reinstallation. Some typical jigs and gauges are shown in Fig. 11.7.

Fig. 11.7. *Typical jigs and gauges used in VHS alignment.*

Reel table heights are critical in avoiding tape edge damage. A reel table that is too high or too low will force the tape out of its normal traveling path and may cause the tape to be damaged at the top or bottom edge. Precise alignment of the cassette tape guides keeps the video tape in the correct path. Tolerances for cassette tape guides and reel table heights are measured within thousandths of an inch.

For the older, top-loading style VHS machines, the position of the cassette tray assembly is adjusted using a cassette tray alignment block. If the cassette tray requires removal for servicing the VCR, the cassette tray alignment block is used to properly align the tray position during its reinstallation. With the alignment block inserted, the cassette tray is placed down into normal operating position. The block allows the tray assembly to be positioned for securing. Once in position, the screws holding the cassette tray assembly in place are tightened. This is simple, but important for proper cassette and tape path alignment.

Alignment Tapes

A factory alignment tape is an essential part of any major video repair tool kit. The alignment tape serves as both the mechanical and electrical standard for the VCR to ensure proper processing for both the audio and video signals and to confirm proper tape path alignment. Factory alignment tapes are produced according to extremely tight standards set by the manufacturer. The VCRs that record alignment signals on these tapes are far more accurate than standard commercial machines. Strict control ensures that the test signals adhere to design specifications. Figure 11.8 shows several alignment tape cassettes.

Alignment tapes are expensive and should be used selectively, never in a machine whose mechanical condition is unknown. When you are testing a unit or making a coarse electronic or mechanical adjustment, use good quality

Fig. 11.8. *Alignment tapes are required to maintain factory specifications.*

tape with a known good recording rather than the alignment tape. Once the machine is aligned close to its proper mechanical and electrical specification, use the alignment tape to fine tune the system to exact specifications.

Some alignment tapes contain several test signals including multifrequency tones for calibrating the VHS linear track audio head assembly and the VHS linear track audio playback electronics. In this case, high-frequency audio signals are used to adjust the linear track audio head azimuth position, while low-frequency audio signals are used to align the head height and playback electronics for that head.

The alignment tape also includes special video signals for calibrating the playback video circuitry and video head preamplifier circuitry. Some tapes include special RF signals that are used only for tape path alignment. Alignment tapes that don't contain the RF signals may still be used to align the tape path. In this case, any of the video test patterns on the tape can be used to adjust the tape path enabling the electronics to produce the proper RF envelope.

Alignment tapes may also include audio FM signals to align VCRs with AFM/Hi-Fi stereo. Each service manual should contain a part number for ordering the alignment tape unique to that VCR. Most VCRs need different alignment tapes. Use only the alignment tape(s) specified for your machine.

The content of these tapes will vary depending on the video head core size of the target VCR and the particular audio option designed into the unit.

A specification sheet comes with some alignment tapes to describe the calibration value of the particular tape. This calibration sheet is important and should be kept with the alignment tape. It describes the amount of inherent error in the tape speed and frequency of the recorded test tones. For example, if the specification indicates an alignment tape speed error of plus 0.1 percent, you should expect this error to be present in the playback speed of a machine under test that has been correctly calibrated. A factory alignment tape is a must for completely maintaining your own machine.

When performing a VCR alignment, many manufacturers instruct the technician to calibrate the playback circuitry before any adjustments are made to the recording section. Playback alignment is typically done first because the record alignment procedure involves playback of the recordings made during alignment. If the machine does not play back properly, playback checks of the recording alignment would be invalid. Once the playback circuitry is properly aligned, the record adjustments are then made.

COMPONENTS AND HOW THEY FAIL

While the use of troubleshooting equipment makes the analysis and isolation of VCR problems much easier, many failures can be found without expensive equipment. In fact, an understanding of how electronic components fail can make troubleshooting and repair relatively simple.

Electronic failures generally occur in the circuits that are used or stressed the most. This includes the motor drive transistors and ICs and power supply transistors and ICs. The microprocessor in the system control circuitry is highly reliable and doesn't fail very often. Most failures involve the other components which require soldering and are not as easy to replace.

Let's look at the kinds of components you'll be analyzing and how these components can fail.

A chip or integrated circuit is fabricated from silicon with tiny particles of metal (impurities) embedded in specific positions in the silicon. By positioning the metals in certain ways, the manufacturing process forms tiny transistors. Applying a voltage to specific places on the chip causes the device to invert a voltage level (+5 volts-logic 1, to 0 volts-logic 0 and vice versa) and to enable all sorts of logic gates (AND, NAND, OR, NOR, etc.) to function. These chips can be made with silicon/metal junctions so tiny that today millions of transistors can be placed on one chip. A memory chip the size of a fingernail can hold over two million transistors.

The problem for chip manufacturers is how to apply voltages and get signals into and out of such a tiny chip. Very thin wires are glued or bonded to tiny pads and used as inputs and outputs to the chip. The other end of each wire is bonded to a larger pad on a supporting material (the big part of what we call the integrated circuit as shown in Fig. 11.9). The supporting structure includes the pins we solder into the printed circuit boards.

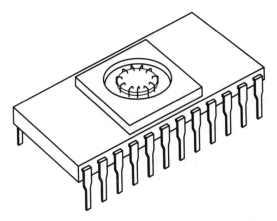

Fig. 11.9. *The tiny leads from the chip to the pins of the chip package are clearly seen.*

These tiny silicon and metal chips are placed in environments that really put them under a lot of stress. They heat up when you use the VCR, cool down when you turn off the machine, then heat up again. This hot-cold-hot effect, called *thermal stress*, affects those tiny strands of wire, or leads, going between the chip and the supporting structure (which includes the large pins that are inserted into the boards). Over time, thermal stress can cause the bonding of the wire lead to break away from the pad on the chip. This disconnect causes an input or output to become an open circuit, and chip replacement is required.

Another chip failure is caused by a phenomenon called *metal migration*. The chip can be compared to an ocean of atoms. Some tiny particles of metal float about in this sea, migrating in directions perpendicular to the electrical current flowing through the chip. Problems occur when these metal particles begin to collect in parts of the chip. If they concentrate in the middle of one of those microelectronic transistors, they cause the transistor to operate differently or not at all. If the resistance of these collected metals gets high enough, it causes the device to operate intermittently or to simply refuse to work. Since the failing transistor is part of a logic gate, the gate malfunctions and the output may become "stuck at 1" or "stuck at 0" regardless of the input signal.

Theoretically, a wearout failure won't occur until after several hundred years of use. However, we shorten the life span of chips by placing them in high-temperature, high-voltage, or power cycling environments.

Other problems occur outside the chip, between the chip leads and the support structure pin leads, the inputs, and outputs of the device. These types of failures include inputs, or outputs shorted to ground, pins shorted to the +5 V supply, pins shorted together, open pins, and connectors with intermittent defects. The most common problem (assuming power is available) is opens or shorts to ground. Under normal use, chips finally fail with an input or output shorted to ground.

Capacitors

Understanding the way a standard capacitor is constructed will help you understand how these devices fail.

In Chapter 2, you discovered that there are several types of capacitors on the VCR boards. The capacitor is constructed of two separated plates. A voltage is placed across the plates and for an instant, current flows across the gap. But electrons immediately build up on one plate and cause the current flow to slow and then stop, leaving the capacitor charged to some voltage potential. In addition to storing a charge, capacitors are used to filter unwanted signal spikes (sharp, quick peaks of voltage) to ground.

The *electrolytic capacitor* is constructed as shown in Fig. 11.10. Two aluminum foils or plates are separated by a layer of porous paper soaked with electrolyte solution, a conductive liquid. On one plate (the positive plate) a thin layer of aluminum oxide is deposited. This is called the *dielectric*. A capacitor has an anode (the positive plate) and a cathode (the electrolyte). Electrons build up on one plate causing it to become so negative that it prevents further current flow (remember that electrons have a negative charge).

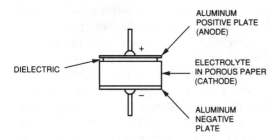

Fig. 11.10. The electrolytic capacitor.

Another type of capacitor is the *film* capacitor. It is constructed of alternating layers of aluminum foil and a plastic (usually polystyrene) insulation. The metal pieces of foil act as the plates and the plastic insulation acts as the dielectric between the plates. A film capacitor is shown in Fig. 11.11. Film capacitors are coated with epoxy and have tinned copper leads.

Capacitors fail open or shorted depending on operating conditions and age. Electrolytic ca-

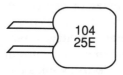

Fig. 11.11. The film capacitor.

pacitors are especially susceptible to the aging process, one effect of which is the drying out of the electrolyte insulator. As the insulator dries out, the capacitance value decreases, and circuit performance decreases. Finally the capacitance value drops dramatically as the plates fold toward each other, and shorting occurs.

Another kind of failure occurs when some of the dielectric oxide dissolves into the moist electrolyte, causing the thickness of the dielectric to shrink. This deforming usually occurs when the electrolytic capacitor sets for a long time without applied voltage. In this case, the capacitance value increases but a large leakage of electrons occurs across the plates, making the capacitor useless.

Electrolytic capacitors may also leak electrolyte as they fail. As electrolyte seeps out of the capacitor and onto the circuit board, unwanted conductive paths may form across board traces shorting out signals and causing operational failure. Copper traces may also be destroyed by the acidic effect of the leaking electrolyte.

The leads of any type of capacitor can physically detach causing an open in the circuit. Also, the plates can short together when a large area of one plate is stripped of its dielectric oxide layer by the application of too much voltage.

Despite the potential for several kinds of failure, capacitors in VCR circuitry seldom fail. However, capacitors can be "fried" by a momentary power surge through the power supply, or damaged by excessive temperatures.

Resistors

These current-limiting, voltage-dropping devices are quite reliable and should function properly for the life of your VCR. However, the same factors that shorten the useful life of the chips also act to reduce the operational life of resistors. High temperatures, high voltages, and power cycling all affect the materials of which the resistors are made. These stresses cause breaks in the carbon, resistive paste, or resistive layers and produce an open conduction path in the circuit. Excessively high voltage can produce electrical current so large that it actually chars resistors to burnt ash. This is rare, especially in a circuit where the highest voltage is 12 volts (usually between 5 volts and 12 volts) and the currents are very tiny (milliamps).

Resistor failures are almost always associated with catastrophic malfunction of some other circuit component.

Diodes and Transistors

The diodes and transistors on the VCR printed circuit boards are made of solid material and act much alike. In fact, the transistor can be considered to be made up, in part, of two diodes.

Diodes are one-way valves for electric current, allowing current flow in only one direction. Diodes are usually made of either silicon or germanium. They are used in power supplies as rectifiers and in some circuits to maintain a constant voltage level. Other diodes are made of gallium arsenide and react by giving off light when biased in a certain way. These are called *light-emitting diodes* or *LEDs*.

Transistors are used as amplifiers or electronic switches in various places in VCR circuitry. Diodes and transistors fail in the same ways and for the same reasons as chips.

USING TOOLS AND TEST EQUIPMENT TO FIND FAILED COMPONENTS

Once the defect has been isolated to the failure of an electronic component, it's time to break out the schematic and block diagrams and turn on the test equipment. Because this guide was produced to cover home VCRs, the block diagrams and troubleshooting flowcharts found in this manual will help you understand and localize defects only to a particular circuit. The schematic diagrams and block diagrams found in VCR service manuals can help you isolate a failure to a specific defective part.

Troubleshooting the electronic circuits can involve tracing signals with a signal generator. These signals are injected into the audio or video circuitry, depending on the nature of the defect, and then traced to the failure point.

A TV monitor/receiver, an oscilloscope, a DVM, an NTSC video generator, an audio signal generator, a new blank video tape, and a prerecorded test signal video tape are some of the basic tools you will need to find the problem. By comparing the symptoms with those found in the troubleshooting flowcharts in Chapter 4 and the troubleshooting tips of Chapters 6 through 10 of this book, you should be able to localize the defect to a particular circuit area. Much troubleshooting time will be wasted if the problem is not localized to a specific circuit. Don't spend time looking for a problem in the audio circuit if your unit has no color. Locating failed components involves checking each of the input and output signals for proper DC voltages, peak-to-peak waveform levels, and frequency (if it applies).

The following example will illustrate how to troubleshoot a brightness problem in the luminance processing circuitry. The first step is to decide if the defect occurs in playback, record, or both. Assume that the defect occurs only in the record mode. You can verify that the VCR plays back properly by putting a video tape containing a good recorded signal in the machine and playing it back. If the signal plays back properly, insert a blank tape and attempt to record. Place the VCR in the E-to-E record mode and attempt to monitor the signal coming from the VCR on your TV receiver.

In this example the program that you have tuned in on the VCR appears on the monitor/receiver, so you conclude that the VCR tuner, and audio and video demodulation circuitry are functioning correctly.

Next, connect the NTSC video generator to the video line input on the VCR and set the VCR to record the NTSC signal. Connect an oscilloscope by placing the clip lead of its signal probe at chassis ground. Take the measuring tip of the probe, place it on the video input line connector, and adjust the oscilloscope display to reveal a few horizontal lines of signal from the NTSC video generator at proper amplitude and frequency.

Check the block diagram in your service manual and compare that with the schematic diagram tracing the signal flow. Key test points will be called out on the block diagram or the schematic. Compare the waveforms observed on your oscilloscope with the waveforms shown in the schematic. Continue tracing the signal through the circuitry from input to output until an abnormality is found.

Eventually you will come to a point where the signal is no longer present or is deformed in amplitude or shape as compared to the schematic. This will be at or near the area of the defective component.

If waveform measurements do not isolate the defective components, use your voltmeter and check components in the area of the location of failure.

Measure and record the DC voltages around the ICs and transistors in the suspected area. Compare the measurements you made with those listed in the service manual. Individually check the components in the area for cracks and open connections along solder traces and connections.

Open connections prevent voltage levels from being transferred and prevent the affected circuit from being able to respond. For example, if one input of a two-input NAND gate in the video record switching logic circuitry has an open input as shown in Fig. 11.12, all but one of the four possible input combinations will be correct. With this type of failure, a gate where only half of the inputs are good could operate correctly most of the time. The failure would be intermittent.

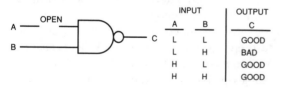

Fig. 11.12. *An open at the input to a NAND gate is only a problem in one of four logic-state cases.*

As shown in Fig. 11.13, if the device being tested is a NOR logic gate, the output would be a logic 1 or "high" only when both inputs are at logic 0 or "low." Should one of the inputs become open, it would float to logic 1 and cause none of the input conditions to produce a logic 1 or "high" output. Thus, the output would be low all the time—just as though the output were shorted to ground.

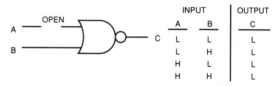

Fig. 11.13. *An open at the input to a NOR gate will prevent the output from ever changing state or going HIGH.*

If the chip has an open pin at its output, it cannot pass a logic 1 or 0 to the next gate. You can measure a voltage at the input to the next gate since it is providing the potential, a logic 1 or "high" level (something around +5 volts). The key here is that at any time an input to a TTL gate opens (a condition we call *floating*), the gate will act as though a logic 1 were constantly applied to that input. The voltage on this floating input will drift between the high supply voltage +5 volts and a level (about 1.5 volts) somewhere between a valid "high" and a valid "low." (A valid "high" is usually above +2.4 volts; a valid "low" is below +0.4 volt.)

A voltmeter reading of about +1.7 volts at the output pin of a gate on a chip is a clue that the output is floating open and the voltage is actually being provided by the next chip or following gate.

Since the VCR boards are flexible at certain points, replacing parts or depressing the boards without supporting them from beneath could cause a break to occur, opening a trace on a circuit board. A hairline crack such as this is often difficult to find, but looking at the board with a magnifying glass and a strong light (or a magnifying lamp) can sometimes reveal a suspected failure. A resistance test can be conducted with a VOM or VTVM by placing a probe at either side of the suspected bad trace, as shown in Fig. 11.14, and observing whether a zero ohm reading is measured. Another way to ascertain if an open trace is present is to compare the voltage levels at both ends of the trace.

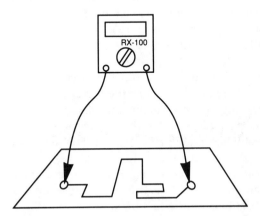

Fig. 11.14. A trace can be tested for an open using an ohmmeter to check the electrical resistance from one end to the other.

Remember, when you are testing for individual shorted or open parts in the VCR board circuitry that more than one part may use the same input or output lines to or from another component. When studying the circuitry, remember that the failure could be located at the other end of the board. One long trace from the end of the board to the part you are looking at may be shorted or open at the distant end. Use of the schematics in the VCR service manual will help here.

Other Troubleshooting Techniques

Here are some interesting tricks that you can use to aid in finding chip failures.

Use your senses

Look, smell, and feel. Sometimes failed components become discolored or develop bub-

bles or charred spots. Blown devices can produce some distinctive smells—a ruptured electrolytic capacitor, for example. Finally, shorted chips can get really hot. By using a "calibrated finger," you can pick out the hot spots on your board.

Heat it, cool it

Heating and then cooling is a fast technique for locating the cause of some intermittent failures. Occasionally, as an aging device warms up under normal operation, it becomes marginal and then intermittently quits working. If you heat the energized area where a suspected bad component is located until the intermittent failures begin, then methodically cool each device with a short blast of canned coolant spray, you can quickly cause a marginally defective chip to function again. By alternately heating, cooling, heating, and cooling, you can pinpoint the trouble in short order.

You can heat the area with a hair dryer or a focused warm air blower designed for electronic testing. Be careful using this technique, because the thermal stress you place on the chip being tested can shorten the life of good components. A one- or two-second spray of freeze coolant is all you should ever need to get a heat-sensitive component working again.

Most coolant sprays come with a focus applicator tube. Use it to pinpoint the spray. Be careful not to spray your own skin. You could get a severe frost burn.

The Easter egg approach

Quite often we can quickly locate a fault to a couple of parts but need further testing to determine which one is the culprit. When time is of essence, take an "Easter egg" approach. Just as a youngster used to pick up and examine Easter eggs one at time to see if his/her name was marked on it, you can try replacing the components one at a time to determine whether the part replaced was causing the problem. You have a 50–50 chance of selecting the right part the first time. If that doesn't work, replace the other component.

If the parts involved are inexpensive, why not replace them both? For 30 cents or more, go ahead and splurge. If the problem's gone, but you're still curious, you can always go back later and test each part individually.

A typical problem occurs in oscillator circuits where a failure can be isolated to one of several capacitors. A fast way to restore proper signal is to replace all of the suspected capacitors.

Testing capacitors

How do you check out a capacitor that you believe has failed? If the device has shorted, resulting in severe leakage of current, you can spot this easily by placing an ohmmeter across the capacitor and reading the resistance. At first you'll notice a low reading, because the capacitor acts as a short until it charges; but then, if the capacitor is working properly, it will charge, and the resistance will rise to a nominally high value. If the device is shorted, the initial low resistance reading continues and the capacitor won't charge.

Should the component be open, you'll not see the instantaneous short at time T0, the moment charge starts to build. An open circuit has infinite resistance. An in-circuit capacitance tester is helpful here.

Total failure as a short or open is pretty easy to find. But how about the device whose leakage depends on temperature or whose dielectric has weakened, changing the capacitance value? To test this capacitor requires a different level of analysis.

Capacitance measuring

If you have an analog ohmmeter whose face has the number 10 in the middle of the scale, you can easily use it to approximate the capacitance of a device. Use the time constant formula $T = RC$, where T equals the time in seconds for a capacitor to charge to 63.2 percent of the supply voltage, R equals the resistance in ohms, and C equals the capacitance in farads. Using a 22 mF (.000022 farad or 22 microfarad) capacitor and a 1 MW (1,000,000

ohm or 1 megohm) resistor, the charge time for one time constant is .000022 × 1,000,000 = 22 seconds. Transposing the formula to read:

$$C = T/R$$

we can determine the value of capacitance by knowing the resistance and counting the seconds required for the charge to cause the ohmmeter needle to reach 63.2 percent of full scale (infinite resistance). This point is at about 17 on the meter.

To do this procedure, disconnect one end of a capacitor from the circuit, turn on your meter. Zero adjust the meter's ohm scale reading. Then estimate the ohm scale multiplier needed to let the capacitor charge in some acceptable time period. For microfarad capacitors use the X 100K scale because this will let the capacitor charge in less than a minute. The 17 on the scale represents 1.7 megohms on the X 100K scale.

Short a low-ohm-value resistor across the two capacitor leads for several seconds to thoroughly drain off any charge. Then connect the ground lead from the meter to the negative side of the capacitor (this can be either lead if the capacitor is not an electrolytic), and touch the positive meter probe to the other side of the capacitor. Use a stop watch to count seconds and tenths of a second, watch the face of the ohmmeter as the capacitor charges and the resistance needle moves up. When the needle gets to 17 on the scale, stop the clock and read the time. This will give you the capacitance value in microfarads. This technique provides a close enough approximation of the capacitance value to determine if the device is good or should be replaced.

Some capacitor failures only occur when a voltage of about 50 percent of their rated working voltage is applied because they might fail only under operational voltage and temperature loads. Yet when they're removed from the circuit and tested as described, they appear to be good. If they are resoldered back in the circuit, voltage measurements or scope presentations of signals in the circuit again point to the capacitors as likely problem candidates. If you think a capacitor is defective and it tests good out of circuit, substitute it with a new capacitor to determine the condition of the original part. Do this for each suspect capacitor.

Replacing capacitors

Always try to use the same type and value capacitor as the one being replaced. Keep the leads as short as possible and solder the capacitor into the solder connector holes using the proper iron. The solder process should not exceed 1.5 seconds per lead; otherwise, heat damage to the components may result.

Testing diodes

If you have a digital multimeter (DMM) with a diode test capability, you can quickly determine whether a suspected diode is bad or good. Placing the meter on the ohmmeter setting and positioning the probes across the diode causes the meter to apply a small amount of current through the diode if the diode is forward biased. The voltage drop across a diode is normally 0.2–0.3 volt for germanium diodes and 0.6–0.7 volt for silicon diodes. Reversing the leads should result in no current flow, so a higher resistance reading should be observed. A low-resistance reading when the diode is biased in either direction indicates that the device is leaking or shorted. A high-resistance reading in both directions indicates the diode bond has opened. In either case, replace the diode immediately.

Diodes can also be tested in-circuit using the ohmmeter to check the resistance across a diode in both directions. With one polarity of the meter probes, you should get a reading that is several hundred ohms different from that obtained when the probes are reversed. For example, in the forward-biased direction, you could read 50–80 ohms; in the reverse biased direction, 300 kW. The difference in readings is called DE for *diode effect* and is useful for evaluating both diodes and transistors.

When diode readings in both directions show low resistance, you can be sure the leaky short is present.

In order for the DE test to work, the ohmmeter must provide enough voltage and current at the test probes to bias the semiconductor junction on in the forward bias connection. Some meters do not provide enough potential in their higher resistance settings to cause the semiconductor to "turn on" in the forward biased configuration. If this happens, the meter will always give a high resistance reading, leading you to believe that the component is bad when it is fine.

Some meters provide a "diode test" feature that causes the meter to read out voltage drops across the semiconductor junction. If your meter gives you this option, use it to test for DE in forward and reverse biased meter lead configurations. The diode test causes the meter to display a DE voltage reading rather than a junction resistance. The diode setting biases the meter probes for sufficient current flow and voltage potential to check most diodes and transistor junctions.

For conventional silicon diodes junctions, 0.6–0.7 volts indicates a good component when the device is forward biased and something above 1 volt (typically 1.4 volts or higher) indicates a good component when the meter leads are reversed. If your readings are something other that these, check to see if the component you are testing is a single junction silicon diode (for example, not a germanium diode or a silicon diode with several diodes in one package). You can check this by referring to the VCR service manual's parts list or a replacement semiconductor cross reference specification sheet for the component you are testing. If it is a conventional silicon device, and the readings are outside of the ranges given above, replace the part.

Testing transistors

It's no fun to desolder a transistor, test it for failure, find it good, and solder it or a new device back into the circuit board.

Fortunately, there is a way to determine the quality of most silicon transistors without removing them from the circuit. In 90 percent of the tests, this procedure will accurately determine whether a device is bad.

A transistor acts like a configuration of diodes as shown in Fig. 11.15. PNP and NPN transistors have opposite facing diodes. The transistor functions by biasing certain pins and applying a signal to one of the leads (usually base) while taking an output off the collector or emitter.

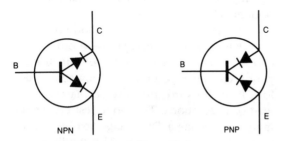

Fig. 11.15. A transistor acts like a pair of diodes.

The following tests apply to both PNP and NPN transistors. If an ohmmeter is placed between the collector (C) and emitter (E) as shown in Fig. 11.16, it effectively bridges a two-diode combination in which the diodes are opposing. You should get a high resistance reading with the leads applied both ways. (It's possible to wire the transistor in a circuit that makes the transistor collector-emitter junction act like a single diode. In this case you could get a DE. Both results are OK.)

Typical C-E resistance readings for germanium transistors are as follows:

Forward biased = 80 ohms
Reverse biased = 8000 ohms (8 KΩ)

For silicon transistors you might read:

Forward biased = 22 megohms
Reverse biased = 190 megohms

The high/low ratio is evident and is about the same for both types.

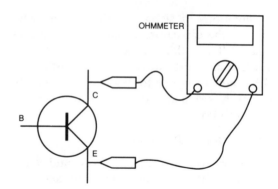

Fig. 11.16. A transistor can be tested using an ohmmeter placed across the collector-to-emitter junction.

Place the probes across the collector-to-base junction leads. Reverse the probes. You should observe a low reading in one case, and a high reading with the test probe leads reversed (the diode effect). Remember that in order for the DE test to work, the ohmmeter must provide enough voltage and current at the test probes to bias the semiconductor junction on in the forward bias connection. Some meters don't provide enough potential in their higher resistance settings to cause the semiconductor to "turn on" in the forward biased configuration. If this happens, the meter will always give a high resistance reading, leading you to believe that the component is bad when it is fine.

Try the same technique on the base-to-emitter (B-E) junction lead (Fig. 11.17). Look for the DE. If the DE is not present in all the preceding steps and you are certain the transistor is a conventional device (shortly you will read that there are hybrid transistors that will give you unusual DE readings) and that your meter is capable of forward biasing the transistor junction, you can be certain the transistor is bad and needs to be replaced.

If your meter provides the diode test feature described earlier, use it to test for DE on a transistor junction. The diode test causes the meter to display a DE voltage reading rather than a junction resistance just as it does for

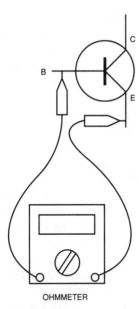

Fig. 11.17. *Check the base-to-emitter junction for diode effect (DE).*

OHMMETER

diode tests. The diode setting biases the meter probes for sufficient current flow and voltage potential to check most diodes and transistor junctions.

Like the diode discussed earlier, a measurement of 0.6–0.7 volts across a conventional silicon transistor junction when the device is forward biased indicates a good component and something above 1 volt (typically 1.4 volts or higher) indicates a good component when the meter leads are reversed. If your readings are something other than these, check to see if the component you are testing is a conventional transistor and not a hybrid transistor or multitransistor device. You can check this by referring to the VCR service manual's parts list or a replacement semiconductor cross reference specification sheet for the component you are testing. If it is a conventional silicon transistor, and the readings are outside of the ranges given above, replace the part.

Another way to evaluate a transistor is to measure the bias voltage from base to emitter

in an energized circuit and then compare your findings with the information shown in Table 11.1. Be certain to confirm the correct supply voltage first because power supply problems have been known to trick troubleshooters into thinking a certain component has failed.

The B-E forward bias for silicon transistors should be between 0.6 and 0.7 volt DC. If the reading is below 0.5 volt, confirm that the transistor is a conventional silicon type from the parts list and, if it is, replace it. The diode junction is leaking too much current. If the reading is almost a volt, the junction is probably open and the device should be replaced.

Although in some isolated cases some other failure could cause the low reading, the most common cause of low bias voltage is failure in the transistor itself.

If the previous tests are inconclusive, there is something else you can try. Measure the voltage across the collector-to-emitter (C-E) junction. If the reading is the same as the source supply and you notice on the schematic that there's plenty of resistance in the collector/base circuit, the junction is probably open. Replace the device.

If your reading is close to 0 volts, take a small length of wire and short the base to the emitter, removing all the transistor bias. The C-E meter reading should rise instantly. If it doesn't, the transistor is shorting internally and should be replaced. If C-E voltage does rise, it suggests a failure in the bias circuitry— perhaps a leaky coupling capacitor.

Recent designs integrate transistor-resistor circuits into a single transistor package. These devices are the same size as the single

Table 11.1. B-E Forward Bias Voltages for Conventional Silicon Transistors

B-E Voltage (Forward Biased)	Action
0.5 V	Replace
0.6–0.7 V	Good, keep
0.9 V	Replace

transistor and look identical. The internal wiring configuration is shown in Fig. 11.18. These components are normally used in switching logic circuits and contain internal resistors to provide bias and decrease switching transition time.

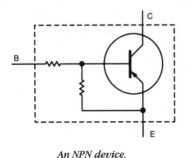

An NPN device.

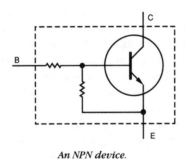

An NPN device.

Fig. 11.18. *Switching transistors may contain resistors inside the case. (A) A PNP device. (B) An NPN device.*

Some manufacturers label these components on schematic diagrams with a different symbol to distinguish them as a hybrid and not a traditional transistor. QR is a typical label designation—Q indicates a transistor, R indicates a resistor. Not all manufacturers segregate hybrids from standard transistors by separate symbols or even identify internal configurations on their schematics. However, most schematics do outline these devices in a rectangular outline as shown in Fig. 11.18, while conventional transistors are drawn inside a circle. If the technical documentation does not specify what's inside the transistor case, and an ohmmeter test doesn't indicate normal transistor behavior, check the specification book to determine its characteristics.

Checking the QR transistor is more difficult with an ohmmeter because the resistor in series with the base may not allow the ohmmeter to bias the base-emitter junction. The results listed in Table 11.1 will likely not apply when checking the QR device. If you suspect that a QR transistor has failed, the most accurate results can be obtained by monitoring the collector-to-emitter junction voltage while applying and removing bias to the base. If the base-to-emitter voltage reads 0.7 volt or more (the nominal range is between 0.7 and 2.0 volts), you can remove the bias by shorting between base and emitter just as you do when checking conventional transistors. For example, when checking an NPN device, the collector-to-emitter voltage should rise when the base-to-emitter junction is shorted, and the collector voltage should be close to zero volts when a bias of 0.7 volt or greater is present across the base-to-emitter junction.

Other combination hybrid devices, such as those shown in Fig. 11.19, are used in VCR circuits. These components contain multiple transistors and may also be identified as QR on schematics. The device in Fig. 11.19A could be used in an amplifier or logic switch. It's frequently labeled with a Q on the schematic. The device in Fig. 11.19B is used only in logic switching circuitry. It's usually labeled QR.

The devices in Fig. 11.19 can be checked with an ohmmeter as long as the meter leads provide 1.4 volts or more potential between them. This minimum potential is required to forward bias the two base-to-emitter junctions. If the devices are silicon-based transistors, each base-to-emitter should drop between 0.6 and 0.7 volt. Table 11.1 can be used if you remember to double the B-E forward bias voltages listed in the table when checking multitransistor packages.

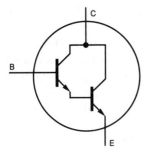

(A) This device may be used as a logic switch or an amplifier.

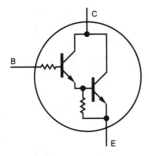

(B) This component functions as a logic switch.

Fig. 11.19. *Some transistor packages contain more than one transistor in a single case.*

However, Table 11.1 is of limited help if a dual transistor with resistor QR device like that in Fig. 11.19B is measured using an ohmmeter. The resistor in the base line will make readings confusing. The dual transistor QR device is most accurately checked in the same manner as that described for checking the single transistor QR component. Monitor the collector-to-emitter voltage while removing and returning the base-to-emitter bias voltage.

TROUBLESHOOTING VIDEO HEAD PROBLEMS AND VHS HI-FI HEAD PROBLEMS

Troubleshooting video head and VHS Hi-Fi head related failures can be challenging and time consuming. The challenge is in isolating the problem to a possible defect in the RF preamplifier stage, a wiring or connection problem between the video head and the VHS Hi-Fi head drum assembly and the preamplifier stage, a defective rotary transformer, a clogged or brown-stained video head, or defective heads on the upper drum. A component failure or bad connection on any one of these items can cause the same symptom—snowy video—making the heads appear worn out.

Devices are available from some test equipment manufacturers for checking the video heads and RF amplification stages. Some of this equipment is expensive, making its purchase impractical for some troubleshooters. This test equipment includes a special purpose inductance meter that checks the electronic condition of the video head core, an RF signal substitution and RF injection signal generator that allows the technician to bypass the video head and inject RF signals to test the RF amplifiers, and a video head and VHS Hi-Fi head tip protrusion gauge that measures the distance of head tip projection from the head disk.

Many technicians rely solely on the oscilloscope presentation of the playback RF waveform for analyzing the 8mm and VHS video head and VHS Hi-Fi head condition. The oscilloscope display of the playback RF waveform is very useful for diagnosing worn video heads and VHS Hi-Fi heads. But when an RF signal from one or both of the video heads or VHS Hi-Fi heads disappears because of either a defective head or an electronic amplification problem, the oscilloscope display is difficult to interpret. This is because the oscilloscope input needs an amplified form of the weak RF signal from the playback head before it can display an acceptable RF waveform. The oscilloscope's test probe must therefore be attached to the output of one of the RF amplifiers. Thus, when either the video head, VHS Hi-Fi head, or the RF amplification electronics are defective, the oscilloscope will display a noisy pattern that is difficult to diagnose.

The oscilloscope and other available test instruments provide different approaches for analyzing worn heads and defects in the video head, the VHS Hi-Fi head, and associated RF amplification stages of a VCR.

The following are suggestions for troubleshooting defective VHS video heads, VHS Hi-Fi heads, rotary transformers, connections between the heads, the RF amplifier circuits, and the RF amplifiers using bench-proven methods of diagnosis and a minimum amount of test equipment. (Although a VHS machine is used in this example, most of these principles also apply to troubleshooting corresponding symptoms in 8mm machines. The major difference in troubleshooting this portion of an 8mm VCR is found in the AFM RF processing section. Since 8mm uses the same pair of heads to write and read video and AFM sound, the playback RF processing circuits are simplified and only one head switching pulse is required.)

Assume that you have an oscilloscope display showing a video RF waveform that is missing one head signal. Figure 7.17 (Chapter 7) shows the oscilloscope display.

The first step in troubleshooting any video head, VHS Hi-Fi head, or video RF amplification problem involves cleaning the heads and the upper and lower drum assemblies. You must eliminate the possibility of a contaminated video head or VHS Hi-Fi head surface from your list of potential causes of the defect. Chapter 3 explains how to clean heads. Remember that if you are servicing an 8mm machine, you must consider the possibility of a brown stain clog on the video head tip. This type of head tip contamination will also cause a lack of RF playback from a video head. Chapters 3 and 5 address this issue and how to solve it.

Once you are sure that the head tip surface is clean, begin your troubleshooting procedure. Your tools and test equipment should include a dual trace oscilloscope, a prerecorded video tape with a known good recording on it, a volt/ohmmeter, a television/monitor, some miscellaneous hand tools (screw drivers, etc.),

the service manual for the model being repaired, and this book.

Anytime you troubleshoot the video head or VHS Hi-Fi head and associated RF amplification circuitry, the presence of the video head or VHS Hi-Fi head RF switching pulse and the correct voltage supply at the RF amplification and the head switching circuitry should be two of the first things confirmed.

The RF section's supply voltages should be checked with a voltmeter at measurement points near the RF circuitry. Compare the measurements you make with the stated values in the service manual. If the supply voltages are missing, find out why. If these voltages are missing or wrong, you have discovered the likely cause of the snowy video. You must troubleshoot the voltage problem.

The presence of the video head switching pulse is confirmed by placing your oscilloscope probe at the head switching pulse input to the video RF head switching circuit. You will need to locate the point of the video head or VHS Hi-Fi head RF switching pulse input to the appropriate head switching circuit in the service manual schematic and place the test probe at this location to check for the presence of this pulse. Comparing Fig. 7.18 in Chapter 7 with the service schematic for the model you are servicing can help you identify the video head switching pulse input to the video head switching circuitry in the preamplification stage.

It's important to check the switching pulse at the input to the head switching circuitry and not somewhere else on the circuit board. This pulse could be present at numerous places on the circuit board but not at the switching circuit input because of a bad connection on the board. Always check for the presence of any signal at its termination point first rather than at some other convenient location.

If the video RF signal pulse is missing, the TV will show a snowy playback picture. Also, the output of the RF amplifier (after the video head switching circuit) will have either no RF output or only one head signal output if this pulse is absent. (See Fig. 7.17 in Chapter 7 for oscilloscope display of just one head output signal from the video head switching circuitry.) Figure 7.16 (in Chapter 7) shows the relationship between the video head switching pulse and each head's video head RF signal. Typically, a missing head switching pulse will cause the RF switching circuitry to permanently latch to only one of the video head input RF channels causing the oscilloscope display shown in Figure 7.17.

For a VHS VCR, a missing Hi-Fi head switching pulse will cause the unit to revert automatically to the longitudinal track audio because of the missing Hi-Fi RF. The playback picture will be fine. Chapter 9 described why VHS Hi-Fi playback audio programs automatically switch when this occurs. For an 8mm machine, a missing head switching pulse will force the AFM sound to mute (because there is no normal audio track to which these units can revert). The playback picture may also be snowy or muted. If the head switching pulse is present in the 8mm video playback processing circuit, but missing in the AFM circuit, the playback picture will be fine: only the playback sound will be muted.

If the video or Hi-Fi head switching pulse is missing, troubleshoot the servo and video head PG circuitry. This is most likely the cause of the missing RF signal and the snowy picture (or poor Hi-Fi operation).

Assuming that head cleaning did not restore the missing head signal and you have the correct supply voltages and head switching pulses present at the respective preamplifier circuit, next consider the possibility of defective RF amplification in one of the RF channels, a defective connection between the rotary transformer and the preamplifier, a defective rotary transformer, or a defective video head.

Here is an effective way to check the lower drum and both the rotating and stationary halves of the rotary transformer in that assembly and the preamplification circuit without using a lot of expensive test equipment. This description will be limited to the video heads and related RF preamplification circuitry, but this method could certainly apply to trouble-

shooting similar defects in the VHS Hi-Fi RF sections. Video head continuity checks are also explained in these procedures.

First we'll explain how to check the lower drum and the stationary half of the rotary transformer.

1. Turn off the power and remove the connector plug attaching the video heads to the preamplification circuitry in the VCR. This connection is normally part of a multiwired harness extending between the lower video drum assembly and the main circuit board in the VCR. Be sure to disconnect only the cable (or cables) supplying the video head RF signal. The motor drive and FG and PG signals will still need to be attached for troubleshooting steps that follow. The schematic of the VCR will help you identify which connector to remove.

2. Locate the wires going to the stationary section of the rotary transformer (this is located in the lower video drum assembly) on the connector you just unplugged. There are normally three wires going to each channel's section of the stationary half of the rotary transformer: the hot or signal carrying lead, the ground lead, and the cable shield.

3. Use an ohmmeter to confirm continuity between the rotary transformer windings and the hot and ground leads on the connector. This is done by placing one probe from the meter to the hot wire of one video head channel of the connector you just removed and the other lead to the ground lead of the same channel. Be careful not to damage the delicate connectors on the wire. Also be sure that you are not attaching the meter probe to the shield wire. This measurement should read something less than 300 Ω and greater than 5 Ω. A reading of less than 5 Ω indicates a short in the wire or stationary half of the

rotary transformer. A reading of something well above 300 Ω indicates an open connection in the same area. You have just verified connection between the connector and the stationary section of the rotary transformer in the lower drum.

4. Next, keep one ohmmeter test probe on one of the wires just checked in Step 3. This can be either the hot or ground lead—it doesn't matter which one. Then place the other meter probe on the shield wire for the head channel being checked. Your meter should read some very high value or infinity. A low reading indicates a short somewhere between the connector and the stationary half of the rotary transformer in the lower drum unit.

5. Repeat Steps 3 and 4 for the other video head RF channel.

NOTES FOR STEPS 3–5:

a. Any ohmmeter reading for either RF channel other than those stated in Steps 3–5 should be investigated and considered a possible cause for the missing RF signal during playback.

b. If you have one good RF channel (as in our example), the continuity readings between the good and bad channels should be compared to aid in locating the fault. If the good and bad channel readings measure the same, the problem is probably not in the section you are checking.

c. If Steps 3–5 reveal a problem for playback, the defect will also cause one RF signal to be missing during record. This happens because the connecting wire harness and stationary rotary transformer section are common for both record and playback functions.

d. If you feel the lower section of the drum assembly or any of its attached leads and connectors are defective, you

must replace the entire lower drum unit. The lower drum assembly is not considered a serviceable item and must be replaced as a complete assembly.

e. It is only necessary to precede to the next steps if Steps 3–5 check out normal for both video head channels.

Assuming that the checks in Steps 3–5 for the RF channels from both heads fail to isolate the problem, use the next steps to confirm continuity of the video head core windings and the rotating half of the rotary transformer.

6. Unsolder the video head leads on the top of the video head assembly. It is not necessary to remove the head—just unsolder the leads. If the head drum you are checking does not have solder leads but is a plug in type, you must remove the upper drum for this next test.

CAUTION

1. Reinstallation of the video head drum after it was removed requires realignment using an alignment tape. Do not remove the video head unless you are prepared and properly equipped to perform the necessary alignments.

2. Be careful not to damage the video heads or the connecting wires and plugs on the upper drum assembly.

7. Use the ohmmeter to confirm continuity of each video head core winding. Typical continuity readings should be between 5 Ω and 300 Ω. A broken head core or an open winding will measure a very high reading or infinity on the meter scale. A shorted winding will measure as a short.

8. Replace the video head cylinder if the measurements are not within the range given in Step 7. If the head continuity checks O.K. in Step 7, continue with the following steps.

9. Use the ohmmeter to measure the continuity between the separate video head connection points and the rotating section of the rotating transformer. Be sure

that the video head leads are removed for this measurement (see Step 6). Otherwise the resistance of the attached head core windings will prevent you from gathering accurate resistance data from this half of the transformer. Normal readings here should be similar to those obtained for the stationary section of the rotary transformer in Step 3 (somewhere between 5 Ω and 300 Ω). As in previous steps, shorts and open connections are indicated by readings outside of this continuity window.

NOTES FOR STEPS 6–9:

a. Any ohmmeter reading for either RF channel other than those stated in Steps 6–9 should be investigated and considered as pointing to a possible cause of the missing RF signal during playback.

b. If you have one good RF channel (as in our example), the continuity readings between the good and bad channels should be compared to aid in locating the fault. If the good and bad channel readings measure the same, the problem is probably not in the section you are checking.

c. If Steps 6–9 reveal a problem for playback, one RF head signal will also be missing during record. This occurs because the connecting wire harness and stationary rotary transformer section are common for both record and playback functions.

d. If you feel that the rotating transformer section of the drum assembly or any of its attached leads and connectors are defective, you must replace the entire lower drum unit. This item is not considered a serviceable item and must be replaced as a complete assembly.

e. If the video head core continuity tests reveal an open or shorted video head core winding, the entire upper video cylinder must be replaced. The video head cylinder is not considered a serviceable item and must be replaced as a complete assembly.

f. It is only necessary to proceed to the next steps if Steps 3–9 check out normal for both video head channels.

g. If you had to remove the plug-in type of video head for checking Steps 6–9 you must reinstall the head at this time and perform the video head switching point alignment as described in your service literature before continuing with the troubleshooting steps that follow.

Assuming that the tests described in Steps 3–9 for the RF channels from both heads check O.K., proceed to the next section of this troubleshooting procedure to confirm proper operation of the video RF preamplification circuitry.

Generally both channels of the preamplification circuitry do not fail at the same time. One channel normally remains functional when a failure occurs in this area. Remember that you are troubleshooting for one missing RF signal; so use voltage, waveform, and continuity comparison measurements between the good and bad channels to help you isolate preamp problems.

10. Be sure the lower drum video head connector is removed from the main circuit board (see Step 3).

11. Be sure the drum motor and FG and PG sensor wires are still attached to the lower drum (see Step 3).

12. Place the VCR in the service position (this will be similar to the position shown in Fig. 3.13 in Chapter 3—see the service manual for your unit for the best position for your unit) and apply power.

13. Insert a video tape with a known good recording on it and engage the play mode. (*Note:* The snow that you now observe on the test monitor screen may appear worse than the original symptom

or the video may now be muted out. Don't worry about this change in symptoms—it happened because you disconnected all RF that used to be present coming out of the preamplifier—even the good RF channel.)

14. Hook one probe of the oscilloscope to the first video RF test point following the video RF switching circuit. Use the VCR schematic to locate this test point.

15. Hook the second oscilloscope channel probe to the video head RF switching pulse test point in the VCR.

16. Put the oscilloscope in the dual trace mode and trigger the oscilloscope from the head switching pulse.

17. Set the scope's horizontal time so that you can see both the negative and positive durations of the head switching pulse.

18. You should see a head switching pulse similar to the one shown in Fig. 7.17 on one trace and a relatively flat line (with some noise) on the scope's other trace.

19. Use the schematic to locate the hot video head socket pin going to the video preamps for both head channels. (These are the socket pins located on the circuit board, not the pins on the unplugged connector harness going to the lower drum you checked earlier.)

20. Now hold the metal shaft of a small screwdriver or some other conductive metal object in one hand and touch one of the preamp RF hot connector pins located in step 19 with the metal shaft. Be sure that your fingers (or hand) are touching the metal shank as it also touches the hot pin of the connector. The important part of this step is to inject noise from your body (which is acting as an antenna) into the high gain preamplifier for amplification.

21. Adjust the amplitude of oscilloscope trace showing the video head RF switching output waveform so that the injected noise amplitude is viewable on the oscilloscope along with the video head switching pulse displayed on the second channel.

22. Note the amplitude of the injected noise and its relationship to the positive and negative plateaus of the video head switching pulse.

23. Next move the metal shank to the other RF channel's hot pin and observe the amplitude of that amplified noise.

NOTES ON STEPS 10–23:

a. These steps have you check and compare (relative to each other) the amplification factor of each preamplifier circuit. If the lack of playback video RF is caused by a failure in one of the two preamp circuit paths, you will notice a large difference between the amplified noise on each of the two channels. If you notice a big difference, troubleshoot the preamp circuit with the lowest amplitude.

b. Remember that a missing video head switching pulse will prevent one preamplifier from outputting the amplified noise for one of the channels.

c. A defect in the video head preamplifier circuitry will not affect record. If tapes recorded on this unit playback O.K. in another VCR, then the video heads, lower drum, rotary transformer, and lower drum connection cables are probably functional.

d. If both preamp channel noise levels are similar and show lots of gain, the two preamp circuits are functioning.

e. Assuming that you performed the checks in Steps 1–23 and all checks come out good, consider replacing the video head as a final check.

The noise injection method can also be used to check the preamplifiers while the lower drum is still connected to the circuit board. To understand this procedure, assume that there is one good RF signal from one video head channel and one missing RF signal on the other channel. The scope is connected as described in Steps 14–18 and the noise is still injected using the shank of a metal tool while the unit is in playback. Since the head connector from the lower drum is attached for this method, the noise must be injected by touching the metal shank from the hand tool to the hot video connector pin where it is soldered to the main circuit board. The video head switched preamplifier RF outputs are monitored on the scope as previously described.

Since the heads are reconnected for this test method, some interpretation of the switched RF output is required. If the amplified levels of the injected noise coming from both the good and bad channels are the same on the oscilloscope, the preamp section is working and the head switching pulse is toggling the preamplified channels as it should.

If the channel originally missing the video head RF output waveform has a substantially lower amplified injected noise level than the functioning channel, a shorted lower drum wiring, shorted rotary transformer winding, shorted video head core winding, or a defective preamplifier is the problem. Shorted wiring or windings are indicated by the reduced noise level because of the *reflected impedance effect* on the injected noise thus reducing the defective RF channels preamplified output. Reflected impedance is a form of *inductive loading* that can be reflected through a series of inductive devices connected together in a circuit. Inductive loading is like the loading effects of a resistor in a circuit.

You can differentiate between shorted windings and wiring and a defective preamplifier by performing Steps 10–23. If the injected noise level of the defective channel's preamplified output suddenly increases to match that of the good channel after the lower drum

connector is unplugged, the preamp is functioning. At that point you should check the lower drum and video head for shorts by performing Steps 1–9.

Now assume that you are checking the defective channel with the lower drum connectors attached and the injected noise level comparison between the good and bad channels shows a significantly larger noise output for the defective channel than for the good one. In fact the amplified injected noise level forces the amplifier into saturation or voltage *clipping*. Clipping is a point where the amplifier's output level comes near to or equals the supply voltage level. This is the upper limit of voltage amplification for the input signal. Since the injected signal is greater in amplitude than the signal normally coming off the video playback head, the amplifier goes into saturation and clips the injected noise amplification level.

When the bad channel shows a significantly higher amplification level (or a clipped voltage level) than the good channel, an open connection is indicated somewhere between the circuit board connector and the video head. The open connection likely exists in the connecting cable running between the main board socket and the lower drum, but may also be present in the rotary transformer, the video head drum, or any wiring between these points. In this case, the defective channel produces an increased preamplified noise output because the open connection has removed the inductive loading naturally present in the circuit when the proper connection is made. Thus, the injected noise for the bad channel has an increased injected noise input level to its preamplifiers compared to the good channel which still has the inductive load.

To troubleshoot the open connection location, you should perform Steps 1–9 as previously described.

Whether the video RF problem is caused by a missing RF waveform on a single channel, a weak RF level on a single channel (like the one shown in Fig. 7.15) or RF envelopes missing from both heads, Steps 1–23 can be used to verify the cause of the defect.

Remember to clean the video heads first and also to confirm the presence of the video head switching pulse at the head switching circuitry before beginning your troubleshooting.

REMOVING SOLDER

One way to remove residual solder is to use a *solder sucker*—a hand-held vacuum pump with a spring-driven plunger to pull the hot, melted solder off a connector (Fig. 11.20). Heat the old solder until it melts, place the spring-propelled vacuum pump in the hot solder, quickly remove the soldering iron while you release the vacuum pump's spring, and suck the solder up into a storage chamber in the pump.

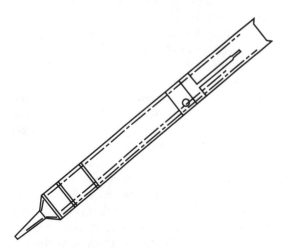

Fig. 11.20. *The spring-driven plunger in the solder vacuum pump is used to pull hot solder off a connection.*

This technique works fine until you try to use it around CMOS chips. Some vacuum pumps produce static electricity, and by now you know what that can do to an MOS or CMOS chip.

A safer way to remove solder is to touch the solder with the end of a strip of braided copper. Then heat the braid just a short distance from the solder (Fig. 11.21). The copper

Fig. 11.21. *Use of a solder wick is another way to remove solder from a connection.*

braid heats quickly, transferring the heat to the solder, which melts and is drawn into the braid by capillary action. Then, cut off the solder-soaked part of the braid and throw it away.

Removing components from double-sided circuit boards is difficult without the use of a special desoldering iron like the one shown in Fig. 11.22. This iron has a hollow point and uses a motorized vacuum pump to heat up and suck the solder into a reservoir. This desoldering tool makes components that have been soldered on both sides of the board much easier to remove. It's also safe for static sensitive parts.

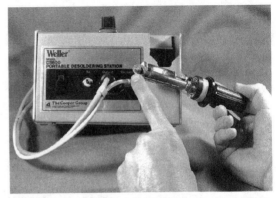

Fig. 11.22. *A hollow-point soldering iron with a motorized vacuum pump to heat up and suck solder into a reservoir.*

If any solder remains in the circuit-board hole, heat the solder and push a toothpick into the hole as the solder cools. The toothpick will keep the hole open so you can easily insert another wire lead for resoldering.

Be careful not to overheat the board during the solder-removal process. Excessive heat can cause part of the circuitry to come away from the board. It can also damage good components nearby.

If you remove the solder from a component and a lead is still stuck on some residual solder, pinch the lead with a pair of needle-nose pliers as you gently wiggle it to break it loose from the solder bond.

The pins of most chips on VCR printed circuit boards are bonded to the circuit board by a process called *wave soldering*. Wave soldering produces an exceptionally good bond without the added manufacturing expense of a socket. This process helps keep the fabrication costs down, but it makes it more difficult for you to replace the chip.

One effective way to remove wave-soldered chips is to cut the chip leads or pins on the component side and remove the bad chip. Then remove the pieces of pin sticking through the board using a soldering iron and solder braid or a vacuum pump.

Some special tools are available to help you remove soldered components. Figure 11.23 shows a desoldering tip that fits over all the leads of a chip or dual-in line package (DIP) device.

Fig. 11.24. *A spring-loaded dual-in-line extractor tool.*

Figure 11.24 is a photograph of a spring-loaded DIP extractor tool. By attaching this device to the surface mounted chip on the printed circuit board and then applying the DIP tip shown in Fig. 11.23 to the soldered connections on the opposite side of the board, you can easily remove the chip. When you press the load button downward and engage the clips, you will cause the extractor to place an upward spring pressure on the chip. When the solder on the reverse side melts enough, the chip will pop up and off the board.

Figure 11.25 is a photograph of a flat pack IC. This low-profile device is mounted di-

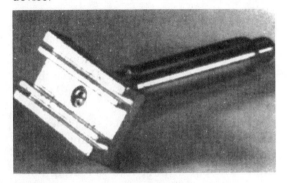

Fig. 11.23. *A desoldering tip for removing chips that are soldered to the circuit board.*

Fig. 11.25. *A flat pack integrated circuit.*

rectly onto the circuit board and removal requires a special desoldering tip to fit over the IC and heat all leads at the same time. With different size ICs, a variety of desoldering tips may be required.

Soldering Tips

No pun intended. Hand soldering is the most misunderstood and most often abused function in electronics repair. Not only do many people use poor soldering techniques, but they also use the wrong soldering irons.

Solder isn't simply an adhesive making two metals stick together. It actually melts and combines with the metals to form a consistent electrical as well as mechanical connection. Time and temperature are critical in this process. The typical hand solder job can be accomplished in 1.5 seconds or less if the soldering iron and tip are properly selected and then properly maintained.

The nominal solder melting temperature is 361°F. Metal combination between the solder and the metals being joined occurs at temperatures between 500°F and 6000°F.

Most soldering jobs join the metals copper and tin, but both of these metals are easily oxidized. Poor or no solder connections are made if the surfaces to be connected are covered by contaminants such as oils, dirt, or even smog, so be sure to use solder with a good cleaning flux. The flux prepares the surfaces for best solder metalization. The flux melts first and flows over the metal surfaces removing oxidation and other contaminants. Then the metal heats so that the solder melts and flows, producing a good, shallow bond.

The key to successful soldering is in the soldering iron tip. Most people selecting their first soldering iron buy a low wattage iron, but this is a mistake. Instead, pick an iron whose tip operating temperature is suited for the circuit board you're going to repair. If the tip temperature is too low, the tip sticks to the surface being soldered. If it's too high, it damages the board surface. The ideal working temperature for soldering on a VCR circuit board is between 600°F and 700°F.

The soldering iron tip is used to transfer the heat generated in the iron out to the soldering surface. The iron should heat the tip quickly, and the tip should be as large as possible yet slightly smaller than any soldering pad on the board.

Tips are made of copper. Copper conducts heat quickly, but it dissolves in contact with tin. Solder is made of tin and lead. To keep the tin from destroying the copper tip, manufacturers plate a thin layer of iron over the soldering tip. The hot iron (now you know where the term *iron* came from) still melts the solder, but now the tip lasts longer. The iron melts above 820°F, so if the heat produced by the iron stays below 700°F, the solder melts but not the iron plating.

The disadvantages of the iron plating are that it doesn't conduct heat as well as copper, and it oxidizes rapidly. To counteract this, you can melt a thin coat of solder over the tip. This is called *tinning*. This solder layer helps the soldering iron heat quickly and also prevents oxidation.

The tip of an old soldering iron is usually black or dirty-brown with oxidation. And it doesn't conduct heat very well. These "burned out" tips can be cleaned with fine emery cloth, retinned, and then used.

Wiping the hot tip with a wet sponge just before returning the iron to its holder is a mistake. This removes the protective coating, exposing the tip surface to atmospheric oxidation. It's much better to add some fresh solder to the tip instead. Keep your iron well tinned.

Figure 11.26 shows the proper way to solder a plug or connector lead. Place the tip of the iron on one side of the lead and the solder on the other side.

As the solder pad heats, the tin-lead solder melts and flows evenly over the wire and the pad. Keep the solder shallow and relatively even. When you think your soldering job is complete, carefully inspect your work. Sometimes, if you aren't careful, you can put too much solder on the joint, so that there's not

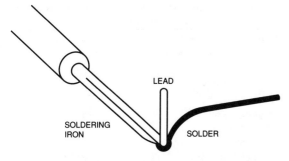

Fig. 11.26. Place the soldering iron on the opposite side of the lead from the solder.

enough solder on the top or bottom of the connection. It's also possible to get internal voids or hollow places inside the solder joint. Large solder balls or mounds invite *cold solder joints* where only partial contact is made. Figure 11.27 shows some examples of inadequate soldering. These kinds of solder joints can be a source for intermittent failure.

Good soldering takes patience, knowledge, and the right tool—a grounded, temperature-controlled soldering iron with a tip temperature maintained in the 500–600°F range for optimum soldering.

Before You Solder It In

A useful thing to do before you solder in a replacement part is to test the device in the circuit. Simply insert the chip or other device

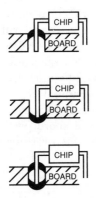

Fig. 11.27. Examples of inadequate soldering on a printed circuit board.

into the solder holes and wedge each lead in its hole with a toothpick. Then energize the circuit and test. After proper function is assured, remove the toothpicks and solder the component into the board.

CIRCUIT BOARD REPAIR

Repairing damaged circuit boards is a lucrative business, and several companies have developed around this activity. For some board failures, you can repair your own circuitry and save some money.

Before soldering in new components, check over the board for any broken traces or pads lifting off the board. If a trace is open and is starting to lift away from the board surface, jumper across the broken spot from one component solder pad to another pad. Use solid #18 or #20 wire tinned at both ends before soldering.

If a pad or trace lifts free, replace it with an adhesive-backed pad or trace overlapping the damaged area. Scrape the coating off the pad on both ends of the trace so the new pad or trace can be soldered firmly to the existing pad or trace. Remove all excess solder and redrill any lead hole that has become covered or plugged with residual solder.

RECOMMENDED TROUBLESHOOTING AND REPAIR EQUIPMENT

If you're planning to tackle failures that usually require service support, you can minimize your investment costs and yet optimize your chance of success by carefully selecting your equipment and tools.

First, get a set of good metric screwdrivers—both Phillips and flat head. VCR screws are metric and require special metric screwdrivers. Get a wide selection of sizes, from the tiny *tweakers* to an 8-inch flat head. You might also find a set of jeweler's screwdrivers quite helpful. You will need standard metric nut drivers and metric hex-nut drivers of assorted sizes.

Then get several sizes of long-nose or needle-nose pliers. Get several sizes of diagonal cutters, or *dykes*, for cutting wire and pins. A good, low-wattage, well-grounded soldering iron whose tip temperature is automatically controlled is a must if you intend to replace components. A simple $3\frac{1}{2}$-digit DVM or DMM is useful for test measurements.

If you expect to do any serious troubleshooting, get a 30–50 MHz oscilloscope with dual trace, delayed sweep, and a time-base range of 200 nanoseconds to a half second. Select a scope with a vertical sensitivity of 10 millivolts per division, or better.

Figure 11.28 is a photograph of typical hand tools used in the maintenance and repair of VHS machines.

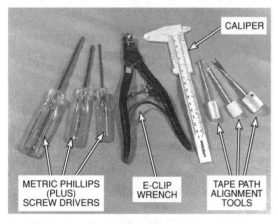

Fig. 11.28. *Standard VHS hand tools.*

Table 11.2 shows an approximate price list for troubleshooting and repair equipment. There are other jigs and adapters that will be required for various models and formats that you service. This list is only a partial guideline to the equipment costs.

Table 11.2. Approximate Prices for Troubleshooting and Repair Equipment

Metric screwdrivers (various sizes and types)	$5 each

Pliers
    $4\frac{1}{2}''$ short nose
    $5\frac{1}{4}''$ long nose
	$15 each

Diagonal cutters
    $4\frac{1}{2}''$ flush
    $4\frac{1}{2}''$ midget
	$15 each

DMM ($3\frac{1}{2}''$ digit)	$150
Tension meter	$450
Torque gauge	$400
Torque tape	$375
Tension tape	$300
Desoldering station	$750
Soldering iron	$125
Oscilloscope	$1800
NTSC generator	$1000
Sweep function oscillator (1 Hz–5 MHz)	$1000
AC voltmeter	$400
Alignment tape	$150 each
Greases and lubricants	$100 for the lot
Service manual	$35 each

SPARE PARTS

Because of the cost involved you will probably want to maintain a minimal stock of repair parts; yet you want to be able to fix a machine quickly when it breaks down.

The most practical backup would limit your spare parts to belts, rubber tires, and torque limiter assemblies. VCRs are much more likely to experience a mechanical failure before an electronic malfunction occurs. These spares represent an investment of under $50 for each model that you are stocking parts for. Several manufacturers sell spare parts packages with belts and tires for their various VCR models.

SUMMARY

There are three possible ways to optimize your VCR's operational life:

1. Buy a highly reliable unit with a good track record of performance.

2. Do preventive maintenance regularly.

3. Become a knowledgeable repair technician yourself.

Armed with the knowledge in this manual, you'll be able to spot downright poor troubleshooting like the "tech" rubbing up and down on the video head using low-grade alcohol to clean it, or the repair person wiping his or her soldering iron on a wet sponge just before putting it in its holder. These are mistakes of poorly trained (or poorly motivated) people working on someone else's machine. You'll also be able to recognize and model your skills after the sharp, highly trained technician who uses the right tools and the right procedures to troubleshoot and repair in minimum time. Then you'll smile to yourself, knowing that you were smart enough to buy this book and do your own repairs the right way.

CONCLUSION

So there you have it! You are now armed with good information and prepared to service and maintain a VCR. In this book you learned about preventive maintenance and how to keep a VCR running at peak performance. You also learned that, when it's time to fix a problem, you have the information available to do the job. And you know the tools and test equipment needed to maintain and service today's VCR. You also learned many of the tricks used by experienced service technicians.

At the back of this book are block diagrams of typical VHS and 8mm electronic signal processing paths for both record and playback. These diagrams will help tie the detailed drawings found in individual chapters together into a complete operating VCR system. These diagrams are another reference tool for quickly isolating and troubleshooting VCR failures.

You've made a good investment in buying and then reading this book. May you experience great success in your VCR servicing efforts.

CHAPTER REVIEW QUESTIONS

1. Why is a pulse cross monitor helpful in troubleshooting VCR problems?

2. Why can't a broadcast video program be used for alignment of video circuits?

3. Connecting the demodulated video output from a television monitor/receiver to the line input of a defective VCR and the VCR's video line output to the television monitor/receiver's line input allows you to bypass what video circuits when troubleshooting the unit?

4. RFI can cause what type of servo-related malfunctions in the VCR during record?

5. What is the purpose of a sweep marker?

6. Why must an isolated AC supply be used when servicing some VCR power supplies?

7. When calibrating VCR electronic circuits, why is it helpful to first align the playback circuits with the factory alignment tape before adjusting the record circuits?

8. What is an LED?

9. What is a typical voltage reading for a good silicon diode that is forward biased?

10. Testing a hybrid transistor with an ohmmeter does not always produce accurate test results. Why?

11. Why must the video head leads be unsoldered from the head disk when checking video head windings for continuity?

Glossary

a Type—Alpha wrap.

ACC—Automatic color control. Used to maintain constant color signal levels.

ACK—Automatic color killer.

A-D—Analog-to-digital. The circuit used to sample an analog signal and generate a string of digital words from the samples taken. (*Also see* Quantization.)

Adaptive closed loop interpolation—A processes of head switching noise reduction used in Hi-Fi audio circuits.

Address Track—A longitudinal track running through the video tracks (usually the vertical blanking interval) which is used for recording cue or time code information.

Adjacent Track—The video track to the immediate right or left of a selected track.

AFC—Automatic frequency control. Used to phase-lock the color circuitry to the recording or playback color signal.

AFM—Audio frequency modulation. A term used to describe Hi-Fi audio recording.

AGC—Automatic gain control. Used to maintain a fixed output signal level in luminance and audio circuits.

Alignment Tape—A prerecorded tape that serves as a reference for electrical and mechanical performance evaluation of a VCR.

Alpha Wrap—The name given to the complete overlapping turn of tape around the video drum as it forms the Greek letter, alpha. This type of wrap allows a single head to be used for recording.

Amplifier—Circuitry which boosts or increases the strength of an input signal.

AND Gate—A logic gate in which all inputs must be high (logic 1) to produce a logic high output.

Annealing—The process of preparing the tape's plastic backing for the magnetic coating by subjecting it to a process of stretching, heating, and cooling.

APC—Automatic phase control. Used to phase-lock color circuits to the recording or playback signal to achieve color signal stability.

APL—Automatic program location. The use of electronic markings on video tapes to enable the locating of the beginning or end of a recording.

APL—Average picture level. A measure of the average luminance (brightness) level of a screen image.

Assemble Edit—Constructing a video recording by adding scenes in sequence.

ATF—Auto Track Find. A signal used in 8mm machines for tracking control.

Audio Dubbing—Adding a new sound track after the original recording has been made. The sound track is erased and rerecorded. The video and sync tracks are unaffected.

Audio Track—The longitudinal portion of the tape on which audio information is recorded using a stationary head.

Automatic Rewind—Electronically rewinding the cassette tape to the beginning.

Azimuth—The left or right tilt of the gap of a recording head that enables extended playing and recording. The angle measured counterclockwise from a perpendicular formed by the head gap and the direction of the track across the tape.

Azimuth Loss—The loss in picture quality caused by the oblique angle relationship of the recording and playback heads.

Balanced Modulator—A circuit that produces an output comprised of the sum or difference between two input signals. The output signal will contain any special characteristics in one of the input signals. No output is produced when the chrominance signals are zero.

Bandwidth—A range of frequencies over which an output remains uniform; measured between the 3 dB power points (point where power drops to half) or 6 dB voltage points.

Bearding—Another name for demodulation noise.

Betamax—A video recording format developed by Sony.

B-H Magnetization Curve—A graph describing the magnetic effect on flux density (B) of an applied magnetic field of intensity (H).

Bias—A high frequency AC signal added to the record longitudinal audio track sound just before it is sent to the recording head. Bias is used to force the longitudinal audio track record head to operate in the linear region of the BH curve.

Binder—The solution used to apply and glue the magnetic particles to the tape backing.

Bi-phase Bit Modulation—The recording scheme used in PCM audio that allows the playback circuit to recognize digital 1's and 0's recorded on the tape.

Blanking—The retrace time when the TV scanning beam shifts from the lower right to the upper left of the screen while the electron beam is deenergized.

Blanking Gap—A vertical interval between two separate video frame signals.

Braking, Magnetic—A servo control using EMF to slow down or speed up a drive shaft.

Brightness—The total energy in the Y luminance signal.

Brown Stain—A form of head tip contamination that occurs when metal video head tips are used to record and playback on a metal tape.

Burst—An 8 to 10 cycle occurrence of the chrominance 3.58 MHz subcarrier signal that appears after the horizontal sync and centers on the blanking portion of the video waveform. Used as the reference for phase comparison during reception and to keep the TV color oscillator locked on a broadcast station.

C—Symbol for chrominance.

Capstan—A small magnetic-tape drive shaft which drives the recording tape to assure positive tape movement.

Capstan Servo—An electromechanical system that maintains capstan motor rotational speed in sync with a comparison signal generated within the electronics of the servo system. A self-correcting speed control system.

Carrier—A waveform containing signal information.

Carrier Leak—A beat effect seen on a display screen as a series of wavy lines. This is a phenomenon that occurs whenever a plain carrier signal appears during playback of a video signal recorded as FM.

Carrier Ripple—An unwanted beat created by the FM modulation/demodulation process when the operating frequency of the FM oscillator and the upper frequencies of the input record video signal are close together. These two high frequencies beat together in the VCR and produce a difference frequency known as carrier ripple.

Channel—A specified range of frequencies used to transmit and receive TV signals.

Chroma—The color part of a video signal. Describes the degree of saturation of color. Dark red is saturated; it becomes less saturated as white light is added to produce light red. On a standard TV, chroma is adjusted with the "COLOR" control knob.

Chroma Bandwidth—The frequency difference between the upper and lower frequency limits of the chroma signal.

Chroma Delay—A time delay produced by restricting the chroma band of frequencies through a low-pass filter.

Chrominance—Another term for chroma.

Chrominance Signal—The composite signal produced by combining hue and chroma. The chrominance subcarrier signal conveys hue and saturation information. (*See* Luminance.)

Clamp—Holding an AC signal at a specified DC level.

Clogging—Contamination on the head's contact surface that prevents detection of the recorded signal during playback. Head clogging may also prevent writing of the signal onto the tape during record.

Closed Loop Interpolation—A processes of head switching noise reduction used in Hi-Fi audio circuits.

Coercive Force—The amount of opposing magnetic intensity that must be applied to a video tape to remove (erase) residual magnetism.

Coercivity—A tape property related to the amount of coercive force required to demagnetize a magnetically saturated portion of tape material.

Color Bar—A color test signal that contains bars of color and black and white video. This set of bars is used to measure and adjust luminance and chrominance performance. From left to right, the bars are white, yellow, cyan, green, magenta, red and blue. Standard color bar test patterns produce 75 percent saturated color bars. The white portion of the pattern may be 77 percent or 100 percent of the peak luminance level depending upon the test pattern used.

Color-Difference Signals—The resulting signals after subtracting the composite luminance Y signal from the primary color signals of red, green, blue, red-Y, blue-Y, and green-Y.

Color Flicker—A 30 Hz change in color saturation caused by video preamp imbalance between two video heads.

Color Signal—A composite signal comprised of luminance, the chrominance subcarriers, and a color sync signal.

Color Sync Signal—Also called Burst and Color Burst. Consists of 8 to 10 cycles of chrominance 3.579545 MHz subcarrier that are added to the end of the TV horizontal sync signal. This signal becomes the reference for phase comparison during reception.

Color Temperature—A way to express color in terms of absolute temperature. A black body heated to high temperature emits light. The relationship between color of light and temperature is constant. Bluish colors have a high color temperature; reddish colors have a low color temperature.

Color-Under—A color-recording technique in which chrominance information is converted (heterodyned) from 3.58 MHz to 629 kHz (VHS) or 743 kHz (8mm).

Comb Filter—An electronic circuit with a frequency pass response resembling a comb. It consists of a 1 H delay line and a summation point that adds (or subtracts) interlaced video to remove the luminance (or chrominance) signal.

Companding—An acronym describing the compression and expansion of an audio signal during the record and playback process for noise reduction processing.

Compatibility—The ability of a tape recording made on one VCR to be played back on another and vice versa.

Composite Sync Signal—The combination of video, blanking, and composite sync signals.

Condensation—Moisture deposits on the head drum will cause the recorder to stop operating until the drum is dry. Operating a VCR when condensation is on the drum can cause the tape to stick to the drum and be ruined. Some recorders have an illuminated moisture indicator.

Control Signal—A special signal recorded on a video tape that is used during playback as a reference for the servo circuits. For VHS machines this pulse is recorded on the control track and marks the beginning of a recorded video track and is used during playback.

Control Track—A special video tape track on VHS machines used to store tape speed synchronization timing control signals.

Converted Subcarrier—The process of shifting the 3.58 MHz color subcarrier and its sidebands down to 629 kHz for VHS and SVHS, 743 kHz for 8mm and Hi 8mm, and 688 kHz for Betamax and SuperBeta.

Counter EMF—An electric field force that is opposite to the field that originated it.

CRCC—Cycle Redundancy Check Code. A form of digital word error detection and correction used in PCM audio.

Crosstalk—Unwanted signals detected by a video head from an adjacent track.

CTL—An abbreviation for control track signal.

CW—Continuous wave.

CW Interference—RF interference in a waveform that is generated by an external carrier signal.

D-A—Digital-to-analog. The circuit used to convert a string of digital words back into an analog signal.

Dark Clip—A circuit used to cut off large negative voltage spikes at an adjustable level to prevent damage to the FM modulation circuitry.

Data Block—The rows and columns of 8-bit digital words recorded onto the tape containing the PCM audio data, the parity words and the CRCC sums.

dB—Decibel. A unit used to express the ratio of voltage, power or sound pressures.

DC Error Voltage—A comparator output proportional in magnitude and polarity with the operational error in a feedback control system.

DDC—Direct drive cylinder. A video head cylinder driven by a self-contained brushless DC motor using no belts or gears. Use of DDC produces increased stability in pictures.

Decoder—A device that converts a coded signal back to its original uncoded condition.

De-emphasis—The process of attenuating previously amplified signals to produce the original signal with less noise.

Definition—A unit of measurement for the sharpness of an image. Denotes the degree of distinguishable picture details. Definition is divided into vertical and horizontal (*Also see* Resolution).

Delay Multivibrator—A monostable multivibrator that generates an output pulse at a predetermined time after being triggered by an input pulse.

Delta Factor—Used to indicate a change in frequency (or phase), such as "wow and flutter" in a playback signal when the frequency off the tape is not stable. The delta frequency is the maximum variation from the center frequency.

Demodulation—The conversion of high-frequency modulated TV signals to audio or video.

Demodulation Noise—Short duration black and white streaks that follow changes in video contrast in a television image.

Depth Multiplexing—The processing of recording the VHS Hi-Fi audio signal deep into the tape's magnetic layer and underneath the video signal that is also recorded on the same track.

Deviation—Used to describe the FM carrier frequency swings during modulation. The excursion of an FM carrier from the original represents the information signal.

Dihedral Error—A time-base error created by video heads that are not mounted 180 degrees apart on the head disk.

Direct Recording—A videotape recording technique in which the complete composite signal is processed and recorded.

Downward Compatible—An SVHS or Hi 8mm VCR that accepts the conventional (lower resolution) tapes for record and playback, as well as the improved tapes for the higher resolution recordings.

Dropout—A momentary absence of FM or color signal on a tape where no video signal was recorded. Dropout is caused by uneven tape oxide, dust on the tape or video heads, or mechanical damage. It distorts the TV picture and appears as tiny black and white dots or as noise streaks.

Dropout Compensator—A device that electronically fills in RF waveform holes produced by dust and tape imperfections which prevent accurate playback. The circuit uses the RF or demodulated video information from the previously stored line.

Drum Servo—The system that controls head rotation by comparing the speed and rotational position (phase) of the head to a reference signal. Any difference is detected as an error. A correction voltage is produced by the drum servo and applied to the drum motor to correct the error.

Duty Cycle—The percent of a waveform that is active during a single cycle of the period.

Dynamic Range—The ratio of the loudest signal that can be recorded on the tape before it reaches magnetic saturation to the lowest level signal that can be detected above the noise threshold.

ED Beta—A metal tape VCR format developed by Sony that offers improved resolution over the Betamax and SuperBeta format VCRs.

Eddy Current—A circular flow of electrons in a conductor exposed to a rapidly changing magnetic field. Eddy currents increase with frequency.

Editing—Adding, deleting, or rearranging sections of audio or video recorded on a tape.

E-E—Electric-to-electric, electronic-to-electronic. A condition during record where the playback output circuit is connected directly to the record input circuit so the video signal can be monitored on a TV set for proper tuning.

E to E—See E-E.

Effective Video Width—The amount of tape width occupied by video recording less the overlap.

EIA System—The TV broadcast standard established by the Electrical Industries Association that uses 525 horizontal scanning lines and 29.97 frames per second to produce a TV screen image.

Eject—Mechanically or electrically opening the cassette housing of a VCR so that a video tape cassette can be removed.

EMF—Electromotive force. The electric force produced by moving electric or magnetic fields.

Emphasis—A process for boosting the level of high-frequency components in a signal.

End Sensor—An automatic stop device using a metal foil tape leader and two contactless oscillator coils for Beta systems, or an LED (or incandescent lamp) and sensing device with clear leader tape for VHS and 8mm systems.

Entrance Guide—A tape guide mounted at the input side of the scanner to control tape angle and height.

EP—Extended play (same as SLP).

Equalizing Pulses—Six serial pulses at twice the horizontal line frequency that occur before and after the vertical sync signal to ensure proper interlaced scanning and to retain horizontal sync during blanking.

Error-Input—A servo feedback signal that corresponds to deviation from a reference standard.

Factor 80—A heuristic relationship between lines of resolution and bandwidth: Bandwidth × 80 = No. lines of resolution.

FG—Frequency generator. Circuit that generates a square or sine wave signal corresponding to the rotational speed of the video head drum, capstan, and reel tables.

Field—The amount of picture that can be described during a vertical scanning period of one sixtieth of a second. It takes 15,750 monochrome horizontal scanning lines (15,734 for color) and 60 fields to draw an image on a screen in a single second (U.S., Canada, and Japan). Two vertical scans will scan 525 lines of horizontal scan so that 30 pictures can be drawn each second. (525 lines = 2 vertical scans = 1 frame × 30 frames per second = 15,750 monochrome horizontal lines/second.)

Field (Flux)—An electric or magnetic force that exists between two unlike poles or around a single pole that can be represented by lines of force flowing out of one pole and into the other.

Field Intensity—A measure of the magnetomotive force per unit length. Expressed in oersteds and symbolized with the letter H.

Field, Magnetic—The force field around a magnetic particle.

Field Rate—The 60 Hz rate at which a field (half screen image) of 262.5 lines appears on the screen.

Field Skip—A system in which only one or two video heads records every other field on tape. During playback, the first head traces a track, and then the second head traces the same track.

Flip-Flop—A collection of logic gates configured such that a logic high input will produce a change of output from high to low or vice versa.

Flutter—A measurement of the fast rate of change in pitch of a constant tone. (*Also see* Wow.) Both wow and flutter generally result from a video tape not being drawn through a VCR at steady speed.

Flux—The summation of magnetic field effects through an arbitrary space. Expressed in Maxwells.

Flux Density—A measure of the flux per unit area. Measured in gauss and symbolized by β.

Flying Erase Head—An erase head mounted on the rotating video drum assembly used for editing a tape.

FM Interleaving—A shift in the FM modulator frequency of 7.867 kHz (_fh) while one of the video heads is recording the FM signal. FM interleaving is used to reduce the visibility of the residual Y FM signal crosstalk not eliminated in the playback head gap by the azimuth offset in some consumer VCR playback speeds.

FM Demodulator—Carrier frequency doublers and pulse density detector circuits which recreate original video from a modulated carrier through use of a low-pass filter.

FM Recording—The modulation of a carrier wave in accordance with the amplitude of a video signal. FM recording permits recording of low frequencies and eliminating level fluctuations.

Footprint—The recorded track pattern on a video tape.

Frame Rate—The 30 Hz rate (29.97 Hz for color) at which a frame (525 lines) appears on a video screen.

Frequency—A measurement of the number of electromagnetic cycles or alternations that occur in one second.

Frequency Interlace—A method to interweave chrominance (color) and luminance (brightness, black and white) information within the same bandwidth.

Frequency Modulation—The variance of a carrier wave frequency in direct relationship to an input signal.

Full Field Recording—Unlike skip field recording, this technique continuously uses both of the two video heads for both recording and playback of every field.

Full Logic Transport Control—Circuitry to enable smooth transition between VCR operating modes. Enables the use of remote control and automatic electronic editing.

FWD Search—Forward search mode.

Gap—The space between the two poles in a magnetic tape head.

Gauss—The CGS unit for measuring flux density.

Ghost—A vague image reflection appearing to the side of a main TV image caused by antenna reflection or impedance mismatch on the antenna producing a reflected wave of signal.

Gilbert—The basic measuring unit for magnetomotive force.

Guard Band—An empty space of tape between two video tracks that provides enough distance between the two tracks to prevent crosstalk. The difference between the video track pitch and the video track width.

HAFC—Horizontal automatic frequency control (system). To automatically maintain the horizontal oscillator frequency within set limits.

Half H shift—Another term for FM interleaving.

Harmonic distortion—The unwanted introduction of extraneous tones into the recording of a pure tone.

HD—High definition audio. A term used to describe Hi-Fi audio recording.

HD—Horizontal drive. A sync pulse used to control the horizontal drive.

Head Cylinder—The metal cylinder which houses the video heads.

Head Drum—A mounting disc containing video heads which rotates to achieve high writing speeds.

Head Gap—The space between the two pole pieces of a ferrite core used as a video head. Typically, this space is filled with glass to maintain the critical spacing dimension.

Head Switching—The process of switching between video heads in a multiple head system.

Head Switching Pulse—A 30 Hz square wave that is applied to the head amplifier to perform head switching.

Helical Scanning—A technique of video recording in which the heads rotate horizontally on a moving drum while the tape is drawn around the drum and over the heads diagonally. This produces diagonal recording tracks on the tape. This recording technique makes mechanical and electronic construction simpler and the system easier to maintain.

Helix Angle—The angle between the plane in which the video heads rotate and the edge of the tape.

High density recording—A process of writing more information within the same track length on a tape than was previously possible. Accomplished by recording higher frequencies (shorter wavelengths) within the same track length boundaries used by older VCR formats.

Horizontal Blanking—The period in which the electron gun is off during horizontal retrace. Horizontal sync is 4.7 μsec; blanking takes 10.9 μsec.

Horizontal Resolution—The number of horizontal pixels that can be distinguished on a horizontal scanning line. Commonly described as "lines of resolution."

H Sync—Horizontal sync.

Hue—The characteristic of a color that makes it unique (e.g., red, yellow, green, blue, etc.). Determined by the dominant wavelength in a color signal regardless of saturation. Represented by the phase angle in NTSC systems.

Hysteresis—The relationship between flux density and magnetic field intensity wherein a lag occurs in the magnetization value due to the changing magnetic force.

Hysteresis Motor—A synchronous motor that is driven by a permanent magnet rotor in a rotating field.

Induction—The phenomenon in which two materials interact electromagnetically due to proximity. The existence of a magnetic field in one material establishes a magnetic field in another adjacent material.

Induction Motor—An AC motor in which current flow in the primary coil induces current flow in a secondary coil wrapped around the armature.

Interchangeability—A description of how well a tape recorded on one VCR will play on another VCR.

Interlace—A scanning system in which the lines of each frame are divided into two fields and half are scanned (odd) then the other half (even) in alternate fashion to generate a 525-line picture or frame in one-thirtieth of a second.

Interlaced Scanning—The scanning of 262.5 odd lines followed by the scanning of 262.5 even lines to produce a complete picture frame. This method reduces flicker without the need to broaden the video signal bandwidth.

Interleaving—A condition in which the harmonics of the chrominance signal occur in between the harmonics of the luminance portion of the video signal. Interleaving of color information reduces interference with luminance information although it is broadcast at the same time.

Intrinsic Coercivity—The property that determines the level of field intensity at which saturation occurs.

Jitter—The instability in a playback signal caused by tape or speed fluctuations. This excessive wow and flutter causes the picture to have a rapid shaking or shivering movement.

Limiter—A circuit that produces a constant amplitude output regardless of varying input. A clipping circuit to strip off amplitude variations in an RF waveform to produce a frequency modulated signal.

Linear Track Audio—The conventional audio record/playback method used in VHS machines. Also known as normal audio or longitudinal track audio.

Loading—The mechanical operation from cassette insertion to completion of tape path setup. In VCR cassette systems, an automatic loading system takes the tape out of the cassette and wraps it around the head drum and capstan so recording and playback can occur.

Lock-Phase Delay—A timing circuit that delays the PG pulse to make it coincident with the on-tape vertical sync signal.

Longitudinal Track Audio—The conventional audio record/playback method used in VHS machines. Also known as normal audio or linear track audio.

LP—Long play.

Luminance—A signal containing brightness information. In black-and-white broadcasts, only the luminance signal is transmitted. Symbol Y.

Magnetic Domains—The intrinsic charge distribution of matter. Charges are divided into balanced pairs of dipoles.

Magnetic Moment—The force exerted at the axis of each magnetic domain or dipole by an external field.

Magnetization—Inducing a residual flux in a material which is proportional to an applied magnetic field intensity.

Magnetomotive Force (mmf)—The force of a magnetic field of one mass particle on another mass particle.

Markers—Dropouts that are placed in a video sweep as reference points for alignment.

Maxwell—The basic unit of measurement for a line of force.

Metal Evaporated Tape (ME)—One of the two types of metal video tapes used in 8mm recording. The metallic layer on this tape does not consist of oxide crystals but is purely metallic in nature.

Metal Particle Tape (MP)—The most common of the two types of metal video tapes used in 8 mm recording. The metallic layer on this tape does not consist of oxide crystals but is purely metallic in nature. MP tape is coated with a metallic alloy such as iron, nickel, and cobalt (Fe-Ni-Co).

MMV—Monostable multivibrator. A circuit that produces a logic high or low output with variable duration in response to an input pulse or voltage level transition.

Modulator—A component or circuit that converts audio and video signals into a television signal by varying the amplitude, frequency, or phase of a carrier as a function of an input signal. Some quality loss in the conversion is unavoidable.

Monitor—A television set without a tuner section.

Monoscope—A video display test pattern used for rough evaluation of resolution and SNR.

Monostable Multivibrator—*See* MMV.

Noise Bar—A rolling horizontal bar on a display screen caused by video heads passing over nonrecorded guard bands.

Noise Canceller Circuit—A circuit designed to remove cross color noise from harmonics of the sync and luminance signals.

Nonsegmented—A video recording system in which a complete field is recorded for each pass of a video head.

Nonsynchronous Motor—A motor with a rotational speed less than the line frequency. The conductor bars of the secondary winding cut the rotating flux field.

Normal Audio—The conventional audio record/playback method used in VHS machines. Also known as longitudinal track audio or linear track audio.

NRZ—Non-return-to-zero. A method of recording digital words on a tape that allows the playback circuitry to recognize strings 1's or 0's that occur in sequence.

NTSC—National Television System Committee. The present color telecasting system in the U.S., Canada, and Japan which uses 525 lines and 60 frames per second to produce a visual image on a TV screen.

Oersted—The unit measure of magnetizing force.

Omega Wrap—A tape threading method in which the tape wrap around the head drum resembles the Greek letter omega.

OR Gate—A logic gate in which any input of a logic high will produce a logic high output.

Overlap—The condition in which the beginning of head B output overlaps slightly with the end of the head A output, eliminating any blank areas between the two tracks. Recorded video is reproduced by the alternating output of the two video heads.

Overmodulation—The condition in which a large signal input to the FM modulation circuit increases the oscillator frequency to such an extent that a black edge-like effect is seen on the right side of white portions of the screen image.

PAL—Phase alternation line. The color television system used in Western Europe (except France) which uses 625 lines and 50 frames per second to produce a visual image on a TV screen.

PAL-SECAM Adaptor—A special attachment to a PAL TV set which permits the receipt and display of SECAM color TV programs.

Parity Check—A form of digital word error detection and correction used in PCM audio.

Pause—Temporarily stopping the tape transport while remaining in the same mode (PLAY or RECORD).

PCM—Pulse code modulation. A signal modification technique used in digital audio recording and playback (Sony Beta and 8 mm).

Permeability—The ability of the magnetic domains in a material to align under the influence of an external magnetic field. Permeability is symbolized by μ.

PG—Pulse generator. The circuitry that produces the rectangular wave pulse signal that is used as a reference during playback and as a comparison signal during recording. A PG pulse is generated each time a video head rotates past a PG coil. With two video heads, a positive pulse is produced on the approach and a negative pulse is produced as the magnet leaves the coil. The two signals are then combined into the rectangular square wave.

PGA,30—A timing pulse generated by head drum rotation and used as feedback in the scanner servo.

PGA Phase—Delay introduced by the 30 PGA and PGB multivibrators.

PGB,30—A second servo timing pulse generated near the head drum for timing video head switching.

Phase Comparator—A circuit that compares phase timing of two events to produce a feedback correction voltage.

Picture Instability—A temporarily unstable display caused by irregular tape travel and its resultant sync signal interference.

Picture Scanning—The method in which a television picture is drawn on a screen using two fields scanned line-by-line, top left to bottom right. The number of lines determines the quality and detail present in the produced image.

Picture Search—Cue and Review, Shuttle Search, Visual Search. A condition in which the recorder operates in fast play without sound so the operator can locate a particular point on the tape by sight rather than by counter.

Pitch—The distance between adjacent video tracks as determined by tape speed.

Pole, Magnetic—One of two regions (dipoles) of a magnetized body at which flux density is concentrated.

Pre-emphasis—The boosting of frequencies to compensate for future frequency roll-off and to reduce noise by increasing the SNR.

Preequalization—The process of boosting the low-frequency components in a video signal before processing to compensate for low-frequency roll-off during playback.

Pulse Generator—A device in the video head servo sync system to produce timing pulses as a head drum rotates.

Quantization Error—A form of digital noise placed upon the signal during the analog-to-digital and digital-to-analog processing.

Quantizing—The process of converting discrete voltage samples taken from an analog signal into a series of digital words that represent the sampled voltages. Quantizing is done by an analog-to-digital converter circuit.

Rabbet—The shelf of the lower video head drum surface which guides the tape around the scanner.

Raster—The 525 horizontal lines which comprise a complete video frame (picture).

Reference Video Head—One of two heads in a helical scanner which is chosen as a frame of reference.

Relative Speed—The speed relationship between the tape and the rotating head. High-frequency video cannot be recorded at audio tape speeds so the head was made to rotate relative to the tape, which greatly increased the effective speed of the system.

Reluctance—The resistance of a material to magnetic influence. This is the inverse of permeability.

Remnant Magnetism—The flux remaining in a material after an external magnetic field has been removed.

Residual Flux—The flux associated with remnant magnetism.

Resolution—The ability to reproduce image detail. When a black and white pattern of lines is produced on a screen, the number of visible lines is used to express the degree of resolution.

Retention—The ability of a magnetic material to hold a residual flux pattern after an external magnetic field has been removed.

REV Search—Reverse search.

RF—Radio frequency. Any frequency at which coherent electromagnetic radiation of energy occurs.

RF Envelope—The curve formed by the peaks of a signal at the video heads or "off-tape" video.

RF Modulator—A device that converts audio and video signals present in a VCR to an appropriate TV broadcast signal. Usually the VCR produces TV signals at television broadcast Channels 3 or 4.

RF Sweep—A 1 field sweep signal containing frequencies from zero to 10 MHz. Dropout markers at 1, 2, 3.58, 4.5, 5.4, and 7 MHz may also be added.

RF Transmission—The method used to transmit video signals by converting the audio and video information to television broadcast signals and passing this information through the air or a cable to the television receiver.

Rotary Chroma—The VHS process that changes the phase of the chrominance signal at 15.734 kHz intervals.

Rotary Erase Head—Another name for flying erase head.

Rotary Transformer—A device that magnetically couples RF signals to and from a spinning video head. This device eliminates the need for brushes in a VCR.

Rotating Drum—The cylinder around which the tape is pulled. In helical scan, the tape traces a diagonal path around the drum forming a helix so that the rotating heads can trace a diagonal path across the tape surface.

Rotating Head—A head (or heads) mounted on a drum cylinder that rotates horizontally as a tape is pulled past on a diagonal path. Over 200 times audio tape speed is achieved by rotating the heads at a high speed while maintaining a slow tape movement. This technique realizes high relative speeds with greatly increased recording area and more efficient use of tape surfaces.

Sample-and-Hold—A comparator circuit process in which a particular signal is measured (sampled) at a specific moment in time and held for later use.

Saturation—The point at which all the magnetic domains in a material are aligned and a further increase in magnetic intensity does not result in greater flux density. Color saturation is the depth of a color or spectral purity as determined by the relative amplitude of the chrominance signals.

Scanner Servo—A feedback control system used to adjust video head timing by controlling head drum speed.

Scanning—The process of structuring a video image into discrete lines of brightness and color (hue and chroma) information and then reconstructing the same image on a television screen. Both horizontal (left to right) and vertical (top and bottom) scanning occur simultaneously.

SECAM—*Sequential Couleur á Memoire*. The 625 line, 50 frames/second color television standard used in France and Eastern Europe.

Segmented—A VCR system in which each video head pass across the tape deposits a fraction of a field and more than a single pass is required to record a complete field.

Servo—An electromechanical device whose mechanical section is constantly measured and regulated by the electrical section so the drive speed or position of the mechanical section closely follows an external reference.

Setup—The difference between the black level seen on the TV screen and the blanking (no video) level. This level is 53.6 millivolts above the blanking level.

Sideband Cutting—The process of limiting bandwidth of one side of a sideband pair by attenuating one sideband to unbalance the pair.

Skew—The change in size or shape of video tracks from the time of recording to the time of playback. When present, this usually results from poor tension caused by misalignment or ambient conditions that affect the tape or tape path. The problem appears as a TV image that is bent out of shape.

Skew Error—An error describing horizontal sync and video signal advance or retard from normal timing.

Slip Ring—Metallic rings that are connected to the coils on the rotating heads and which form the interface with the heads and the electronics of the VCR.

Slip/Stick—Another term for stiction.

Slow Motion—Generating screen images at slower than normal speed.

SLP—Super long play.

S/N (Signal-to-noise)—A measure of the ratio of the desired signal to unwanted noise, expressed in decibels (dB). The higher the ratio, the clearer the screen image and the higher quality the audio. Video S/N is about 40 dB.

SNR—Signal-to-noise ratio.

Sound on Sound—A method of recording new audio material over an existing audio track without erasing the previous recording. During playback the two audio signals mix in the playback audio head to produce a composite audio signal.

SP—Standard play.

Spacing Loss—A reduction, or a complete loss of signal coming from a magnetic playback head caused by an increased distance between the head's contact surface and the surface of the magnetic tape. Contamination on the head's surface (head clog) is one cause of spacing loss.

Standard Tape—A high tolerance reference tape used to evaluate machine performance during design and prototyping of a new VCR.

Stiction—An unwanted condition where the smooth surface of a video tape sticks to the stationary portion of a video head drum assembly, stationary tape guides, and other stationary heads in the tape path. This phenomenon produces a situation where the tape sticks then slips repeatedly to the stationary surfaces as the mechanism attempts to pull the tape smoothly through the tape path.

Subcarrier—The 3.58 MHz continuous wave (CW) signal used to carry color information.

SuperBeta—A VCR format developed by Sony that offers improved resolution over the conventional Betamax format VCR.

Suppressed Carrier—A color recording technique in which the carrier is attenuated or canceled and only color sidebands are produced.

S-Video—Separate video. A means of connecting separate luminance and chrominance signals between pieces of video equipment.

SVHS—Super VHS.

Sync Head—A head used in some broadcast video recording systems that records information only during the vertical blanking periods. Also called a "$1\frac{1}{2}$ head."

Sync Signal—The pulse signals that synchronize the video circuitry. Horizontal sync defines the start of horizontal scanning; vertical sync determines the beginning of vertical scanning.

Sync Tip—The lowest point in a video waveform corresponding to the horizontal sync signal.

Synchronous Motor—A motor whose speed is synchronized to the line frequency.

Tape Guide—A metal cylinder or tapered metal shape that guides the tape properly past the heads, capstan, and along the tape path.

Tape Thickness Loss—A phenomenon resulting from recording a magnetic signal in the direction of the tape's thickness. Tape thickness loss increases as the magnetic signal gets farther away from the playback head. It is one cause of lower output signal from the playback head. Tape thickness loss has a similar effect on the playback signal strength as spacing loss.

Tension Servo—A servo system that maintains constant tension along a video tape between the supply and take-up reel spindles.

Threading—The operation in which a cassette tape is automatically positioned around the tape path.

Time-Base Errors—Mechanically induced timing errors in a VCR caused by incorrect tape tension, tape stretch, etcetera.

Time-Base Stability—Term used to describe how closely sync timing enables the playback video signal to match an external reference video signal. The ability of a VCR to minimize recorded errors due to mechanical transport errors that cause hue errors from frequency and phase deviation of the color signals.

Timer—A device built into the tuner of a video section that enables the recording of different TV broadcasts at pre-selected times and days.

Track—The magnetic pattern imposed into a tape surface by a recording head.

Track Angle—The angle between the video track and the edge of the tape.

Track Length—The tape distance on which video information is recorded.

Track Width—The core piece thickness at the head gap which controls the width of the tape area affected by the magnetic field from the head gap area.

Tracking—The ability of the spinning video heads to accurately pass over the recorded video RF information during playback. Good tracking is indicated by a strong RF signal; poor tracking is indicated by low level RF or noise.

Tracking Adjustment—An electronic adjustment that delays the VHS and SVHS during CTL (Control Track) signal during playback so its effective position matches the position of the CTL head on the recording machine.

Tracking Control—A variable delay MMV used for playback tracking adjustment on VHS and SVHS machines.

Tracking Errors—The visible interference noise bars produced on a screen caused by the inability of video heads to correctly follow the video tracks recorded on a tape. The tracking control adjustment corrects for this error.

TTL—Transistor-transistor-logic.

U-Loading—A tape system in which the tape forms a U shape as it passes around the head drum.

U-Matic—A video system standard using $\frac{3}{4}$ inch wide tape.

Unloading—Replacing the tape back into its original position in the cassette.

Utilization Ratio—The ratio of lines in a raster contributing to vertical resolution and the total number of raster lines. This ratio should be 0.7.

VCO—Voltage controlled oscillator. An oscillator whose frequency is controlled by an external voltage level.

Vertical Blanking—The video electron gun off-period during vertical retrace. Vertical sync is 3 horizontal lines long while blanking is 21 lines long.

VHS—Video home system. A tape recording format developed by JVC in Japan and in use worldwide.

Video Drum—A drum cylinder with opposing video heads mounted 180 degrees apart.

Video Head—An electromagnet wrapped at one end with a tiny coil of wire in which signal current flows producing a magnetic flux field that enables RF information to be stored on a tape.

Video Muting—The electronic blanking of a VCR screen during machine threading and unthreading, start-up, and playback. The muted screen may turn black or blue during muting.

Video Track—The area on a tape in which RF information is magnetically recorded as a recording head passes over.

Visual Color Acuity—The ability of the human eye to discern different shaped colors at varying distances. The ability to resolve small picture shape and color details.

V Sync—Vertical sync.

VTR—Video tape recorder. A collective term for all video recording systems.

V-V—Video-to-Video. The playback screen image produced from a tape.

VXO—Voltage controlled crystal oscillator. An oscillator whose frequency is controlled by an internal quartz crystal.

Weber—A measure of flux. 1 weber = 1×10^8 Maxwells.

White Clip—A circuit used to cut off excessive positive-going spikes (overshoot) from the emphasis circuitry and hold the overshoot to a predetermined, adjustable level.

Wow—Wow, a measurement of the slow rate of change in pitch of a constant audio tone. (*Also see* Flutter.) Both wow and flutter generally result from a video tape not being drawn through a VCR at steady speed.

Wow and Flutter—Tape transport speed fluctuations that can cause regular instability in the screen image and produce a wavering or quivering effect in the sound during recording and playback. Fluctuations below 3 Hz are called *wow*, and short cycles above 3 Hz are called *flutter*.

Wrap Angle—The angle produced by the tape against the scanner surface measured at the center of the scanner.

Writing Speed—The relative speed between the record head and the tape surface.

XTAL—An abbreviation for a crystal used in an oscillator circuit.

Y/C Video—Another name for S-video.

Y Signal—An abbreviation for luminance. It is the portion of a video signal containing black and white information and sync.

Appendix

SELECTING A VCR

Selecting a video tape recorder can be frustrating and confusing. Here are some guidelines to help you make your selection.

First, understand clearly how the VCR will be used. Will it be used during tape exchanges with a neighbor or relative? Will you use it just to play back rented movies on weekends, or to record programs while you are away from home? Consider what your immediate and future needs might be so the VCR you select will be appropriate now and later. Ask friends what they wish their VCR could do and what options will be included on the next machine that they purchase.

Selecting a VCR format is important too. The 8mm format produces wonderful video, but VHS is far more popular. Some feel that the quality of the 8mm picture is superior to that of VHS. However, VHS has been widely marketed in the United States. VHS machines do excel in recording time capability. The sharp, crisp images of Hi8mm and SVHS have given the world new standards of excellence in video reproduction.

In the end, you must decide if you prefer a VHS or 8mm machine. You may wish to base part of the decision on the kind of tapes available for rent at your local video store. Consider what format your friends and relatives are using in case you want to exchange tapes with them.

Do you intend to use the VCR to record home movies? Do you receive TV programs through a cable system or an antenna? If you presently use an antenna, do you intend to get cable in the future? Do you want special effects features such as slow motion, or a fast playback speed that allows you to play the program at two or three times the normal viewing speed in either forward or reverse direction? Or perhaps you'd like a still frame feature that allows you to pause the machine during PLAY so that you have a perfect still picture (similar to a slide) on your TV screen. Then there are the search or cue and review features that allow you to zip through unwanted sections of your program either in forward or reverse at about nine or 10 times the normal playback speed. Cue and review is one of the most popular features for reviewing a particular scene or speeding through commercials in one-tenth the time.

Many machines allow you to pause during playback with some form of a viewable picture on the screen, and to search in forward or reverse direction without offering other, more expensive special effects. Good still frame and slow motion are features generally reserved for the more expensive machines. Selecting special effects generally creates the most confusion during a VCR purchase.

The type and number of programming events that can be set determine if you will be able to record one or more programs while you are away. The key feature is the timer. A VCR timer lets you record a program automatically. The recorder is pre-programmed to turn on to a specified channel at a selected time, to record for a set period of time, and to turn off automatically at the end of this period.

The basic timer can record one program within a 24-hour period. More expensive models let you pre-program two, three, four, or more programs over a period ranging from a few days to several weeks. On some VCRs, you can even pre-program events up to a year in advance.

Some VCR models include a feature that allows you to set up the VCR's timer programming using a device that looks and operates like a bar code scanner. The scanner operates similar to the bar code reader used at check out counters in many grocery and department stores. The VCR bar code reader is a specialized remote control device that is held in your hand as you scan the reader tip across a series of black and white vertical stripes on a sheet of paper. The scanner reads the black and white bar code as it travels across the stripes. The bar code on the paper looks similar to the UPC bar code on packaged merchandise in a store.

The bar code read by the scanner is translated into VCR programming time, date, channel and program length information. After scanning the data into the remote reader, the programming information is sent over an in-frared beam to the VCR's remote control infrared receiver. The VCR communicates the received data to the timer control circuitry and programs the recorder for the future recording.

Many weekly TV program booklets contain the bar code information for programming your VCR, but your VCR must be capable of recognizing the programming data stream being transmitted from the remote bar code scanner to your VCR.

Some remote bar code readers provide a means for manually entering the decimal numbers represented by the bar code directly into the hand held scanner. This feature is necessary because not all TV program listings include printed bar codes. Instead, these booklets contain just the decimal numerical sequence corresponding to the individual bar code stripes. This lets the operator manually enter the numerical equivalent of the bar code into the hand held scanner. Then the data is sent from the remote scanner to the VCR via the infrared beam.

Virtually all home VCRs sold today contain what is known as a cable-ready tuner. If you are considering the purchase of a VCR with a cable-ready tuner, you might find the following general discussion to be helpful.

Standard TV broadcast signals consist of three bands of frequencies—VHF low band, VHF high band, and UHF. Cable companies transmit their signals on the VHF low band, VHF high band, and on special bands known as midband, or superband.

Other bands such as hyperband are also used occasionally, but midband and superband are most commonly used. Cable companies don't use the UHF band. Older VCRs without a cable-ready tuner require a special cable box to receive midband and superband channels. These channels are converted to a VHF low frequency band such as channel 3 so that a non-cable-ready TV or a non-cable-ready VCR can receive the cable programs. The cable boxes receive and tune the midband, superband, and hyperband stations.

Cable-ready VCRs have a tuner that is designed to directly receive most of the cable transmitted midband, superband, and hyperband channels. Be aware, however, that not all VCR manufacturers think and act alike. Some VCRs that vendors claim are cable ready actually only have a cable type connector (commonly called an "F" connector) protruding from the back of the VCR so that it is ready for a coaxial cable cable hook-up. These so called cable-ready VCRs do not have a tuner capable of receiving all cable channels. Fortunately, the Electronic Industries Association, cable TV company operators, and TV and VCR manufacturers developed standards to clarify the cable-ready specifications and to help provide consistent labeling of VCR capabilities.

The presence of a cable-ready tuner in the VCR eliminates the need for a cable converter box. Some cable systems still require a cable converter box if you want to receive scrambled pay-TV cable programs such as "HBO," "Disney Channel," or others.

Just because the VCR that you're thinking of buying is advertised as cable-ready, there is no guarantee that the VCR's tuner will be able to receive all of the channels that your cable company provides. Some cable-ready VCR tuners have gaps in the cable frequency bands they receive making it impossible to view some of the cable programs. Be sure to compare the channel reception specifications of the VCR's tuner with the channels provided by your cable company.

Another factor in the selection of your VCR is the number of motors used to drive various sections of the mechanism. VCRs with all direct drive motors have an advantage over a gear or belt driven mechanism. A direct drive motor VCR means that each essential mechanical, moving part of the VCR is directly driven by its own separate motor without using belts, pulleys, or gears.

Many older model VCRs have several motors. For example, one motor may be required to load and unload the tape while another is used to move some other part of the machine. In a front load machine, a separate motor is used to raise and lower the video cassette into the unit. Separate motors are also used to rotate the head drum, and the capstan and reel tables.

Nearly all consumer VCRs sold today use fewer motors to perform all mechanical functions. This is one reason why selling prices have fallen for new models.

Most VCRs have at least two motors: one to pull the tape and move the mechanism (typically the capstan motor does all of these functions) and one to turn the head drum. VCRs with few motors depend on a series of belts, clutches, gears, pulleys and linkages to transfer energy from a common motor to various sections of the mechanism. In current designs, the capstan motor commonly drives the front load tray movement, controls the threading and unthreading action of the tape around the video head drum, provides the tape take-up reel torque for the play and record modes, and turns the reel tables in the fast forward and rewind modes. All of these actions are in addition to the main function of the capstan motor; turning the capstan shaft to move the tape through the machine at a precise speed.

Current VCRs with multiple direct drive motors are generally reserved for expensive industrial and professional grade equipment because VCRs with direct drive motors provide longer performance life (no belts or worn plastic gears to replace) and much tighter control over tape handling. This provides finer edits, better tape control and, in general, better stability of video recording and playback.

On the other hand, direct drive motors require more electronic circuitry for precise operational control. Another disadvantage is that if these motors fail, the repair can be more expensive.

Original VCRs were driven with a series of belts, pulleys, and idler tires. These machines had only two motors. One motor was responsible for the action of the threading, and another motor controlled all other VCR operations. The latter (main) motor was powered directly from the 115 volt AC line. It rotated at 1800 revolutions per minute (rpm) and was synchronized by the 60 Hz AC input. The head drum was rotated at a precisely-controlled speed as calculated by the diameter of pulleys over which the connecting belts were attached. The capstan shaft was rotated at a precisely-controlled speed by another series of belts and pulleys connected to the same AC motor. Reel tables were also driven by this same AC motor through a series of belts, torque limiter assemblies, tires, and pulleys.

With the introduction of two- and three-speed machines, a different motor system was required to allow for changes in speed. This led to the introduction of the DC-controlled capstan motor. The head drum and reel tables were still controlled by the AC motor. In current consumer VCRs we typically find two-motor belt, pulley, and gear-driven machines. These motors include the capstan motor and drum motor.

Two-motor belt, gear, and pulley-driven cassette recorders are common in all price ranges of today's VCRs. Manufacturers recommend that belts and rubber tires be replaced every one to two thousand hours of use. Belts can break without warning, and tires may begin to slip even if the unit was not used excessively. Rubber will deteriorate with time causing possible tape movement problems. Defective belts and tires can cause the unit to not play, to slow, or to not rewind or fast forward. Such defects can cause excess tape to remain in the machine after the cassette is ejected.

The plastic gears, linkages, and levers in today's models do not require preventive maintenance replacement. When these items fail (and they do fail) the failure comes suddenly and without warning. Failure of these components forces the machine out of mechanical alignment. Unfortunately, no practical preventive maintenance will prevent this from happening. Defects in these components often occur when a cassette tape is forcefully shoved into the front load mechanism, or placed into the tray mouth at a slight angle.

Gears sometimes jump a tooth because of excessive torque applied to the mechanism during a change in mechanical position. This action can cause a gear, lever, or linkage to become defective without warning.

Defective gears, linkages and related mechanical misalignment are inevitable in VCRs sold today. The comparatively lower prices (even for VCRs with full features) accompany less reliable designs, thus reducing the trouble free hours available from these machines.

Before purchasing a VCR, also consider the sound features that you want. VCRs are now available with AFM, Hi-Fi (HD), and PCM sound capability. Hi-Fi, AFM, and PCM audio require special electronics and, in the case of VHS, extra heads that are mounted on the video head drum assembly. These machines reproduce stereo sound with fantastic quality, exceeded only by compact disc players. External amplification and external speakers are required to enjoy the full benefits of these features. These sophisticated VCRs will still play back the high quality audio on a monaural TV, but you would be wasting your money without the proper speakers and amplifier. All 8mm machines use AFM audio. Some models also provide stereo AFM sound. And a special PCM audio feature is available on some 8mm machines. On VHS machines Hi-Fi audio may also be called *high definition (HD) audio*.

Linear track or normal audio is a standard feature on all VHS machines. Current 8mm models do not provide linear track sound. For VHS, the linear track audio record and playback method uses stationary audio heads to record sound with traditional recording techniques. The recording process is identical to methods used on reel-to-reel and audio cassette recorders. Some VCRs offer stereo linear audio. To enjoy stereo reproduction on linear audio heads you must connect an external amplifier and speakers.

Other current VCR options include stereo tuners and audio demodulators that let you record TV programs broadcast in stereo. This option should be considered if you'd like to hear stereo audio. VCRs that don't contain the stereo tuner and demodulator, yet have the stereo audio option, will play back prerecorded stereo programs and record stereo audio through the external audio input jacks.

The number of video heads is yet another option. VCRs have two to seven heads mounted on the video head disk. Only two video heads are used at any given time to record and play back picture information. These two heads are mounted exactly 180 degrees apart. One pair of heads may be used to record and play back at standard speed while a second set of heads can be used to record long play and extended play (super long play). Most four-head VHS machines work this way. Inexpensive VCRs use two heads to record and play back all three speeds.

A VHS VCR with Hi-Fi audio, has a separate pair of heads to record special Hi-Fi audio information. It is possible to have four video heads and two Hi-Fi audio heads mounted on the video head assembly. These machines have been called *six-head* VCRs.

Some machines have a special video head that is used only for improving special effects during playback. This head is used with the normal video heads during playback and is electronically switched on and off to provide optimum video presentation during slow motion and still frame modes. You can find VCRs with three, five, or seven heads mounted on the rotating head disk assembly.

Current 8mm technology does not require as many heads. The 8mm AFM and PCM sound and video are all recorded using the same pair of video heads. It's typical to find 8mm video recorders with AFM and PCM audio and superb playback special effects using only three or four heads.

In VHS machines, the number of video heads can have a direct effect on the quality of playback special effects. VHS recorders with four video heads generally provide better picture quality in Standard Play than machines having only two heads recording in the same speed. This is because the extended play heads must be physically smaller to allow the recording information tracks to be spaced closer together on the tape. While the heads are smaller, the video head gaps are the same size.

For VHS machines with only two video heads, the SP speed playback picture quality suffers slightly because the FM information track recorded onto the tape is not as wide as the tracks recorded by the larger SP heads used in four head models, so the FM signal being read off the tape is not as strong as the signal recorded with the designated SP heads. This reduces overall video playback quality. Four-head VHS machines with one set of heads for SP provide better, cleaner, video quality in this speed. In four video head machines, smaller heads are used for both the EP (SLP) and LP speeds.

VHS units with two video heads must use the smaller extended play head size for all three heads. The SP playback video is not as crisp on these units as it is on VCRs with four heads. Another advantage of the four-head VHS VCR is the option to record and play back in standard speed for better video quality.

However, machines with more than two video heads experience increased replacement cost and potentially increased video tape wear because more heads contact the tape during each head rotation. Some VHS manufacturers recommend the replacement of both lower and upper drum assemblies during a video head replacement on Hi-Fi audio capable VCRs because high precision is required for placement of the upper drum assembly. This creates a substantially higher cost than for the non-Hi-Fi machine. When video heads wear out on a standard VCR, only the upper video head disk needs to be replaced. Video heads cannot be replaced individually. The entire head disk must be replaced at one time.

This is a sampling of the features that you should consider before purchasing a VCR. Other good sources of questions to ask yourself and "food for thought" are the consumer publications. Consumer buying guides often have a section rating VCR brands. By understanding yourself, your desires, the desires of those around you, and available products, you should be able to select a good VCR.

CHOOSING THE BEST VIDEO TAPE

Technicians and engineers are often asked which brand of video tape is best to use in a VCR. From a technical perspective, one would guess that the best tape would have the most durable magnetic coating, the strongest magnetic coating binder, and the best head lubricant. While these factors are indeed important, the cassette housing itself is also a critical aspect in tape selection.

Magnetic tape production has reached such a level of maturity that many companies can successfully produce high quality tape. However, not all can produce high quality cassette housings. That black, hard plastic tape housing that you take so much for granted is produced with unbelievable precision. Should the reel tables not fit exactly into the cassette, the mechanics of the VCR could actually be damaged by the cassette.

Several years ago, *Consumer Reports* and *Video Review* magazines conducted lengthy studies on video cassette tapes and the results are interesting. Of prime importance is their finding that the range of quality between the best and worst tapes made by major, well known video tape manufacturers is not large. Tape quality has improved greatly since the introduction of the first VCR. Today, most major brand video tapes can provide a better picture than a typical VCR can record or play back.

Manufacturers have introduced as many as a half dozen grades of tape with as many price tags, but the difference in video signal quality observed on a TV screen is slight. An-

other finding is that much like the "plain Jane" or "vanilla" floppy disks of the computer industry, the video cassette industry also has its no name, unlicensed tape vendors. If the tape does not meet the minimum quality standards of 8mm or VHS, avoid these discount tapes. Your video tape is actually part of the equipment in your system, so shop carefully. You could spend too much money and get too little quality for the price.

Video Tape Performance

About two dozen companies manufacture video cassette tapes. Tape quality can be determined by analyzing the tape's video and audio performance. Video performance can be partitioned into four categories: (1) dropout, (2) frequency response, (3) noise, and (4) signal retention.

Dropout

This is the most visible (and irritable) problem in videotape. Dropouts appear as intermittent horizontal streaks, tiny spots, glitches, or unwanted lines on the screen image. They are caused by missing bits of magnetic energy that come from scratches and flaws in the tape's magnetic surface, or dust particles that cling to the tape surface. This problem is measured by the number of dropouts that occur per minute of tape play. VHS tapes should not exceed fifty 15-microsecond (or longer) periods of dropout in a minute of play. Most tapes average around 10 dropouts of 15 microseconds (or longer) per minute.

Usually, dropout isn't noticed by the consumer-viewer because most VCRs have built-in circuitry that fills in the image holes when dropout occurs. In addition, vivid colors and large screen objects distract the viewer from noticing minor dropouts. Dropouts are most noticeable during slow-motion and freeze-frame.

A reduction in FM video tape signal level of 16 dB or greater and that lasts for 5 microseconds or longer is considered a dropout. Dropouts tend to be worse at both ends of a videotape. This is why it's a good idea not to record anything on the first and last minute of a cassette tape.

Frequency Response

Frequency response corresponds to picture resolution. Good frequency response produces good picture resolution. The better the frequency response, the better the definition, or detail in the playback picture.

The tape output signal should have the same strength at all frequencies, but the VCR in playback amplifies the middle frequencies more. This isn't usually a problem during playback, except some tapes don't maintain the same signal output level at the highest frequencies and experience some loss in detail.

Video Noise

Video noise exists in two forms: luminance (brightness) noise and chrominance (color) noise. Luminance noise is seen as graininess or snow in the screen image. Chrominance noise can be noticed as amplitude modulation snow along horizontal lines of a color picture or as phase modulation changes in the shade of colors. At its worst, color noise can turn solid areas of color into a confused disarray of colored blips and streaks.

Both luminance and chrominance noise are expressed as signal-to-noise ratios (SNR) and measured in decibels (dB). An SNR difference of 1 dB is perceptible by the viewer. The higher the SNR value, the better the quality of the tape. A luminance SNR of 45 dB or more is quite good. A chrominance SNR of 45 dB or better is very good. A value of 43 dB is noticeable. Most tapes demonstrate little luminance and chrominance noise (SNRs are high). In fact, the chrominance SNR for most videotapes is better than that of the consumer VCR.

Magnetic Signal Retention

This is a measure of the longevity of a recorded program. Magnetic particles tend to forget their magnetic orientation after a number of plays. Also, recorded signal strength deteriorates with time as the tape is stored. This lessening of the composite (RF) signal strength causes a signal loss that degrades the picture image.

Most of the initial degradation in image quality from demagnetization occurs during the first few times a tape is played back. After that, signal strength losses become more gradual with repeated use and long term storage. Signal loss occurring from repeated playback is measured in dB and is based on the loss of signal strength from the initial recording to that remaining after the tenth playback. A value of 6 dB represents a reduction in signal level by one half. The ten playback cycle measurement is used for this rating because most of the playback loss of a recorded tape happens during this 10 playback cycle period. Most video tapes have less than 1 dB loss after 10 replays.

Loss of signal strength resulting from the long term storage of recorded material is a highly debated issue. Although tape manufacturers make time-simulation tests to discover the long term storage capabilities of their tapes, these specifications are rarely revealed to the public.

Audio Performance

Audio performance is determined by analyzing four factors: (1) tape edge regularity; (2) harmonic distortion; (3) dynamic range; and (4) bandwidth.

Tape edge regularity

This factor becomes especially important when you are using tapes in a linear track stereo VHS machine. This type of VCR records two tracks in the same tape space area as is used for a single track of monaural audio. Conventional stereo VCRs record both right and left channel audio into this narrow band. The right channel touches the tape edge so damage near this edge results in a difference in signal strength between the right and left channels.

Video cassette tapes usually have no irregularities along the edges. However, improper storage of the cassette or misalignment in the tape path can damage the tape edge and dramatically affect the stereo playback.

AFM and Hi-Fi VCRs record both sound and video diagonally across the tape so minor tape edge irregularities don't affect the sound quality.

Harmonic distortion

The introduction of extraneous tones into the recording of a pure tone (usually 1 kHz) is called *harmonic distortion*. This specification defines the amount of distortion that the tape itself adds to an audio track. Harmonic distortion is expressed in percent and represents the amount of distortion in an audio signal recorded at 10 dB lower than a maximum (0 dB) reference.

A distortion of one percent from a reference tone level can be detected by most listeners. Usually, no significant differences in harmonic distortion exist between tapes.

Dynamic range

Dynamic range represents the span between the loudest and softest audio signal that can be played from a recorded tape. At the loud end, the signal gets badly distorted; at the soft end, the playback audio becomes lost in the noise or hiss coming from the rape.

Dynamic range is measured in SNR and expressed in dB. Like video SNR, the higher the reading, the better. A value of 50 dB or more is required to qualify as high fidelity.

Most video tapes cannot match the dynamic range of a good audio tape.

Bandwidth

For an audio signal, this is an expression of the range of audio frequencies that a tape can reproduce. Most magnetic tape can record midrange tones, so one way to determine bandwidth is to compare the recording of a high treble tone (*Consumer Reports* uses 11 kHz) to a midrange tone (1 kHz). The best tapes will reproduce the treble tone almost as well as the midrange tone. *Video Review* records 20 Hz to 20 kHz at 10 dB below a reference 0 dB level. During playback, they note the bass and treble frequencies, where signal strength falls off by 3 dB. This is a half-power point where sound intensity loss can be clearly discerned.

Video frequency bandwidth relates to the ability of magnetic tape to retain fine picture detail. Low frequencies correspond to large objects in the reproduced picture. High frequencies are required to reproduce fine details in the video. As the retained signal strength of the high frequency portion of the magnetic signal rolls off, the reproduced picture detail deteriorates making the playback picture appear softer and somewhat blurry.

The magnetic particle composition and size and the density of the particles on the tape account for much of its video bandwidth specifications. High grade video tape typically has improved magnetic formulations that provide better video signal bandwidth than the less expensive tape. Video bandwidth measurements of a high grade tape often exceed the bandwidth capabilities of the VCR. Thus, high grade tapes provide the best resolution reproduction possible.

With all these factors fresh in your mind, the only other thing yet to be decided is the VCR format to choose. Battle lines have again been drawn in the war of video formats. The initial skirmish was between Beta and VHS formats. This ended with VHS the victor. But the format battle is far from over. Some feel that the 8mm format is superior to VHS and sales of 8mm machines are growing. The advantage of 8mm's smaller physical size is evident in home movie applications and the camcorder (camera-recorder).

EDITING AND DUBBING TAPES

In VCR operation, editing and dubbing are the same basic functions. When editing a tape, the VCR is making a dub of the original ("raw") footage recorded earlier. The difference between a plain dub and an edited tape is that the finished (edited) product has unwanted scenes from the original recording cut out of the finished dub or has additional scenes inserted into the final program. The added scenes may not have existed in the original recording. In other words, an edited tape is a dub with specific segments either removed or inserted with a high degree of precision into the final dub.

Two types of edits can be made on a VCR—*insert* and *assemble*. An insert edit involves inserting program material into a length of tape that already has information recorded on it. The inserted information can be audio, video or both audio and video. With stereo audio machines audio inserts may occur on one or both channels of audio. For example, background music or narration can be inserted on one track of a two track recording after the original recording was completed.

For VHS machines, audio (without video) inserts can *only* be performed on the longitudinal audio track. Inserting only audio for VHS Hi-Fi sound is not possible. For 8mm units, audio only editing must be performed in the PCM audio (if available). AFM audio only insert editing cannot be done on 8mm machines. FM audio only inserts are not possible on either format because both the Hi-Fi and AFM signals used in these units are recorded

on the same track as the video FM signal (see Chapter 9). This means that the FM audio and FM video cannot be separated for individualized editing. Therefore Hi-Fi and AFM audio only editing cannot be performed without destroying the previously recorded video. Insert audio edits can be made for both the VHS Hi-Fi and 8mm AFM audio signals but only when video inserts are simultaneously made, thus rewriting the video signal along with the FM audio edit.

In order for an insert edit to work, a good quality continuous recording of existing material must already be on the tape. Prior to performing an insert edit, the tape must be "striped" (referring to writing prerecorded helical tape tracks on the tape) or "blacked" (referring to pre-recording the tape with black video, or sync only signals). Any pre-recorded video signal will suffice for blacking or striping a tape as long as the pre-recording contains a continuous and reliable signal with proper color video sync.

An example of the importance of a continuous and reliable pre-recorded tape for insert edits is found in the VHS format. When performing an insert edit of any type (audio, video or both audio and video), the unit uses the existing CTL pulses to phase lock the servos before, during, and after the period that the insert edit is performed. The VHS format does not rewrite new CTL pulses onto the tape during any type of insert edit. Therefore, if the previously striped tape is not continuous (has sections with no CTL pulses), the insert edits will appear as broken up and highly unstable recordings. This is caused by poor CTL pulse integrity in the edited section of tape. To avoid this, many VHS editing VCRs will abort the edit if the pre-recorded CTL pulses are not suitable.

An assemble edit differs from an insert edit in that it does not require existing recordings on the tape before making an edit. However, the first few seconds on the tape must have previously recorded material on it before starting the first assembly editing sequence. The rest of the tape can be new, and unrecorded.

An assemble edit does not place program material into an existing recording. Instead, it functions to tag new material onto the end of a previous recording. Assemble editing compares to forming a paper link chain, like the one you made in elementary school to decorate your home at holiday time. Each new link of the chain is attached to the link just made until you have formed a long chain. In an assemble edit session, you attach segments of video onto the last segment you just created. This attaching ("tagging") of new segments onto the end of the previous one continues until the final edited program is completely assembled.

A second difference between an insert edit and an assemble edit, is that in an assemble edit the operator does not have the option of selecting audio only, video only, or both audio and video edit options. Assemble edits record audio and video simultaneously.

A third difference between assemble and insert edits is that in the assemble edit mode, VHS machines record new CTL pulses on the tape, even if the pulses are present from a previous recording.

For 8mm machines, the ATF pilot signals used for tracking control are always rewritten whenever video is recorded onto the tape. Thus, ATF signals are rewritten for all video insert and assemble edits and normal record modes. This is necessary because the ATF pilot signal is intertwined with the video FM and audio FM signals and recorded on the same RF track.

All editing VCRs should have a flying erase head to assist in clean edits at the "edit in" (edit start) and "edit out" (edit stop) points. Flying erase heads also ensure good video quality throughout the duration of the edit. Some low priced editing models do not provide flying erase heads. Instead these models increase video head record current during an edit to minimize the appearance of interference occurring from an incomplete erasure of previously recorded material during an edit. (See Chapter 5 for a discussion on this topic.)

Video Editing Controllers for Home Use

Simple edits can be performed on any VCR by using the record/pause button. For example you can remove commercials from an off-air program by pressing the record/pause button each time you want to delete unwanted material from the recording you are making. This form of editing is done on the original, raw footage tape as it is being recorded. Stop and start record timing is not precise for this type of editing, however, precise in and out edit points are generally not important for this type of editing.

When it's important to have accurate edit-in and edit-out points, some form of editing control device is required. An edit control device allows you to mark the exact edit-in and edit-out points on the tape and to perform each edit in a semi-automatic manner. For example, with an edit control device, you can insert your own titles onto your recorded movies or home videos. To edit two tapes together, you need two VCRs; one for recording and one for playing back the source tape.

Some of the higher priced VCRs have an internal edit controller that is part of the VCR's front panel or remote control device. With this feature, you can control both the record and source VCRs from the front panel of the recording unit. An interconnecting cable between the two machines allows you to control some of the source VCR's functions from the record machine's front panel.

Stand-alone (external) edit controllers are provided as optional accessories by some manufacturers. External edit controllers allow the operator to control selected functions on both machines using commands that are passed over a cable between each unit. In this configuration, one VCR is designated the editing (record) VCR, and the other VCR acts as the source (playback) unit. Connecting an external edit controller is only possible for VCRs that will accept operational commands from an external control device.

Whether the edit control is internal or external to your machine, actual editing is performed by marking the edit-in and edit-out points on the record and source tapes. You must also tell the edit controller if it is performing an assemble or insert edit. And, if you are performing an insert edit, you must also select audio only, video only, or both audio and video edits on the controller.

After the edit-in and edit-out points are identified and the appropriate editing choices have been selected on the controller, the edit start button is depressed. This causes the edit controller to take over the operation of both machines. The source and record tapes are pre-rolled back to pre-determined points (usually a minimum of 5 seconds before the marked edit-in point on the tape), and then both tapes begin to roll forward toward the marked edit-in points. The record machine is switched from the pre-roll forward mode to record (edit) at the precise spot on the tape that you set as the edit-in position. When the record starts at the edit-in point on the record machine, the source tape will have also reached the edit-in point that you marked for that tape during the initial edit set up sequence.

Pre-rolling the tape backwards at the start of the edit cycle allows the servos in each unit to lock up and stabilize the tape movement before the edit start point is reached. If the tape is not moving smoothly when an edit is made, the video at the edit point will be unstable on the TV screen as you playback the edited tape.

The accuracy of the edit-in and edit-out points on the tape is maintained by counting RF tracks for the 8mm units, or by counting CTL pulses for VHS models as the record and source tapes roll backwards during the pre-roll period and again as the tape rolls forward toward the edit-in point. The pre-roll and forward counting sequence takes place just after the edit start button is pressed by the operator. Both the record and source units must pre-roll their respective tapes by the same amount so that both tapes reach their respective edit-in points at the same time.

Hook-Up Configurations for Editing and Dubbing

When connecting two VCRs for an edit or dub, always use the audio line inputs and outputs for the sound. If possible, video should be connected using an S-Video cable. The line video jacks are the next best choice for video connection if S-Video is not available on your machines.

If you have an external character generator for creating titles on your edited tapes, special effects generator for producing unique video effects, or an external color and video corrector for improving the color and brightness, you should connect this equipment between the source VCR video output and the record VCR's video input. If available (and for best performance) these devices should also be connected using the S-Video cables.

PREVENTIVE MAINTENANCE RECORD

Use Table A-1 as a guide in conducting preventive maintenance on your VCR.

ROUTINE PERIODIC MAINTENANCE SCHEDULE

Table A-1 describes the suggested periodic maintenance and replacement actions that should be conducted on a VCR to achieve optimum performance and to maximize operational life. In this chart, "C" stands for "clean," "L" stands for "lubricate," "O" stands for "oil," and "R" stands for "replace."

CAUTION
When cleaning, oiling, or lubricating any part of the VCR mechanism, use only the manufacturer's recommended chemicals. Use of other lubricants and cleaning solutions can damage your VCR!

Table A-1. Suggested Periodic Maintenance and Replacement Schedule

Description	Hours of Operating Time					
	500	1,000	1,500	2,000	3,000	5,000
Loading motor	Replace after 20,00 operations					
Capstan shaft	C	C,L				R
Capstan motor					R	
Lower drum and video head drum motor	C					R
Reel motor				R		
Cassette tray motor	Replace after 20,00 operations					
Belts	C	R				
Rubber tires	C	R				
Belt and tire pulleys and shafts		C,O				
Video heads	C		R			
Stationary audio heads (VHS only)	C				R	
Full track erase head	C					
Reel tables	C	C,O				
Brake shoes	C	R				
Forward holdback tension regulator band		R				
Sensor lamp (incandescent type only)		R				
Torque limiter assemblies	C,O		R			
Mechanical linkages				C,L		
Loading post slider base			C,L			
Pinch roller	C			R		
Tape guides	C				R	

ANSWERS TO CHAPTER REVIEW QUESTIONS

Chapter 1 Answers

1. In 1965, Sony introduced the CV series reel-to-reel video tape recorder.

2. The Electronic Industries Association of Japan (EIAJ) formed in 1968.

3. The five disadvantages of early skip field recorders are: (1) no interchangeability with other models; (2) requires special modified monitor/receiver to overcome severe bending or flagging of vertical objects in the picture; (3) some loss of picture detail; (4) slight vertical jitter; and (5) discontinuity in the picture's horizontal motion.

4. VHS is an acronym for Video Home System.

Chapter 2 Answers

1. A notch or groove is located at one end of the IC to identify where pin 1 can be found. Some ICs also have a dot printed on the case near pin 1.

2. The colors brown, black, red, and silver indicate that this resistor has a value of 1k ohm with a 10% tolerance rating.

3. A component marked 1S1519 is a diode.

4. Leakage means that current is allowed to travel in the opposite direction to normal current flow.

Chapter 3 Answers

1. Several reasons to prevent dust buildup in a VCR include (1) Dust can form an insulating layer on electrical components causing heat to build up in the components and not escape (convect) into the space inside the unit; (2) Dust can cause mechanical parts inside the unit to wear out faster; and (3) Dust also causes video heads to wear faster.

2. Video heads must be cleaned on a regular basis to prevent buildup of the tape's magnetic coating on the head's surface. Buildup affects the head's performance and can cause a tape head clog. Severe head clogs can damage the tape's magnetic surface.

3. Storing video tapes horizontally can cause the tape on the spools to shift downward damaging the edge of the tape.

4. Move the cleaning instrument opposite to the plane of tape travel (moving up and down) can break the tiny head chips in a VCR.

5. Use only manufacturer recommended grease to lubricate a VCR because petroleum-based or other types of lubricants may react with the nylon and other synthetic materials in the gears and tape mechanism. This can cause them to operate sluggishly or to bind.

6. Two negative effects of having increased tape play tensions in a VCR are:
(1) shorter tape life due to the tape surface wearing out faster; and
(2) increased head and tape guide wear due to greater friction.

Chapter 4 Answers

(There are no review questions in Chapter 4.)

Chapter 5 Answers

1. Audio bias is mixed with the normal record audio signal to force the recording head to operate in the linear region of the BH curve. Its purpose is to eliminate distortions that would otherwise be present.

2. 60 to 100 kHz.

3. The bias oscillation is also used by the full track erase and audio erase heads to perform the erase function.

4. FM deviation is a change in frequency from the sync tip frequency. In a VCR, the maximum deviation frequency represents the peak white video level.

5. A) +/– 10 degrees
   B) +/– 6 degrees

6. Storing a video cassette vertically helps the tape to remain properly packed in the spools during storage.

7. Dropout is a brief loss of playback video resulting from a loss of magnetic energy from the tape. It is generally caused by a head-to-tape contact malfunction or a defect on the magnetic surface of the tape.

8. Brown stain is an unwanted deposit that forms on the surface of metal video heads. It can only be removed by an abrasive type of cleaning tape.

9. A metal head pole tip is capable of producing a much stronger magnetic field than a ferrite head before it becomes saturated. The maximum field energy developed by a ferrite head is insufficient for the higher coercivity metal tape.

10. The tape performs a slight scrubbing action on the head to polish and clean the head surface. This action creates some wear on the head surface.

11. Here are some symptoms of a defective tape:
    1) High levels of drop out are observed on that one tape.
    2) Stiction.
    3) Video and/or audio heads require frequent cleaning each time that tape is used.
    4) Poor normal audio reproduction from this tape.
    5) Poor tracking each time this tape is used.
    6) Poor picture quality reproduction.
    7) Poor HiFi audio from this tape.

12. In a VHS VCR, the control track head is used to write and read control track pulses onto a longitudinal track at the bottom edge of the tape. These pulses are used during playback to identify the location of the RF tracks on the tape and to control the tape's playback speed. They are also required for tape playback interchangeability. The 8mm format uses an ATF pilot signal that is imbedded in the RF track to perform these same functions. Since the 8mm video heads write and read the ATF signal along with the video FM signals, a separate control track head is not required.

Chapter 6 Answers

1. Down conversion is required for two reasons. First, the 3.58 MHz sub-carrier will beat against the FM luminance signal if not converted to a lower frequency. This will produce an unwanted beat pattern on the screen. Second, the complete record and playback color-under process provides an economical means of time-base-stabilization during the playback up-conversion operation.

2. APC maintains precise control over the phase of the chroma frequency processing circuits. Without this control, the color hue would drift.

3. VHS AFC corrects playback frequency errors and un-rotates the phase rotated color-under signal.

4. During playback the APC ID corrects for large frequency errors in the 8mm format.

5. Horizontal sync and color burst alignment between adjacent tracks produce increased beat patterns if the VHS LP playback picture is emphasized during record.

6. Chroma burst is emphasized to aid in reducing the noise present in the playback burst signal. Reducing noise in the burst signal increases the accuracy of the chroma playback processing.

7. Without playback ACC, chroma levels will fluctuate because of head to tape contact errors, gain differences between playback heads, and tape surface defects.

8. Color burst provides the only stable reference for monitoring the chroma levels.

9. ACK monitors the status of the APC section.

10. The phase alternates 180 degrees to reduce the visible appearance of the chroma subcarrier on the TV screen.

11. Electronic color-under crosstalk cancellation is required because the color-under frequency on the tape is too low for the head gap azimuth angle offset between the heads to cancel the chroma crosstalk.

12. Your list should include: 1) playback horizontal sync, 2) video head switching, 3) burst flag (also known as burst gate), 4) 3.58 MHz crystal reference oscillation, and 5) playback chroma.

Chapter 7 Answers

1. Sync tip or free run frequency are two other names for the FM carrier frequency.

2. The carrier or free run frequency.

3. Equalization is required to flatten out the 6dB per octave response of the video head. If the signal is not equalized, beat patterns may appear in the reproduced video caused by an amplitude modulated FM carrier.

4. The FM deviation frequencies represent changes in brightness.

5. White and dark clip circuits limit the pre-emphasis applied to the modulated video.

6. Interleaving the FM aids the VCR in reducing the remaining adjacent track crosstalk not canceled by the video head gap azimuth.

7. The lower sideband contains the picture detail. Without recovery of this portion of the bandwidth, video resolution would suffer.

8. Clamping the input video signal is required to establish the sync tip frequency and maintain that reference when the DC content of the video changes.

9. Sub-emphasis reduces the appearance of noise generated in the playback process.

10. The drop out compensator masks the snow that would otherwise be present when the video head encounters a portion of tape without an RF signal on it. DOC action replaces the missing information with the last good line of video.

11. Possible causes include; 1) worn video heads; 2) RF pre-amp failure; 3) playback equalization misalignment;

4) demodulator balance misalignment;
5) limiter balance misalignment;
6) a failure or misalignment of the rotary transformer; 7) low playback RF levels.

Chapter 8 Answers

1. Control track inductance (where L is the symbol for inductance).

2. The source of the record CTL signal is separated vertical sync coming from the record video input.

3. In the 8mm format, the ATF pilot frequencies supersede the need for a CTL signal. They provide the playback tracking feedback data to the capstan playback phase (ATF) control.

4. Yes. The capstan phase is controlled by a digital PWM signal in record. ATF is only used for phase control in the playback mode.

5. The FG signal provides the drum and capstan speed servos with speed feedback information.

6. The drum record phase control assures that the video head is contacting the video tape $10\frac{1}{2}$ H before the start of the record vertical sync signal.

7. The record capstan phase control is a fine tape speed control.

8. During record, the source of the capstan PG signal is counted down capstan FG.

9. A) The tape speed would be too fast.
   B) The drum motor would rotate too fast.

   NOTE: If the microprocessor monitors the drum and capstan rotation speeds, the unit will shut down a few seconds after the motor went into the run away condition. Therefore, operation shut down may be another symptom of the missing motor FG signal.

10. 1) Rear lock detection
    2) Automatic playback speed selection.

11. A 3.58 MHz crystal oscillator counted down to 30 Hz.

12. 1) The head switching pulse will drift vertically through the playback picture.
    2) The VCR will shut down after play is engaged if the head switching pulse is also missing at the system control head switching input line.

Chapter 9 Answers

1. AFM recording is the standard audio recording means for 8mm.

2. The left channel is 1.3 MHz and right channel is 1.7 MHz.

3. The 8mm machines with stereo AFM, multiplex the stereo audio during record so that all monaural-only 8mm VCRs will be able to play back audio from both the recorded left and right stereo channels and output that mixed audio on the single monaural channel. Multiplexing also makes it possible for stereo AFM machines to output monaural sound on both the left and right channels when a monaural recording is played back.

4. Audio bias is also used as the erase signal for the full track erase and audio erase heads.

5. Wow and flutter are generated by any defect that causes the tape speed to change as it passes across the stationary audio head. This in turn causes the pitch of the recorded sound to vary.

6. Companding serves two basic functions: 1) It compresses the dynamic range of the input signal so that a narrower deviation range can be used for FM audio recordings and fewer bits are required in PCM audio; 2) It is also used for noise reduction and dynamic range extension over and above the generated noise level.

7. The amount of deviation (swing above and below) from the carrier frequency produces the volume of the FM audio signal.

8. The frequency (rate) at which the deviation sweeps back and forth across the carrier frequency determines the reproduced audio frequency.

9. Quantization error is noise placed on the input signal resulting from inter-polation errors in the A-D circuit for record and in the D-A circuit for playback.

10. For record, quantization error occurs in the A-D conversion.

11. It's 2.9 MHz for logic 0 and 5.8 MHz for logic 1.

Chapter 10 Answers

1. Mechanical gear, linkage and mode sensing switch alignments, sensors, connections, and peripheral electronic components supporting the micro-processor should all be checked before suspecting a defective microprocessor IC.

2. System Control issues the power-on command after the operator presses the POWER-ON button. System Control issues the power-on command only after checking the initial state of the VCR.

3. Handshaking is a term used to describe serial or parallel data communications between individual microprocessors that share responsibility in the System Control.

4. The VCR/TV button selects the VCR output provided at the RF output jack on the back of the VCR. During record, it allows the operator to monitor either an RF modulated form of the E-to-E audio and video (VCR mode) or to view an antenna through-put of all channels supplied to the antenna input to the VCR during record (TV mode).

5. The VCR RF modulator modulates either the record E-to-E audio/video signal or the playback audio/video signal to Channel 3 or Channel 4 broadcast frequencies for transmission to the television set tuner.

6. The preferred order of VCR hookup to the television are: (For video) 1) S-Video (if available), 2) Line video, 3) RF modulator output; (For audio) 1) Line audio, 2) RF modulator output.

7. The mode position sensing switch tells the System Control microprocessor the particular status (position) of the mechanism. This is done by identifying (by a series of contact closures) the position of certain arms and levers during various modes of operation.

8. Most VCRs cause the loading mechanism to shift position slightly at power on so that the initial state of the mechanism can be identified by the System Control microprocessor when the VCR first wakes up. This is an important part of power on initialization.

Chapter 11 Answers

1. The pulse cross monitor enables viewing of important horizontal sync and vertical sync signals and the head switching interval which are not normally observable on a television display. The observation of these hidden portions of the picture assists technicians in assessing the correct operation of the servo, tape path alignment, and video head condition. Unwanted noise that causes vertical synchronization instabilities (picture jumping) can also be observed with a pulse cross monitor.

2. The broadcast picture does not provide fully saturated color signals, or constant color hues for aligning the chroma processing circuits. Also, the broadcast TV signal rarely provides 100% peak white video levels required for aligning the luminance circuitry.

3. Applying the demodulated video input from a television monitor/receiver to the VCR's video line input allows the servicing technician to bypass the VCR's RF tuner and RF demodulation stages. Record input defects can be quickly isolated to the tuner or demodulator if the VCR suddenly starts recording when these circuits are bypassed. Connecting the video from the VCR's line video output connector to a monitor/ receiver allows the service technician to bypass the channel 3/channel 4 RF modulator for playback defect fault isolation.

4. RFI may cause the VCR servos to become unstable. The playback video may jump vertically or the playback speed may change erratically.

5. Sweep marker identification uses an electronic marker superimposed on the output of the frequency generator to identify specific frequencies. Sweep marker generators can sweep a selected range and identify the position of certain frequencies within that range using markers.

6. Most VCRs have a hot chassis power supply that has one side of the AC line connected directly to the chassis. Test instruments (such as oscilloscopes) with ground probes that connect to the AC

power line cannot be connected to machines with hot chassis power supplies without an isolation transformer because you could destroy both units. If you blindly connect test equipment to machines that have these types of power supplies, there is great potential for damage to the VCR, the test equipment and you.

7. Playback alignment is typically done first because the record alignment procedure involves playback of the recordings made during alignment. If the machine does not play back properly, playback checks of the recording alignment would be invalid. Once the playback circuitry is properly aligned, the record adjustments are then made.

8. Some diodes are made of gallium arsenide and react by giving off light when biased in a certain way. These are called light emitting diodes or LEDs.

9. A typical forward biased silicon diode will read between 0.6 and 0.7 volts.

10. Checking hybrid transistors with an ohmmeter often reveals incorrect results because the device is made up of resistors and a semiconductor. Sometimes a hybrid transistor has more than one transistor which will also throw off the ohmmeter reading results.

11. The rotary transformer must be disconnected from the video head lead wires to produce an accurate continuity measurement. When the rotary transformer is left connected to the video head, the ohmmeter will not reveal an open head winding. This is because the head winding and the rotary transformer winding are in parallel to the meter probes if left connected.

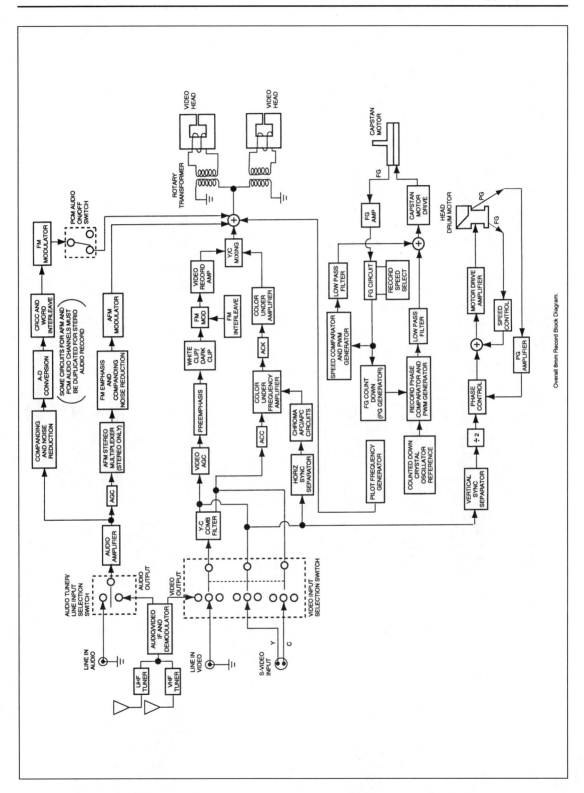

Overall 8mm Record Block Diagram.

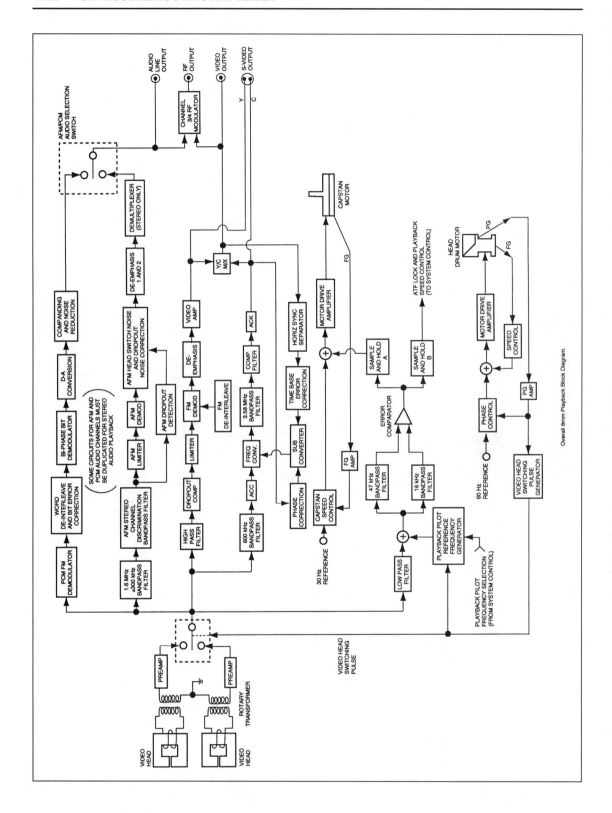

Overall 8mm Playback Block Diagram.

Maintenance Record

Date: _____

INITIAL CHECK

Playback _____ Fast Forward _____

Record _____ Rewind _____

Channel Tune _____ Threading _____

INTERMEDIATE CHECK AND REPAIR

CLEAN:

Audio Heads _____

Video Heads _____

Control Track Head _____

Scanner _____

Tape Path _____

Pulleys _____

Pinch Roller _____

Capstan _____

Reel Tables _____

Brake Pads _____

Tension Bands _____

CLEAN OR REPLACE:

Belts _____

Tires _____

Pinch Roller _____

LUBRICATE:

Reel Tables _____

Idler Tires _____

Pulley Shafts _____

Linkages/Levers _____

FINAL CHECK

Brake Torque _____ Record _____

FFWD. Torque _____ Playback _____

RWD. Torque _____ Tracking _____

Play Torque _____ RF Envelope _____

Hold Back Tension _____

Notes:

Index

Page numbers in *italics* denote figures; "t" denote tables

415